Springer Series in
OPTICAL SCIENCES 80

founded by H.K.V. Lotsch

Springer
Berlin
Heidelberg
New York
Barcelona
Hong Kong
London
Milan
Paris
Singapore
Tokyo

Springer Series in

OPTICAL SCIENCES

The Springer Series in Optical Sciences, under the leadership of Editor-in-Chief *William T. Rhodes*, Georgia Institute of Technology, USA, and Georgia Tech Lorraine, France, provides an expanding selection of research monographs in all major areas of optics: lasers and quantum optics, ultrafast phenomena, optical spectroscopy techniques, optoelectronics, quantum information, information optics, applied laser technology, industrial applications, and other topics of contemporary interest.
With this broad coverage of topics, the series is of use to all research scientists and engineers who need up-to-date reference books.

The editors encourage prospective authors to correspond with them in advance of submitting a manuscript. Submission of manuscripts should be made to the Editor-in-Chief or one of the Editors. See also http://www.springer.de/phys/books/optical_science/

Kazuaki Sakoda

Optical Properties of Photonic Crystals

With 95 Figures and 28 Tables

Springer

Professor Kazuaki Sakoda
Hokkaido University
Research Institute for Electronic Science
North 12 West 6, Kita-ku
Sapporo 060-0812
Japan

ISSN 0342-4111

ISBN 3-540-41199-2 Springer-Verlag Berlin Heidelberg New York

Library of Congress Cataloging-in-Publication Data

Sakoda, Kazuaki, 1957-
Optical properties of photonic crystals / Kazuaki Sakoda.
p.cm. – (Springer series in optical sciences, ISSN 0342-4111 ; 80)
Includes bibliographical references and index.
ISBN 3-540411992 (acid-free paper)
1. Photons. 2. Crystal optics. I. Title. II. Springer series in optical sciences ; v. 80.
QC793.5.P427 S24 2001   548'.9–dc21   00-066118

Springer-Verlag Berlin Heidelberg New York
a member of BertelsmannSpringer Science+Business Media GmbH

http://www.springer.de

Camera-ready by the author using a Springer TEX macropackage
Cover concept by eStudio Calamar Steinen using a background picture from The Optics Project. Courtesy of John T. Foley, Professor, Department of Physics and Astronomy, Mississippi State University, USA.
Cover production: *design & production* GmbH, Heidelberg

Printed on acid-free paper   SPIN 10771912   57/3141/di   5 4 3 2 1 0

To Fumi and Midori

Preface

The interaction between the radiation field and matter is the most fundamental source of dynamics in nature. It brings about the absorption and emission of photons, elastic and inelastic light scattering, the radiative lifetime of electronic excited states, and so on. The huge amount of energy carried from the sun by photons is the source of all activities of creatures on the earth. The absorption of photons by chlorophylls and the successive electronic excitation initiate a series of chemical reactions that are known as photosynthesis, which support all life on the earth. Radiative energy is also the main source of all meteorological phenomena.

The fundamentals of the radiation field and its interaction with matter were clarified by classical electromagnetism and quantum electrodynamics. These theories, we believe, explain all electromagnetic phenomena. They not only provide a firm basis for contemporary physics but also generate a vast range of technological applications. These include television, radar, optical and microwave telecommunications, lasers, light-emitting diodes, solar cells, etc.

Now, the interaction between the radiation field and matter is so fundamental that it may seem universal and invariant. But in fact it is *controllable*. This discovery has been the great motivating force of intensive investigations in optical physics during the last three decades. In this book, I will show how it is controlled using *photonic crystals*, a remarkable invention realized by the combination of optical physics and contemporary microfabrication techniques. I will also show how the controlled radiation field alters the optical properties of atoms and molecules embedded in the photonic crystals and what kinds of new phenomena and new physics are expected to manifest themselves.

This book was written to serve as a comprehensive textbook covering the optical properties of photonic crystals. It deals not only with the properties of the radiation field inside the photonic crystals but also their peculiar optical response to external fields. Only an elementary knowledge of electromagnetism, quantum mechanics, solid-state physics, and complex analysis is required of the reader. Therefore, undergraduate students in physics, applied physics, optics, electronics, and electrical engineering in the final year should be able to read this book without difficulty. Since the main recent developments, such as the enhancement of stimulated emission, second harmonic generation, and quadrature-phase squeezing, are also treated in a detailed and understandable manner, this book also provides important ideas for graduate students and researchers in this field.

I would like to thank Professor Eiichi Hanamura and Professor Kuon Inoue who gave me a wonderful introduction to this exciting field. I would also like to thank Professor Kazuo Ohtaka and Professor Joseph W. Haus who gave me many suggestions on important problems. I am grateful to Professor Sajeev John for giving me the opportunity to visit his laboratory in Toronto University in the summer of 1998. During that period, I learned much about the quantum optics of photonic crystals. I am also grateful to Dr. Kurt Busch who helped me to confirm that the group theory worked well for the fcc structure. I acknowledge Professor Takao Koda for his continuous encouragement since I was an undergraduate student of Tokyo University. I deeply acknowledge Professor Toshimitsu Asakura and Dr. Claus Ascheron for giving me the opportunity to write this book. I am grateful to many of my graduate students for their efforts in numerical calculations. In particular, Ms. Hitomi Shiroma-Hirata, Ms. Noriko Kawai, and Mr. Takunori Ito did many of the calculations that are shown in this book. I am also grateful to Dr. Tetsuyuki Ochiai who made the main contribution to the study of photonic crystal slabs presented in Chapter 8.

Sapporo
February 2001

Kazuaki Sakoda

Contents

1. Introduction

As is well known, there is the following relation between the frequency ν, the velocity c, and the wavelength λ_0, of the radiation field in free space:

$$c = \lambda_0 \nu. \tag{1.1}$$

When we define the wave number k by,

$$k = \frac{2\pi}{\lambda_0}, \tag{1.2}$$

we obtain the relation between the angular frequency ω and k:

$$\omega = ck. \tag{1.3}$$

This equation is called the dispersion relation of the radiation field. If one thinks of the radiation field in a uniform material with refractive index η, one can obtain its dispersion relation by replacing c by $v = c/\eta$ and λ_0 by $\lambda = \lambda_0/\eta$ in (1.2) and (1.3). The density of states of the radiation field in the volume V of free space, $D(\omega)$, is proportional to ω^2 (Fig. 1.1):

$$D(\omega) = \frac{\omega^2 V}{\pi^2 c^3}. \tag{1.4}$$

The density of states in the uniform material is obtained by replacing c by v in this equation.[1] The optical properties of atoms and molecules strongly depend on $D(\omega)$. As an example, let us consider the spontaneous emission of a photon from an electronic excited state of an atom or a molecule. Quantum mechanics tells us that the rate of the spontaneous emission is proportional to $\omega D(\omega)$. Since the spontaneous emission is an origin of the energy dissipation and the fluctuation of the radiation field, it suppresses the occurrence of laser oscillations. This suppression is marked in the high frequency region since $D(\omega)$ is proportional to ω^2. This is one of the reasons why the laser oscillation is difficult to realize at high frequencies.

Now, if we can design and modify $D(\omega)$, we can substantially change the optical properties of atoms and molecules [1]. This is a key idea of contemporary optical physics, and it is possible. One method is to use optical microcavities while another is to use *photonic crystals*. In this book, it will be shown how the characteristics of the radiation field are modified in photonic

[1] The derivation of (1.4) will be given in Sect. 5.2.

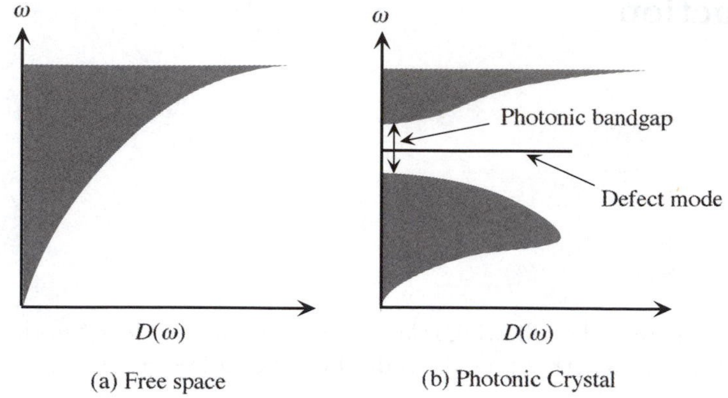

Fig. 1.1. Schematic illustration of the density of states of the radiation field (**a**) in free space and (**b**) in a photonic crystal. In the illustration for the photonic crystal, a photonic bandgap and a localized defect mode with a delta-function like density is included (see text)

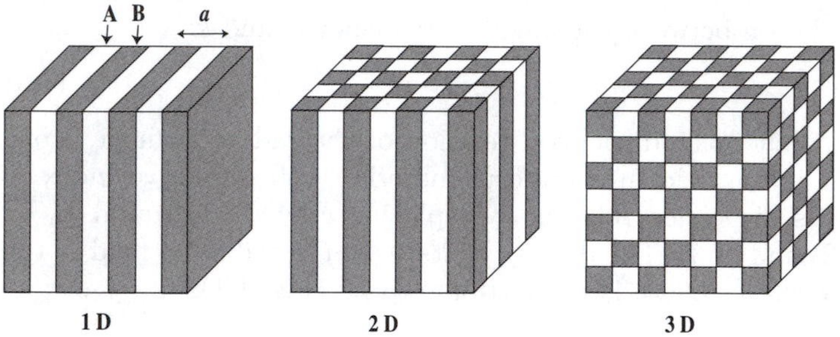

Fig. 1.2. Schematic illustration of one-dimensional (1D), two-dimensional (2D), and three-dimensional (3D) photonic crystals. a is the lattice constant

crystals and how the optical properties of atoms and molecules embedded in them are altered.

Photonic crystals are regular arrays of materials with different rafractive indices. Figure 1.2 shows the simplest case in which two materials denoted by A and B are stacked alternately. The spatial period of the stack is called the *lattice constant*, since it corresponds to the lattice constant of ordinary crystals composed of a regular array of atoms. Actually, many basic ideas are common to both crystals and they will be utilized to build fundamental theories of the photonic crystals, as will be shown in the following chapters. However, one big difference between them is the scale of the lattice constant. In the case of ordinary crystals, the lattice constant is on the order of angstroms. On the other hand, it is on the order of the wavelength of the relevant electromagnetic waves for the photonic crystals. For example, it is about 1 mm or less for visible light, and is about 1 cm for microwaves.

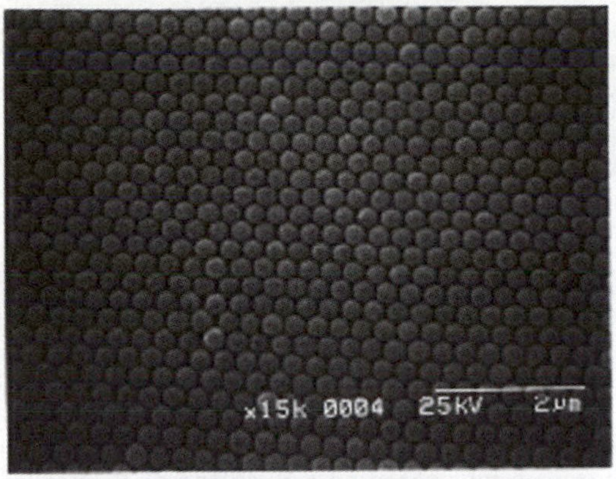

Fig. 1.3. SEM image of a 3D photonic crystal composed of an fcc array of SiO_2 spheres with a diameter of 300 nm. (After [2])

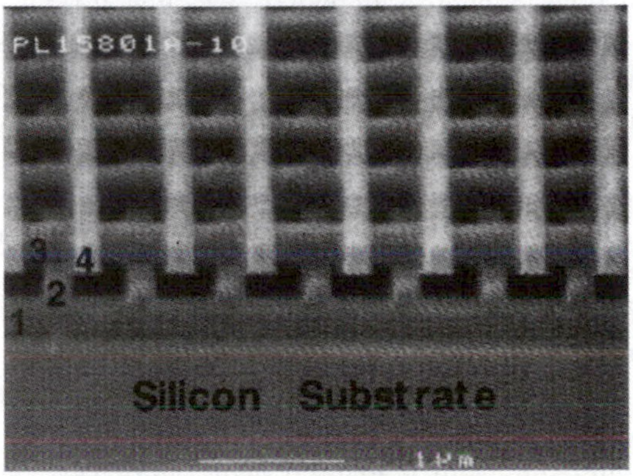

Fig. 1.4. SEM image of a 3D photonic crystal made of silicon. (After [3])

Photonic crystals are classified mainly into three categories, that is, one-dimensional (1D), two-dimensional (2D), and three-dimensional (3D) crystals according to the dimensionality of the stack (see Fig. 1.2). The photonic crystals that work in the microwave and far-infrared regions are relatively easy to fabricate. Those that work in the visible region, especially 3D ones are difficult to fabricate because of their small lattice constants. However, various technologies have been developed and applied to their fabrication in the last ten years, and many good crystals with a lattice constant less than 1 mm are now available. For example, Fig. 1.3 shows an SEM (scanning electron micrograph) image of an fcc (face-centered-cubic) lattice composed of silica spheres [2]. The diameter of the spheres is 0.3 mm. On the other hand, Fig. 1.4 is an SEM image of a 3D structure made of Si [3].

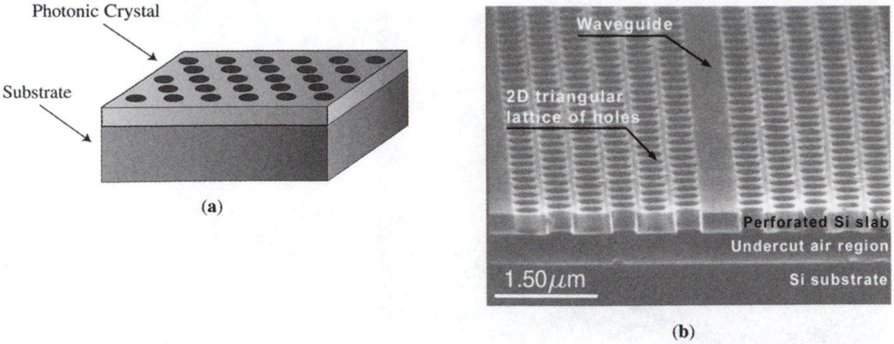

Fig. 1.5. (**a**) Schematic illustration of a photonic crystal slab and (**b**) the scanning electron micrograph of an actual specimen fabricated on a Si substrate. (After [4])

If we design a 3D photonic crystal appropriately, there appears a frequency range where no electromagnetic eigenmode exists. Frequency ranges of this kind are called *photonic bandgaps*, since they correspond to bandgaps of electronic eigenstates in ordinary crystals. Moreover, if we introduce a disorder into the regular dielectric structure of the photonic crystal, we may obtain midgap modes whose eigenfunctions are strongly localized around the disorder. These modes are called *localized defect modes*. The density of states for a 3D photonic crystal with a photonic bandgap and a localized defect mode is schematically illustrated in Fig. 1.1(b). If the emission frequency of an atom or a molecule embedded in the photonic crystal lies just in the photonic bandgap, the spontaneous emission of a photon from its electronic excited state is completely forbidden, since there exists no photon in the gap. On the other hand, if the emission frequency coincides with the eigenfrequncy of the localized mode and the atom is located near the defect, the spontaneous emission is accelerated.

Another class of photonic crystals known as *photonic crystal slabs* [4–13] is illustrated in Fig. 1.5. Photonic crystals of this type are usually fabricated on a substrate made of a semiconductor or an insulator. They have been investigated energetically in recent years, because many sophisticated technologies such as electron beam lithography and thin-layer formation developed in the field of electronics and opto-electronics can be applied to their fabrication. Monolayers made of polymer micro-spheres [14–16] may also be regarded as photonic crystal slabs.

In order to gain an intuitive understanding of the photonic bands and bandgaps, we examine 1D crystals in some detail. The 1D photonic crystals are traditionally called dielectric multilayers and their optical properties are well-known [17]. We take the x axis in the direction perpendicular to the surface of the dielectric layers as shown in Fig. 1.6. We only deal with electromagnetic waves propagated in the x direction and polarized linearly here. We take the y axis in the direction of the polarization. The electric field of the

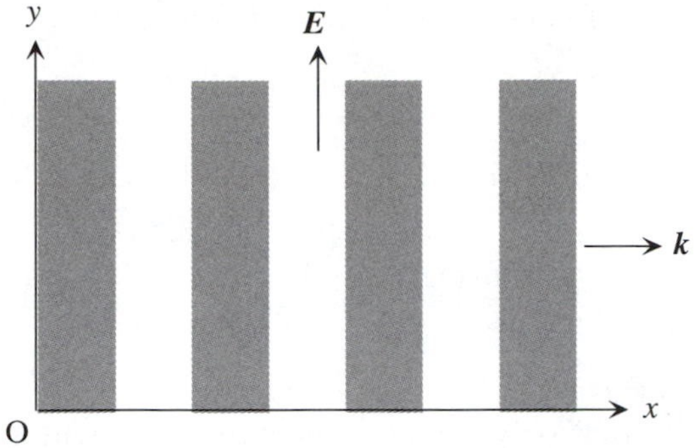

Fig. 1.6. Geometry of the calculation of the photonic band structure, or the dispersion relation, of a 1D photonic crystal

propagated wave is denoted by a complex function $E(x,t)$ for convenience. The actual electric field is, of course, a real quantity. It is given by the real part of $E(x,t)$.

Now, the wave equation for $E(x,t)$ is given by

$$\frac{c^2}{\varepsilon(x)}\frac{\partial^2 E}{\partial x^2} = \frac{\partial^2 E}{\partial t^2}, \tag{1.5}$$

where $\varepsilon(x)$ denotes the position-dependent relative dielectric constant of the 1D photonic crystal, which will be called the *dielectric function* hereafter.[2] In (1.5), we assumed that the magnetic permeability of the photonic crystal is equal to that in free space, since we do not treat magnetic materials in this book. Because $\varepsilon(x)$ is a periodic function of x,

$$\varepsilon(x+a) = \varepsilon(x), \tag{1.6}$$

$\varepsilon^{-1}(x)$ is also periodic and can be expanded in a Fourier series:

$$\varepsilon^{-1}(x) = \sum_{m=-\infty}^{\infty} \kappa_m \exp\left(\mathrm{i}\frac{2\pi m}{a}x\right), \tag{1.7}$$

where m is an integer and $\{\kappa_m\}$ are the Fourier coefficients. Since we assume that $\varepsilon(x)$ is real in this chapter, $\kappa_{-m} = \kappa_m^*$. It is well-known that Bloch's theorem holds for the electronic eigenstates in ordinary crystals because of the spatial periodicity of the potential energy that an electron feels due to the regular array of atomic nuclei [18]. The same theorem holds for electromagenetic waves in photonic crystals.[3] Any eigenmode in the 1D crystal is thus characterized by a wave number k and expressed as follows:

$$E(x,t) \equiv E_k(x,t) = u_k(x)\exp\{\mathrm{i}(kx - \omega_k t)\}, \tag{1.8}$$

[2] The derivation of the wave equation from Maxwell's equations will be described in Chap. 2.

[3] The proof will be given in Sect. 2.1.

where ω_k denotes the eigen-angular frequency and $u_k(x)$ is a periodic function:

$$u_k(x+a) = u_k(x). \tag{1.9}$$

Hence, it can also be expanded in a Fourier series. As a result, (1.8) is modified to

$$E_k(x,t) = \sum_{m=-\infty}^{\infty} E_m \exp\left\{\mathrm{i}\left(k + \frac{2\pi m}{a}\right)x - \mathrm{i}\omega_k t\right\}, \tag{1.10}$$

where $\{E_m\}$ are the Fourier coefficients.

Now, we assume for simplicity that only components with $m = 0$ and ± 1 are dominant in the expansion (1.7):

$$\varepsilon^{-1}(x) \approx \kappa_0 + \kappa_1 \exp\left(\mathrm{i}\frac{2\pi}{a}x\right) + \kappa_{-1}\exp\left(-\mathrm{i}\frac{2\pi}{a}x\right). \tag{1.11}$$

When we substitute (1.10) and (1.11) into the wave equation (1.5), we obtain

$$\begin{aligned}&\kappa_1\left\{k + \frac{2(m-1)\pi}{a}\right\}^2 E_{m-1} + \kappa_{-1}\left\{k + \frac{2(m+1)\pi}{a}\right\}^2 E_{m+1}\\ &\approx \left\{\frac{\omega_k^2}{c^2} - \kappa_0\left(k + \frac{2m\pi}{a}\right)^2\right\} E_m.\end{aligned} \tag{1.12}$$

For $m = 0$,

$$E_0 \approx \frac{c^2}{\omega_k^2 - \kappa_0 c^2 k^2}\left\{\kappa_1\left(k - \frac{2\pi}{a}\right)^2 E_{-1} + \kappa_{-1}\left(k + \frac{2\pi}{a}\right)^2 E_1\right\}. \tag{1.13}$$

For $m = -1$,

$$E_{-1} \approx \frac{c^2}{\omega_k^2 - \kappa_0 c^2 (k - 2\pi/a)^2}\left\{\kappa_1\left(k - \frac{4\pi}{a}\right)^2 E_{-2} + \kappa_{-1}k^2 E_0\right\}. \tag{1.14}$$

Therefore, if $k \approx |k - 2\pi/a|$ (i.e., $k \approx \pi/a$), and if $\omega_k^2 \approx \kappa_0 c^2 k^2$, E_0 and E_{-1} are dominant in the expansion (1.10). In this case, we neglect all other terms and obtain the following coupled equations:

$$\left(\omega_k^2 - \kappa_0 c^2 k^2\right) E_0 - \kappa_1 c^2 \left(k - \frac{2\pi}{a}\right)^2 E_{-1} = 0, \tag{1.15}$$

$$-\kappa_{-1} c^2 k^2 E_0 + \left\{\omega_k^2 - \kappa_0 c^2\left(k - \frac{2\pi}{a}\right)^2\right\} E_{-1} = 0. \tag{1.16}$$

These linear equations have a nontrivial solution when the determinant of coefficients vanishes:

$$\begin{vmatrix} \omega_k^2 - \kappa_0 c^2 k^2, & -\kappa_1 c^2\left(k - \frac{2\pi}{a}\right)^2 \\ -\kappa_{-1} c^2 k^2, & \omega_k^2 - \kappa_0 c^2\left(k - \frac{2\pi}{a}\right)^2 \end{vmatrix} = 0. \tag{1.17}$$

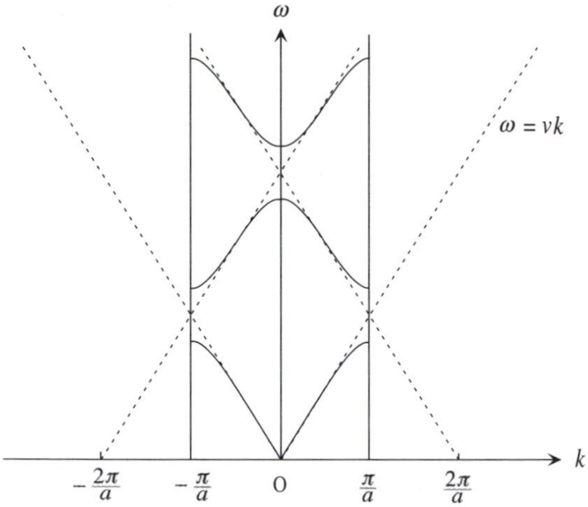

Fig. 1.7. Dispersion relation for a 1D photonic crystal (*solid lines*). The boundary of the first Brillouin zone is denoted by two vertical lines. The dispersion lines in the uniform material are denoted by *dashed lines*. They are folded into the first Brillouin zone taking into account the identity of the wave numbers which differ from each other by a multiple of $2\pi/a$. When two dispersion lines cross, they repel each other and a photonic bandgap appears

If we introduce $h = k - \pi/a$, the solutions are given by

$$\omega_{\pm} \approx \frac{\pi c}{a}\sqrt{\kappa_0 \pm |\kappa_1|} \pm \frac{ac}{\pi|\kappa_1|}\left(\kappa_0^2 - \frac{|\kappa_1|^2}{2}\right)h^2, \tag{1.18}$$

as far as $|h| \ll \pi/a$. So, there is no mode in the interval

$$\frac{\pi c}{a}\sqrt{\kappa_0 - |\kappa_1|} < \omega < \frac{\pi c}{a}\sqrt{\kappa_0 + |\kappa_1|}. \tag{1.19}$$

This gap disappears when $\kappa_1 = 0$. This result can be interpreted that the modes with $k \approx \pi/a$ and $k \approx -\pi/a$ were mixed with each other in the presence of the periodic modulation of the dielectric constant and this mixing led to a frequency splitting.

In general, those wave vectors which differ from each other by a multiple of $2\pi/a$ should be regarded as the same because of the presence of the periodic spatial modulation of the dielectric constant. When the spatial modulation is small, the dispersion relation in the photonic crystal is not so far from $\omega = vk$, but it should thus be expressed with the wave vector in the first Brillouin zone, $[-\pi/a, \pi/a]$.[4] In addition, if two dispersion lines cross each other, a frequency gap appears. All these things are schematically illustrated in Fig. 1.7. There are an infinite number of frequency gaps in the spectrum. However, we should note that this is true only as far as we deal with optical

[4] This is the so-called reduced zone scheme.

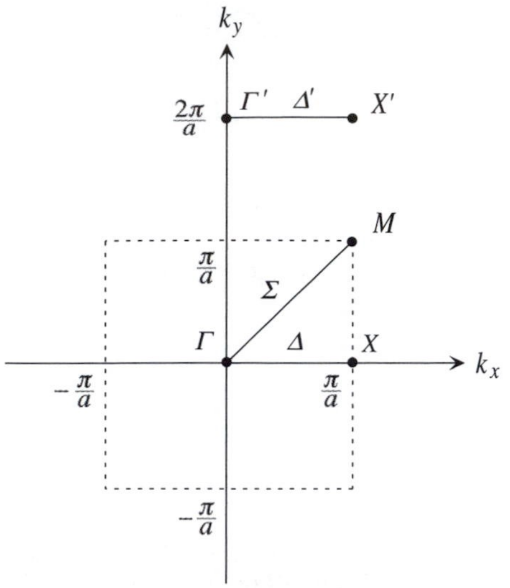

Fig. 1.8. Reciprocal lattice space of the 2D square photonic crystal with the lattice constant a. The first Brillouin zone is surrounded by a *dashed line*

waves travelling in the x direction, and that there is no gap when we take into consideration those modes travelling in other directions.

1D photonic crystals, or dielectric multilayers, have many important applications. For example, they are utilized as high-reflection mirrors and anti-reflection coatings. However, most investigations related to the control of the radiation field in recent years have been concentrated on 2D and 3D photonic crystals, because they offer a rich variety of new physics of the radiation field, as will be described in this book. One example is the photonic bandgaps that may be realized in a certain class of 3D crystals. Another example is the *group-velocity anomaly* that is realized in all 2D and 3D crystals. We consider this point in some detail here.

Like the 1D crystals, the dispersion relation of a 2D crystal with a small spatial variation of the dielectric constant can be well-represented by the folding of the dispersion line of the uniform material, $\omega = vk$, into the first Brillouin zone. The fact that all wave vectors are not parallel to each other in two dimensions, however, makes a large difference. As an example, Fig. 1.8 shows the reciprocal lattice space of a 2D square lattice. The lattice constant is denoted by a as before. The first Brillouin zone is surrounded by a dashed line in this figure. The elementary lattice vectors are

$$\boldsymbol{a}_1 = (a, 0) \qquad \text{and} \qquad \boldsymbol{a}_2 = (0, a), \tag{1.20}$$

and the elementary reciprocal lattice vectors are

$$\boldsymbol{b}_1 = \left(\frac{2\pi}{a}, 0\right) \qquad \text{and} \qquad \boldsymbol{b}_2 = \left(0, \frac{2\pi}{a}\right). \tag{1.21}$$

Those wave vectors that differ from each other by the sum of a multiple of $\boldsymbol{b}_1$ and that of $\boldsymbol{b}_2$ should be regarded as the same. There are three highly symmetric points in the first Brillouin zone, i.e., the Γ point, $(0,0)$, the X point, $(\pi/a, 0)$, and the M point, $(\pi/a, \pi/a)$. Γ' and X' shown in Fig. 1.8 are equivalent to Γ and X, respectively. The wave vectors between the Γ and the X points are called the Δ point. On the other hand, that between the Γ and the M points is called the Σ point.

Now, the anomalous group velocity takes place when the original wave vector in the uniform material is not parallel to either $\boldsymbol{b}_1$ or $\boldsymbol{b}_2$. We examine this point for the wave vectors between Γ' and X', which are denoted by Δ' in Fig. 1.8. Those wave vectors are

$$\boldsymbol{k}' = \left(k_x, \frac{2\pi}{a}\right), \qquad 0 \leq k_x \leq \frac{\pi}{a}. \tag{1.22}$$

The original dispersion relation is given by

$$\omega = v|\boldsymbol{k}'| = v\sqrt{k_x^2 + \left(\frac{2\pi}{a}\right)^2}, \tag{1.23}$$

and this should also be regarded as a good approximation to the rigourous dispersion relation of the photonic crystal for the corresponding wave vectors $\boldsymbol{k}$ in the first Brillouin zone:

$$\boldsymbol{k} = (k_x, 0). \tag{1.24}$$

The dispersion relation for the square lattice with an infinitesimally small modulation of the dielectric constant thus obtained is shown in Fig. 1.9. The averaged refractive index is assumed to be 1.28 in this figure. The band that we have just discussed is the third lowest one on the Δ point. This band has a very small slope compared with the rest. This is a consequence of the fact that the original wave vector is not parallel to the elementary reciprocal lattice vectors as was mentioned above.

The group velocity $\boldsymbol{v}_\mathrm{g}$ of the radiational eigenmode is given by the slope of the dispersion curve:

$$\boldsymbol{v}_\mathrm{g} = \frac{\partial \omega}{\partial \boldsymbol{k}}. \tag{1.25}$$

Hence, the third lowest band has a small group velocity over its entire frequency range. We refer to the small group velocity of this type as the *group-velocity anomaly.* Because we can prove that the group velocity is equal to the energy velocity in the photonic crystal,[5] the small group velocity implies that the interaction time between the radiational mode and the matter system is long [19]. This leads to a large effective coupling between them, and various optical processes are enhanced. I will show later the enhancement of stimulated emission, sum-frequency generation (Chap. 5), and quadrature-phase

[5] See Sect. 2.5.

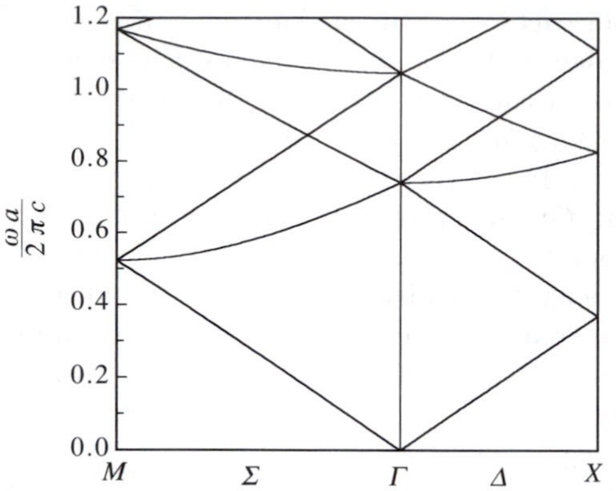

Fig. 1.9. Dispersion relation for a 2D photonic crystal with an infinitesimally small spatial variation of the dielectric constant. The abscissa represents the wave vector in the first Brillouin zone of the 2D square lattice. The ordinate is the normalized frequency where a and c stand for the lattice constant and the light velocity in free space

squeezing (Chap. 10). Light absorption is also enhanced. As for nonlinear optical processes such as sum-frequency generation, the phase-matching condition, which is an important character to obtain a large output, is given by the conservation of the crystalline momentum (see Sect. 5.4):

$$\boldsymbol{k}_{\mathrm{f}} = \boldsymbol{k}_{\mathrm{i}} + \boldsymbol{G}, \tag{1.26}$$

where $\boldsymbol{k}_{\mathrm{i}}$ ($\boldsymbol{k}_{\mathrm{f}}$) denotes the initial (final) wave vector of the radiation field, whereas $\boldsymbol{G}$ denotes a reciprocal lattice vector. As for the uniform material,

$$\boldsymbol{k}_{\mathrm{f}} = \boldsymbol{k}_{\mathrm{i}}. \tag{1.27}$$

This equation may be regarded as a special case of (1.26) for which $\boldsymbol{G} = 0$. Because $\boldsymbol{G}$ is an arbitrary reciprocal lattice vector for the photonic crystals, the phase-matching condition is relaxed compared with the uniform materials. The group-velocity anomaly and the relaxed phase-matching condition are very important aspects of nonlinear optical properties of the photonic crystals.

This book is organized as follows. The eigenvalue problems for the radiation field in the photonic crystals will be formulated in Chap. 2. An Hermitian operator related to the electric field will be introduced, and the properties of its eigenfunctions will be studied in detail. The retarded Green's function of this Hermitian operator will be derived for later use. In Chap. 3, group theory will be formulated to analyze the symmetry of the eigenmodes, and will be applied to some examples. The uncoupled modes, that is, those eigenmodes which cannot be excited by external plane waves due to the mismatching of the symmetry property, will be identified by group theory. In Chap. 4, a numerical method to calculate the transmittance and the Bragg reflectance will be formulated and applied to some examples. The precise coincidence between the calculated spectra and the band structure will be shown. In Chap. 5, a general formula for the description of the optical response of the photonic crystals will be derived based on the method of Green's function. Four examples of its application, i.e., dipole radiation, stimulated emission, sum-frequency generation, and free induction decay will be presented. The phase-matching condition and the additional selection rule for nonlinear optical processes peculiar to photonic crystals will be given there. An example of the numerical analysis of the sum-frequency generation will also be presented. In Chap. 6, an efficient numerical method to calculate the localized defect modes will be derived based on the general formula to describe the dipole radiation in the photonic crystals given in Chap. 5. An excellent agreement between the calculated eigenfrequencies and the experimental observation will be shown. In Chap. 7, I will present a general and accurate numerical method to calculate the band structure with frequency-dependent dielectric constants. This method will be applied to a photonic crystal with metallic components. The presence of extremely flat bands related to surface plasmon polaritons will be shown. In Chap. 8, the optical properties of photonic crystal slabs will be examined. The low-threshould lasing due to the enhanced stimulated emission caused by the group-velocity anomaly will be discussed in Chap. 9. In addition to the analytical formula describing the onset of lasing, the lasing threshould will be analyzed numerically. The radiation field will be quantized in Chap. 10. Two quantum optical properties, i.e., quadrature-phase squeezing due to optical parametric amplification and Lamb shift of atoms in the photonic crystal will be examined. The enhancement of the former due to the anomalous group velocity and that of the latter due to the anomalous density of states of the radiation field will be shown.

2. Eigenmodes of Photonic Crystals

As the first step of the analysis of the radiation field in a photonic crystal, we will formulate the eigenvalue problem of the wave equation and give a general numerical method to solve it. Based on the complete set of the eigenfunctions, we will also derive the expression for the retarded Green's function related to the electric field. In addition to the three-dimensional (3D) case, we will treat the two-dimensional (2D) crystals, for which the vectorial wave equation reduces to two independent scalar equations and the relevant expressions are simplified.

2.1 Wave Equations and Eigenvalue Problems

We begin with Maxwell's equations. Because we are interested in the eigenmodes of the radiation field, and the interaction between the field and matter will be discussed in later chapters, we assume here that free charges and the electric current are absent. In this case, Maxwell's equations in the most general form are given in MKS units as follows.

$$\boldsymbol{\nabla} \cdot \boldsymbol{D}(\boldsymbol{r}, t) = 0, \tag{2.1}$$

$$\boldsymbol{\nabla} \cdot \boldsymbol{B}(\boldsymbol{r}, t) = 0, \tag{2.2}$$

$$\boldsymbol{\nabla} \times \boldsymbol{E}(\boldsymbol{r}, t) = -\frac{\partial}{\partial t}\boldsymbol{B}(\boldsymbol{r}, t), \tag{2.3}$$

$$\boldsymbol{\nabla} \times \boldsymbol{H}(\boldsymbol{r}, t) = \frac{\partial}{\partial t}\boldsymbol{D}(\boldsymbol{r}, t). \tag{2.4}$$

The standard notations for the electric field ($\boldsymbol{E}$), the magnetic field ($\boldsymbol{H}$), the electric displacement ($\boldsymbol{D}$), and the magnetic induction ($\boldsymbol{B}$) are used in these equations.

In order to solve the wave equations derived from Maxwell's equations, we need so-called constitutive equations that relate $\boldsymbol{D}$ to $\boldsymbol{E}$ and $\boldsymbol{B}$ to $\boldsymbol{H}$. Since we do not deal with magnetic materials in this book, we assume that the magnetic permeability of the photonic crystal is equal to that in free space, μ_0:

$$\boldsymbol{B}(\boldsymbol{r}, t) = \mu_0 \boldsymbol{H}(\boldsymbol{r}, t). \tag{2.5}$$

As for the dielectric constant, we assume in this chapter that it is real, isotropic, perfectly periodic with respect to the spatial coordinate $\boldsymbol{r}$, and does not depend on frequency. Those cases in which the dielectric constant is complex, has a disorder, or depends on frequency, which are very interesting and important, will be dealt with in Chaps. 5–9. Readers who are interested in the treatment of anisotropic dielectric constants may consult [20]. We denote the dielectric constant of free space by ε_0 and the relative dielectric constant of the photonic crystal by $\varepsilon(\boldsymbol{r})$. The electric displacement is thus given by

$$\boldsymbol{D}(\boldsymbol{r},t) = \varepsilon_0\varepsilon(\boldsymbol{r})\boldsymbol{E}(\boldsymbol{r},t). \tag{2.6}$$

The periodicity of $\varepsilon(\boldsymbol{r})$ implies

$$\varepsilon(\boldsymbol{r}+\boldsymbol{a}_i) = \varepsilon(\boldsymbol{r}) \qquad (i = 1,2,3), \tag{2.7}$$

where $\{\boldsymbol{a}_i\}$ are the elementary lattice vectors of the photonic crystal. Because of this spatial periodicity, we can expand $\varepsilon^{-1}(\boldsymbol{r})$ in a Fourier series. For this, we introduce the elementary reciprocal lattice vectors $\{\boldsymbol{b}_i; i = 1,2,3\}$ and the reciprocal lattice vectors $\{\boldsymbol{G}\}$:

$$\boldsymbol{a}_i \cdot \boldsymbol{b}_j = 2\pi\delta_{ij}, \tag{2.8}$$

$$\boldsymbol{G} = l_1\boldsymbol{b}_1 + l_2\boldsymbol{b}_2 + l_3\boldsymbol{b}_3, \tag{2.9}$$

where $\{l_i\}$ are arbitrary integers and δ_{ij} is Kronecker's delta. $\varepsilon^{-1}(\boldsymbol{r})$ is expressed as

$$\frac{1}{\varepsilon(\boldsymbol{r})} = \sum_{\boldsymbol{G}} \kappa(\boldsymbol{G}) \exp\left(\mathrm{i}\boldsymbol{G}\cdot\boldsymbol{r}\right). \tag{2.10}$$

Because we assumed that the dielectric function is real, $\kappa(-\boldsymbol{G}) = \kappa^*(\boldsymbol{G})$. When we substitute (2.5) and (2.6) into (2.1)–(2.4), we obtain

$$\boldsymbol{\nabla}\cdot\{\varepsilon(\boldsymbol{r})\boldsymbol{E}(\boldsymbol{r},t)\} = 0, \tag{2.11}$$

$$\boldsymbol{\nabla}\cdot\boldsymbol{H}(\boldsymbol{r},t) = 0, \tag{2.12}$$

$$\boldsymbol{\nabla}\times\boldsymbol{E}(\boldsymbol{r},t) = -\mu_0\frac{\partial}{\partial t}\boldsymbol{H}(\boldsymbol{r},t), \tag{2.13}$$

$$\boldsymbol{\nabla}\times\boldsymbol{H}(\boldsymbol{r},t) = \varepsilon_0\varepsilon(\boldsymbol{r})\frac{\partial}{\partial t}\boldsymbol{E}(\boldsymbol{r},t). \tag{2.14}$$

When we eliminate $\boldsymbol{E}(\boldsymbol{r},t)$ or $\boldsymbol{H}(\boldsymbol{r},t)$ in (2.13) and (2.14), we obtain the following wave equations:

$$\frac{1}{\varepsilon(\boldsymbol{r})}\boldsymbol{\nabla}\times\{\boldsymbol{\nabla}\times\boldsymbol{E}(\boldsymbol{r},t)\} = -\frac{1}{c^2}\frac{\partial^2}{\partial t^2}\boldsymbol{E}(\boldsymbol{r},t), \tag{2.15}$$

$$\boldsymbol{\nabla}\times\left\{\frac{1}{\varepsilon(\boldsymbol{r})}\boldsymbol{\nabla}\times\boldsymbol{H}(\boldsymbol{r},t)\right\} = -\frac{1}{c^2}\frac{\partial^2}{\partial t^2}\boldsymbol{H}(\boldsymbol{r},t), \tag{2.16}$$

where c stands for the light velocity in free space:

$$c = \frac{1}{\sqrt{\varepsilon_0 \mu_0}}. \tag{2.17}$$

Now we seek the solutions of (2.15) and (2.16) of the form

$$\boldsymbol{E}(\boldsymbol{r}, t) = \boldsymbol{E}(\boldsymbol{r})\mathrm{e}^{-\mathrm{i}\omega t}, \tag{2.18}$$

$$\boldsymbol{H}(\boldsymbol{r}, t) = \boldsymbol{H}(\boldsymbol{r})\mathrm{e}^{-\mathrm{i}\omega t}, \tag{2.19}$$

where ω is the eigen-angular frequency, and $\boldsymbol{E}(\boldsymbol{r})$ and $\boldsymbol{H}(\boldsymbol{r})$ are the eigenfunctions of the wave equations. These eigenfunctions should thus satisfy the next eigenvalue equations.

$$\mathcal{L}_E \boldsymbol{E}(\boldsymbol{r}) \equiv \frac{1}{\varepsilon(\boldsymbol{r})} \boldsymbol{\nabla} \times \{\boldsymbol{\nabla} \times \boldsymbol{E}(\boldsymbol{r})\} = \frac{\omega^2}{c^2} \boldsymbol{E}(\boldsymbol{r}), \tag{2.20}$$

$$\mathcal{L}_H \boldsymbol{H}(\boldsymbol{r}) \equiv \boldsymbol{\nabla} \times \left\{ \frac{1}{\varepsilon(\boldsymbol{r})} \boldsymbol{\nabla} \times \boldsymbol{H}(\boldsymbol{r}) \right\} = \frac{\omega^2}{c^2} \boldsymbol{H}(\boldsymbol{r}), \tag{2.21}$$

where the two differential operators $\mathcal{L}_E$ and $\mathcal{L}_H$ are defined by the first equality in each of the above equations.

Because ε is a periodic function of the spatial coordinate $\boldsymbol{r}$, we can apply Bloch's theorem to (2.20) and (2.21) as in the case of the electronic wave equation in ordinary crystals with a periodic potential due to the regular array of atoms.[1] $\boldsymbol{E}(\boldsymbol{r})$ and $\boldsymbol{H}(\boldsymbol{r})$ are thus characterized by a wave vector $\boldsymbol{k}$ in the first Brillouin zone and a band index n and expressed as

$$\boldsymbol{E}(\boldsymbol{r}) = \boldsymbol{E}_{\boldsymbol{k}n}(\boldsymbol{r}) = \boldsymbol{u}_{\boldsymbol{k}n}(\boldsymbol{r})\mathrm{e}^{\mathrm{i}\boldsymbol{k}\cdot\boldsymbol{r}}, \tag{2.22}$$

$$\boldsymbol{H}(\boldsymbol{r}) = \boldsymbol{H}_{\boldsymbol{k}n}(\boldsymbol{r}) = \boldsymbol{v}_{\boldsymbol{k}n}(\boldsymbol{r})\mathrm{e}^{\mathrm{i}\boldsymbol{k}\cdot\boldsymbol{r}}, \tag{2.23}$$

where $\boldsymbol{u}_{\boldsymbol{k}n}(\boldsymbol{r})$ and $\boldsymbol{v}_{\boldsymbol{k}n}(\boldsymbol{r})$ are periodic vectorial functions that satisfy the following relations:

$$\boldsymbol{u}_{\boldsymbol{k}n}(\boldsymbol{r} + \boldsymbol{a}_i) = \boldsymbol{u}_{\boldsymbol{k}n}(\boldsymbol{r}), \tag{2.24}$$

$$\boldsymbol{v}_{\boldsymbol{k}n}(\boldsymbol{r} + \boldsymbol{a}_i) = \boldsymbol{v}_{\boldsymbol{k}n}(\boldsymbol{r}), \qquad \text{for } i = 1, 2, 3. \tag{2.25}$$

Because of the spatial periodicity of these functions, they can be expanded in Fourier series like $\varepsilon^{-1}(\boldsymbol{r})$ in (2.10). This Fourier expansion leads to the following form of the eigenfunctions.

$$\boldsymbol{E}_{\boldsymbol{k}n}(\boldsymbol{r}) = \sum_{\boldsymbol{G}} \boldsymbol{E}_{\boldsymbol{k}n}(\boldsymbol{G}) \exp\{\mathrm{i}(\boldsymbol{k} + \boldsymbol{G}) \cdot \boldsymbol{r}\}, \tag{2.26}$$

$$\boldsymbol{H}_{\boldsymbol{k}n}(\boldsymbol{r}) = \sum_{\boldsymbol{G}} \boldsymbol{H}_{\boldsymbol{k}n}(\boldsymbol{G}) \exp\{\mathrm{i}(\boldsymbol{k} + \boldsymbol{G}) \cdot \boldsymbol{r}\}. \tag{2.27}$$

[1] The proof will be given at the end of this section.

The expansion coefficients in reciprocal lattice space, i.e., $\boldsymbol{E}_{\boldsymbol{k}n}(\boldsymbol{G})$ and $\boldsymbol{H}_{\boldsymbol{k}n}(\boldsymbol{G})$ are denoted by the same symbols as the original ones in real space. Substituting (2.10), (2.26), and (2.27) into (2.20) and (2.21), we obtain the following eigenvalue equations for the expansion coefficients $\{\boldsymbol{E}_{\boldsymbol{k}n}(\boldsymbol{G})\}$ and $\{\boldsymbol{H}_{\boldsymbol{k}n}(\boldsymbol{G})\}$:

$$-\sum_{\boldsymbol{G}'} \kappa\left(\boldsymbol{G}-\boldsymbol{G}'\right)\left(\boldsymbol{k}+\boldsymbol{G}'\right) \times \left\{\left(\boldsymbol{k}+\boldsymbol{G}'\right) \times \boldsymbol{E}_{\boldsymbol{k}n}(\boldsymbol{G}')\right\} = \frac{\omega_{\boldsymbol{k}n}^2}{c^2} \boldsymbol{E}_{\boldsymbol{k}n}(\boldsymbol{G}), \tag{2.28}$$

$$-\sum_{\boldsymbol{G}'} \kappa\left(\boldsymbol{G}-\boldsymbol{G}'\right)\left(\boldsymbol{k}+\boldsymbol{G}\right) \times \left\{\left(\boldsymbol{k}+\boldsymbol{G}'\right) \times \boldsymbol{H}_{\boldsymbol{k}n}(\boldsymbol{G}')\right\} = \frac{\omega_{\boldsymbol{k}n}^2}{c^2} \boldsymbol{H}_{\boldsymbol{k}n}(\boldsymbol{G}), \tag{2.29}$$

where $\omega_{\boldsymbol{k}n}$ denotes the eigen-angular frequency of $\boldsymbol{E}_{\boldsymbol{k}n}(\boldsymbol{r})$ and $\boldsymbol{H}_{\boldsymbol{k}n}(\boldsymbol{r})$. By solving one of these two sets of equations numerically, we can obtain the dispersion relation of the eigenmodes, or the photonic band structure [21, 22].

This numerical method, which is based on the Fourier expansion of the electromagnetic field and the dielectric function, is called the plane-wave expansion method. In the actual numerical calculation of the photonic bands, the summation in (2.28) or (2.29) is calculated up to a sufficiently large number N of $\boldsymbol{G}'$, and an eigenvalue problem for each $\boldsymbol{k}$ is solved, which is equivalent to the diagonalization of the matrix defined by the left-hand side of (2.28) or (2.29). The dimension of the matrix that should be diagonalized is $3N$ for $\{\boldsymbol{E}_{\boldsymbol{k}n}(\boldsymbol{G}) : \boldsymbol{G}\}$. On the other hand, it is $2N$ for $\{\boldsymbol{H}_{\boldsymbol{k}n}(\boldsymbol{G}) : \boldsymbol{G}\}$, since $\boldsymbol{H}_{\boldsymbol{k}n}(\boldsymbol{G})$ should be perpendicular to $\boldsymbol{k}+\boldsymbol{G}$ according to (2.12) and (2.27), and its degree of freedom is two. The CPU time that is necessary for the diagnalization is usually proportional to the cube of its dimension. Hence, the CPU time for the photonic band calculation by means of the plane-wave expansion method is proportional to N^3. This fact sometimes leads to a serious constraint of the accuracy of the calculation. In fact, the convergence of the plane-wave expansion method is not good when the amplitude of the spatial variation of the dielectric constant is large, and numerical error exceeds 5% in certain cases even though we take N greater than 3000. The numerical error also depends on the dimensionality of the crystal. The convergence of the plane-wave expansion method has been discussed by several authors. Readers who are interested in this problem should consult [23–26].

The poor convergence of the plane-wave expansion method has been improved by several means. We may use spherical waves instead of plane waves as a basis set if the photonic crystal is composed of dielectric spheres or circular cylinders. This method is called the spherical-wave expansion method or the vector KKR (Koringa–Kohn–Rostker) method. Although it can treat only spherical and cylindrical dielectric structures, the convergence of the numerical calculation is usually much better than the plane-wave expansion method. The reader may consult [27–30].

We should note that $\boldsymbol{E}_{\boldsymbol{k}n}(\boldsymbol{r})$ and $\boldsymbol{H}_{\boldsymbol{k}n}(\boldsymbol{r})$ have, of course, the same eigenfrequency, since they are related to each other by (2.13) and (2.14). There is a difference in the number of the eigenvalues between $\boldsymbol{E}_{\boldsymbol{k}n}(\boldsymbol{r})$ and $\boldsymbol{H}_{\boldsymbol{k}n}(\boldsymbol{r})$ since the dimension of the matrix that should be diagonalized is $3N$ for the former and $2N$ for the latter. Actually N of $3N$ eigenvalues of $\boldsymbol{E}_{\boldsymbol{k}n}(\boldsymbol{r})$ are equal to zero. This point will be discussed again in Sect. 2.7 in connection with the quasi-longitudinal modes of the radiation field.

Because $\boldsymbol{H}_{\boldsymbol{k}n}(\boldsymbol{G})$ is perpendicular to $\boldsymbol{k}+\boldsymbol{G}$, it can be expressed by a linear combination of two orthogonal normal vectors, $\boldsymbol{e}_{\boldsymbol{G}1}$ and $\boldsymbol{e}_{\boldsymbol{G}2}$:

$$\boldsymbol{H}_{\boldsymbol{k}n}(\boldsymbol{G}) = h_{\boldsymbol{k}n}^{\boldsymbol{G}1}\boldsymbol{e}_{\boldsymbol{G}1} + h_{\boldsymbol{k}n}^{\boldsymbol{G}2}\boldsymbol{e}_{\boldsymbol{G}2}. \tag{2.30}$$

We assume without the loss of generality that

$$\left\{\boldsymbol{e}_{\boldsymbol{G}1},\ \boldsymbol{e}_{\boldsymbol{G}2},\ \frac{\boldsymbol{k}+\boldsymbol{G}}{|\boldsymbol{k}+\boldsymbol{G}|}\right\} \tag{2.31}$$

constitute a right-hand system. It is easy to derive the following equation from (2.29).

$$\sum_{\boldsymbol{G}'}\sum_{j=1}^{2} M_{\boldsymbol{k}}^{ij}(\boldsymbol{G},\boldsymbol{G}')h_{\boldsymbol{k}n}^{\boldsymbol{G}'j} = \frac{\omega_{\boldsymbol{k}n}^2}{c^2}h_{\boldsymbol{k}n}^{\boldsymbol{G}i}, \tag{2.32}$$

where $\mathrm{M}_{\boldsymbol{k}}(\boldsymbol{G},\boldsymbol{G}')$ is given by

$$\mathrm{M}_{\boldsymbol{k}}(\boldsymbol{G},\boldsymbol{G}') = |\boldsymbol{k}+\boldsymbol{G}||\boldsymbol{k}+\boldsymbol{G}'|\kappa(\boldsymbol{G}-\boldsymbol{G}') \times \begin{pmatrix} \boldsymbol{e}_{\boldsymbol{G}2}\cdot\boldsymbol{e}_{\boldsymbol{G}'2}, & -\boldsymbol{e}_{\boldsymbol{G}2}\cdot\boldsymbol{e}_{\boldsymbol{G}'1} \\ -\boldsymbol{e}_{\boldsymbol{G}1}\cdot\boldsymbol{e}_{\boldsymbol{G}'2}, & \boldsymbol{e}_{\boldsymbol{G}1}\cdot\boldsymbol{e}_{\boldsymbol{G}'1} \end{pmatrix}, \tag{2.33}$$

which is an Hermitian matrix:

$$M_{\boldsymbol{k}}^{ij}(\boldsymbol{G},\boldsymbol{G}') = M_{\boldsymbol{k}}^{ji*}(\boldsymbol{G}',\boldsymbol{G}). \tag{2.34}$$

The eigenvectors $\{\boldsymbol{h}_{\boldsymbol{k}n}\}$ are thus orthogonal to each other:

$$\sum_{\boldsymbol{G}}\sum_{i=1}^{2} h_{\boldsymbol{k}n}^{\boldsymbol{G}i*}\cdot h_{\boldsymbol{k}n'}^{\boldsymbol{G}i} = \delta_{nn'}. \tag{2.35}$$

From this equation, we obtain

$$\begin{aligned}
&\int_V \mathrm{d}\boldsymbol{r}\boldsymbol{H}_{\boldsymbol{k}n}^*(\boldsymbol{r})\cdot\boldsymbol{H}_{\boldsymbol{k}n'}(\boldsymbol{r}) \\
&= \int_V \mathrm{d}\boldsymbol{r}\sum_{\boldsymbol{G}}\sum_{\boldsymbol{G}'}\boldsymbol{H}_{\boldsymbol{k}n}^*(\boldsymbol{G})\cdot\boldsymbol{H}_{\boldsymbol{k}n'}(\boldsymbol{G}')\mathrm{e}^{\mathrm{i}(\boldsymbol{G}'-\boldsymbol{G})\cdot\boldsymbol{r}} \\
&= V\sum_{\boldsymbol{G}}\boldsymbol{H}_{\boldsymbol{k}n}^*(\boldsymbol{G})\cdot\boldsymbol{H}_{\boldsymbol{k}n'}(\boldsymbol{G}) \\
&= V\delta_{nn'},
\end{aligned} \tag{2.36}$$

where V denotes the volume of the photonic crystal. On the other hand, if $\boldsymbol{k}\neq\boldsymbol{k}'$,

$$\int_V \mathrm{d}\boldsymbol{r}\boldsymbol{H}^*_{\boldsymbol{k}n}(\boldsymbol{r})\cdot\boldsymbol{H}^*_{\boldsymbol{k}'n'}(\boldsymbol{r}) = \int_V \mathrm{d}\boldsymbol{r}\boldsymbol{v}^*_{\boldsymbol{k}n}(\boldsymbol{r})\cdot\boldsymbol{v}_{\boldsymbol{k}'n'}(\boldsymbol{r})\mathrm{e}^{\mathrm{i}(\boldsymbol{k}'-\boldsymbol{k})\cdot\boldsymbol{r}}. \tag{2.37}$$

Because both $\boldsymbol{k}$ and $\boldsymbol{k}'$ are wave vectors in the first Brillouin zone, $\boldsymbol{k}' - \boldsymbol{k}$ cannot coincide with a reciprocal lattice vector. The periodicity of the function $\boldsymbol{v}^*_{\boldsymbol{k}n}(\boldsymbol{r})\cdot\boldsymbol{v}_{\boldsymbol{k}'n'}(\boldsymbol{r})$ thus leads to the above integral vanishing. Hence, we finally obtain

$$\int_V \mathrm{d}\boldsymbol{r}\boldsymbol{H}^*_{\boldsymbol{k}n}(\boldsymbol{r})\cdot\boldsymbol{H}_{\boldsymbol{k}'n'}(\boldsymbol{r}) = V\delta_{\boldsymbol{k}\boldsymbol{k}'}\delta_{nn'}. \tag{2.38}$$

This orthogonality is a direct consequence of the fact that the matrix $\mathrm{M}_{\boldsymbol{k}}$ defined by (2.33) is Hermitian. The latter is, on the other hand, a consequence of the fact that $\mathcal{L}_H$ defined by (2.25) is an Hermitian operator. As for $\mathcal{L}_E$, it is not Hermitian, and so its eigenfunctions are not necessarily orthogonal to each other. This point will be described in more detail in Sect. 2.7.

Proof of Bloch's Theorem

Bloch's theorem holds for the eigenmodes of the regular photonic crystals; here we will prove (2.22). Equation (2.23) can be proved in a similar manner. First, we express the eigenfunction of the electric field by a Fourier integral:

$$\boldsymbol{E}(\boldsymbol{r}) = \int \mathrm{d}\boldsymbol{k}\boldsymbol{A}(\boldsymbol{k})\mathrm{e}^{\mathrm{i}\boldsymbol{k}\cdot\boldsymbol{r}}. \tag{2.39}$$

We write (2.20) in a slightly different way:

$$\boldsymbol{\nabla}\times\{\boldsymbol{\nabla}\times\boldsymbol{E}(\boldsymbol{r})\} = \frac{\omega^2}{c^2}\varepsilon(\boldsymbol{r})\boldsymbol{E}(\boldsymbol{r}). \tag{2.40}$$

Then we expand the periodic dielectric function in a Fourier series:

$$\varepsilon(\boldsymbol{r}) = \sum_{\boldsymbol{G}}\varepsilon(\boldsymbol{G})\mathrm{e}^{\mathrm{i}\boldsymbol{G}\cdot\boldsymbol{r}}. \tag{2.41}$$

Here we used the same notation for the expansion coefficients as the original dielectric function. When we substitute (2.39) and (2.41) into (2.40), we obtain

$$\int \mathrm{d}\boldsymbol{k}\ \boldsymbol{k}\times\{\boldsymbol{k}\times\boldsymbol{A}(\boldsymbol{k})\}\mathrm{e}^{\mathrm{i}\boldsymbol{k}\cdot\boldsymbol{r}} + \frac{\omega^2}{c^2}\sum_{\boldsymbol{G}}\int \mathrm{d}\boldsymbol{k}\varepsilon(\boldsymbol{G})\boldsymbol{A}(\boldsymbol{k}-\boldsymbol{G})\mathrm{e}^{\mathrm{i}\boldsymbol{k}\cdot\boldsymbol{r}} = 0. \tag{2.42}$$

Since this equation holds for all $\boldsymbol{r}$, the integrand should vanish:

$$\boldsymbol{k}\times\{\boldsymbol{k}\times\boldsymbol{A}(\boldsymbol{k})\} + \frac{\omega^2}{c^2}\sum_{\boldsymbol{G}}\varepsilon(\boldsymbol{G})\boldsymbol{A}(\boldsymbol{k}-\boldsymbol{G}) = 0. \tag{2.43}$$

This equation implies that only those Fourier components that are related by the reciprocal lattice vectors constitute the eigenvalue problem, that is, a set of linear eigenvalue equations. Hence, only those Fourier components are necessary to express the eigenfunction in (2.39):

$$\boldsymbol{E}_{\boldsymbol{k}}(\boldsymbol{r}) = \sum_{\boldsymbol{G}} \boldsymbol{A}(\boldsymbol{k}-\boldsymbol{G})\mathrm{e}^{\mathrm{i}(\boldsymbol{k}-\boldsymbol{G})\cdot\boldsymbol{r}}. \tag{2.44}$$

When we define $\boldsymbol{u}_{\boldsymbol{k}}(\boldsymbol{r})$ as

$$\boldsymbol{u}_{\boldsymbol{k}}(\boldsymbol{r}) = \sum_{\boldsymbol{G}} \boldsymbol{A}(\boldsymbol{k}-\boldsymbol{G})\mathrm{e}^{-\mathrm{i}\boldsymbol{G}\cdot\boldsymbol{r}}, \tag{2.45}$$

it is periodic, i.e., it satisfies (2.24). $\boldsymbol{E}_{\boldsymbol{k}}(\boldsymbol{r})$ is given by

$$\boldsymbol{E}_{\boldsymbol{k}}(\boldsymbol{r}) = \boldsymbol{u}_{\boldsymbol{k}}(\boldsymbol{r})\mathrm{e}^{\mathrm{i}\boldsymbol{k}\cdot\boldsymbol{r}}. \tag{2.46}$$

Since (2.43) generally has an infinite number of eigenvalues and eigenfunctions, we distinguish them by a subscript n. Hence, we obtain (2.22). ■

2.2 Eigenvalue Problems in Two-Dimensional Crystals

For two-dimensional (2D) crystals, the eigenvalue equations are much simplified if the $\boldsymbol{k}$ vector is parallel to the 2D plane. We examine this case here. In the 2D crystal, the dielectric structure is uniform in the z direction (see Fig. 1.2). The electromagnetic waves travel in the x-y plane and are also uniform in the z direction. Hence, $\varepsilon(\boldsymbol{r})$, $\boldsymbol{E}(\boldsymbol{r})$, and $\boldsymbol{H}(\boldsymbol{r})$ are independent of the z coordinate in (2.13) and (2.14). In this case, these vectorial equations are decoupled to two independent sets of equations. The first is

$$\frac{\partial}{\partial y}E_z(\boldsymbol{r}_{/\!/},t) = -\mu_0\frac{\partial}{\partial t}H_x(\boldsymbol{r}_{/\!/},t), \tag{2.47}$$

$$\frac{\partial}{\partial x}E_z(\boldsymbol{r}_{/\!/},t) = \mu_0\frac{\partial}{\partial t}H_y(\boldsymbol{r}_{/\!/},t), \tag{2.48}$$

$$\frac{\partial}{\partial x}H_y(\boldsymbol{r}_{/\!/},t) - \frac{\partial}{\partial y}H_x(\boldsymbol{r}_{/\!/},t) = \varepsilon_0\varepsilon(\boldsymbol{r}_{/\!/})\frac{\partial}{\partial t}E_z(\boldsymbol{r}_{/\!/},t), \tag{2.49}$$

and the second is

$$\frac{\partial}{\partial y}H_z(\boldsymbol{r}_{/\!/},t) = \varepsilon_0\varepsilon(\boldsymbol{r}_{/\!/})\frac{\partial}{\partial t}E_x(\boldsymbol{r}_{/\!/},t), \tag{2.50}$$

$$\frac{\partial}{\partial x}H_z(\boldsymbol{r}_{/\!/},t) = -\varepsilon_0\varepsilon(\boldsymbol{r}_{/\!/})\frac{\partial}{\partial t}E_y(\boldsymbol{r}_{/\!/},t), \tag{2.51}$$

$$\frac{\partial}{\partial x}E_y(\boldsymbol{r}_{/\!/},t) - \frac{\partial}{\partial y}E_x(\boldsymbol{r}_{/\!/},t) = -\mu_0\frac{\partial}{\partial t}H_z(\boldsymbol{r}_{/\!/},t). \tag{2.52}$$

Here, $\boldsymbol{r}_{/\!/}$ denotes the 2D position vector (x,y). From the first set of these equations, we obtain the following wave equation by eliminating $H_x(\boldsymbol{r}_{/\!/},t)$ and $H_y(\boldsymbol{r}_{/\!/},t)$.

$$\frac{1}{\varepsilon(\boldsymbol{r}_{/\!/})}\left\{\frac{\partial^2}{\partial x^2}+\frac{\partial^2}{\partial y^2}\right\}E_z(\boldsymbol{r}_{/\!/},t)=\frac{1}{c^2}\frac{\partial^2}{\partial t^2}E_z(\boldsymbol{r}_{/\!/},t). \tag{2.53}$$

From the second set, we obtain the wave equation for $H_z(\boldsymbol{r}_{/\!/},t)$:

$$\left\{\frac{\partial}{\partial x}\frac{1}{\varepsilon(\boldsymbol{r}_{/\!/})}\frac{\partial}{\partial x}+\frac{\partial}{\partial y}\frac{1}{\varepsilon(\boldsymbol{r}_{/\!/})}\frac{\partial}{\partial y}\right\}H_z(\boldsymbol{r}_{/\!/},t)=\frac{1}{c^2}\frac{\partial^2}{\partial t^2}H_z(\boldsymbol{r}_{/\!/},t). \tag{2.54}$$

We seek, as before, the solutions of these equations of the form

$$E_z(\boldsymbol{r}_{/\!/},t)=E_z(\boldsymbol{r}_{/\!/})\mathrm{e}^{-\mathrm{i}\omega t}, \tag{2.55}$$

$$H_z(\boldsymbol{r}_{/\!/},t)=H_z(\boldsymbol{r}_{/\!/})\mathrm{e}^{-\mathrm{i}\omega t}. \tag{2.56}$$

The eigenvalue equations are thus given by

$$\mathcal{L}_E^{(2)}E_z(\boldsymbol{r}_{/\!/})\equiv-\frac{1}{\varepsilon(\boldsymbol{r}_{/\!/})}\left\{\frac{\partial^2}{\partial x^2}+\frac{\partial^2}{\partial y^2}\right\}E_z(\boldsymbol{r}_{/\!/})=\frac{\omega^2}{c^2}E_z(\boldsymbol{r}_{/\!/}), \tag{2.57}$$

$$\mathcal{L}_H^{(2)}H_z(\boldsymbol{r}_{/\!/})\equiv-\left\{\frac{\partial}{\partial x}\frac{1}{\varepsilon(\boldsymbol{r}_{/\!/})}\frac{\partial}{\partial x}+\frac{\partial}{\partial y}\frac{1}{\varepsilon(\boldsymbol{r}_{/\!/})}\frac{\partial}{\partial y}\right\}H_z(\boldsymbol{r}_{/\!/})=\frac{\omega^2}{c^2}H_z(\boldsymbol{r}_{/\!/}), \tag{2.58}$$

where the two differential operators $\mathcal{L}_E^{(2)}$ and $\mathcal{L}_H^{(2)}$ for the 2D case are defined by the first equality in each of the above two equations. These two kinds of eigenfunctions represent two independent polarizations; one is called the E polarization for which the electric field is parallel to the z axis, and the other is called the H polarization for which the magnetic field is parallel to the z axis.

When we apply Bloch's theorem as before, we can express $E_z(\boldsymbol{r}_{/\!/})$ and $H_z(\boldsymbol{r}_{/\!/})$ as

$$E_z(\boldsymbol{r}_{/\!/})=E_{z,\boldsymbol{k}_{/\!/}n}(\boldsymbol{r}_{/\!/})=\sum_{\boldsymbol{G}_{/\!/}}E_{z,\boldsymbol{k}_{/\!/}n}(\boldsymbol{G}_{/\!/})\exp\{\mathrm{i}(\boldsymbol{k}_{/\!/}+\boldsymbol{G}_{/\!/})\cdot\boldsymbol{r}_{/\!/}\}, \tag{2.59}$$

$$H_z(\boldsymbol{r}_{/\!/})=H_{z,\boldsymbol{k}_{/\!/}n}(\boldsymbol{r}_{/\!/})=\sum_{\boldsymbol{G}_{/\!/}}H_{z,\boldsymbol{k}_{/\!/}n}(\boldsymbol{G}_{/\!/})\exp\{\mathrm{i}(\boldsymbol{k}_{/\!/}+\boldsymbol{G}_{/\!/})\cdot\boldsymbol{r}_{/\!/}\}, \tag{2.60}$$

where $\boldsymbol{k}_{/\!/}$ and $\boldsymbol{G}_{/\!/}$ are the wave vector and the reciprocal lattice vector in two dimensions. Substituting (2.59) and (2.60) into (2.57) and (2.58), we obtain the following eigenvalue equations for the expansion coefficients:

$$\sum_{\boldsymbol{G}'_{/\!/}}\kappa\left(\boldsymbol{G}_{/\!/}-\boldsymbol{G}'_{/\!/}\right)\left|\boldsymbol{k}_{/\!/}+\boldsymbol{G}'_{/\!/}\right|^2E_{z,\boldsymbol{k}_{/\!/}n}(\boldsymbol{G}'_{/\!/})=\frac{\omega^{(E)2}_{\boldsymbol{k}_{/\!/}n}}{c^2}E_{z,\boldsymbol{k}_{/\!/}n}(\boldsymbol{G}_{/\!/}), \tag{2.61}$$

$$\sum_{G'_{/\!/}} \kappa\left(G_{/\!/} - G'_{/\!/}\right)\left(k_{/\!/} + G_{/\!/}\right) \cdot \left(k_{/\!/} + G'_{/\!/}\right) H_{z,k_{/\!/}n}(G'_{/\!/})$$
$$= \frac{\omega^{(H)2}_{k_{/\!/}n}}{c^2} H_{z,k_{/\!/}n}(G_{/\!/}), \tag{2.62}$$

where $\omega^{(E)}_{k_{/\!/}n}$ and $\omega^{(H)}_{k_{/\!/}n}$ denote the eigen-angular frequencies of $E_{z,k_{/\!/}n}(r_{/\!/})$ and $H_{z,k_{/\!/}n}(r_{/\!/})$, respectively. When we define a matrix $\mathrm{M}_{k_{/\!/}}$ by

$$M_{k_{/\!/}}(G_{/\!/},\ G'_{/\!/}) = \kappa(G_{/\!/} - G'_{/\!/})(k_{/\!/} + G_{/\!/}) \cdot (k_{/\!/} + G'_{/\!/}), \tag{2.63}$$

it is Hermitian:

$$M_{k_{/\!/}}(G_{/\!/},\ G'_{/\!/}) = M^*_{k_{/\!/}}(G'_{/\!/},\ G_{/\!/}). \tag{2.64}$$

The eigenvalue equation, (2.62), is thus expressed as

$$\sum_{G'_{/\!/}} M_{k_{/\!/}}(G_{/\!/},\ G'_{/\!/}) H_{z,k_{/\!/}n}(G'_{/\!/}) = \frac{\omega^{(H)2}_{k_{/\!/}n}}{c^2} H_{z,k_{/\!/}n}(G_{/\!/}). \tag{2.65}$$

As we derived the orthogonality of $\boldsymbol{H}_{kn}(\boldsymbol{r})$ in (2.38), we can prove that

$$\int_{V^{(2)}} \mathrm{d}r_{/\!/} H^*_{z,k_{/\!/}n}(r_{/\!/}) H_{z,k'_{/\!/}n'}(r_{/\!/}) = V^{(2)} \delta_{k_{/\!/}k'_{/\!/}} \delta_{nn'}, \tag{2.66}$$

where $V^{(2)}$ denotes the 2D volume of the photonic crystal. This orthogonality relation is a consequence of the fact that $\mathcal{L}^{(2)}_H$ is an Hermitian operator. On the other hand, $\mathcal{L}^{(2)}_E$ is not Hermitian, and so its eigenfunctions are not necessarily orthogonal to each other. This point will also be described in more detail in Sect. 2.7.

For the detailed numerical method, see [31] and [32]. Also, see [33] for the off-plane dispersion of the 2D crystals.

2.3 Scaling Law and Time Reversal Symmetry

There are two useful properties of the photonic bands. One property is the scaling law and the other is the time reversal symmetry of the wave equation. The scaling law tells us that two photonic crystals which are similar to each other essentially have the same photonic band structure, that is, the difference between the two band structures is simply the scales of frequency and the wave vector. On the other hand, the time reversal symmetry tells us that any photonic band structure has inversion symmetry even though the crystal structure does not have inversion symmetry. The proof of these properties is given in the following.

First, the following scale transformation to (2.15) is performed:

$$\frac{1}{a}\boldsymbol{r} = \boldsymbol{r}' \qquad \text{and} \qquad \frac{c}{a}t = t'. \tag{2.67}$$

The new variables $\boldsymbol{r}'$ and t' are dimensionless. When we define a new dielectric function $\varepsilon_{\mathrm{sc}}$ and a new vector field $\boldsymbol{E}_{\mathrm{sc}}$ by

$$\varepsilon_{\mathrm{sc}}(\boldsymbol{r}') = \varepsilon(\boldsymbol{r}), \qquad \text{and} \tag{2.68}$$

$$\boldsymbol{E}_{\mathrm{sc}}(\boldsymbol{r}', t') = \boldsymbol{E}(\boldsymbol{r}, t), \tag{2.69}$$

$\boldsymbol{E}_{\mathrm{sc}}$ satisfies the following wave equation.

$$\frac{1}{\varepsilon_{\mathrm{sc}}(\boldsymbol{r}')}\boldsymbol{\nabla}' \times \{\boldsymbol{\nabla}' \times \boldsymbol{E}_{\mathrm{sc}}(\boldsymbol{r}', t')\} = -\frac{\partial^2}{\partial t'^2}\boldsymbol{E}_{\mathrm{sc}}(\boldsymbol{r}', t'), \tag{2.70}$$

where $\boldsymbol{\nabla}'$ stands for the differentiation with respect to $\boldsymbol{r}'$. Hence, if the structures of two photonic crystals are similar to each other and their difference is simply the scale of the length, i.e., the lattice constant, then their wave equations are attributed to the same dimensionless wave equation by the scale transformation.

Now, we denote the dimensionless wave vector and the dimensionless eigen-angular frequency in the $(\boldsymbol{r}', t')$ space by $\boldsymbol{k}'$ and $\omega'_{\boldsymbol{k}'n}$, respectively. Because $\boldsymbol{k}$ and ω have the inverse dimensions of $\boldsymbol{r}$ and t, respectively, the following transformations are necesary to return to real space:

$$\boldsymbol{k}' = \frac{a}{2\pi}\boldsymbol{k} \qquad \text{and} \qquad \omega' = \frac{a}{2\pi c}\omega, \tag{2.71}$$

where we included the factor $1/2\pi$ according to the convention in this field. Therefore, if we measure the wave vector in units of $2\pi/a$ and the angular frequency in units of $2\pi c/a$, all dispersion curves are the same for those crystals which have similar dielectric functions. Thanks to this scaling law, we can conduct simulation experiments for crystals with a lattice constant of about 1 mm, for example, using specimens with a lattice constant of 1 cm. Since the fabrication of specimens with smal lattice constants is usually a difficult task, the scaling law is very usuful to accelerate experimental studies and to confirm theoretical predictions.

As for the time reversal symmetry, we should note that the wave equation, (2.15), is invariant when we change the sign of the time variable. When we define a new variable t' and a new vector field $\boldsymbol{E}_{\mathrm{tr}}$ by

$$t' = -t \qquad \text{and} \tag{2.72}$$

$$\boldsymbol{E}_{\mathrm{tr}}(\boldsymbol{r}, t') = \boldsymbol{E}(\boldsymbol{r}, -t), \tag{2.73}$$

$\boldsymbol{E}_{\mathrm{tr}}$ satisfies the same wave equation as $\boldsymbol{E}$ does:

$$\frac{1}{\varepsilon(\boldsymbol{r})}\boldsymbol{\nabla} \times \{\boldsymbol{\nabla} \times \boldsymbol{E}_{\mathrm{tr}}(\boldsymbol{r}, t')\} = -\frac{1}{c^2}\frac{\partial^2}{\partial t'^2}\boldsymbol{E}_{\mathrm{tr}}(\boldsymbol{r}, t'). \tag{2.74}$$

The eigenfrequency and the eigenfunction are thus given by

$$\omega = \omega_{\boldsymbol{k}n} \qquad \text{and} \tag{2.75}$$

$$\boldsymbol{E}_{tr}(\boldsymbol{r},t') = \boldsymbol{u}_{\boldsymbol{k}n}(\boldsymbol{r}) \exp\{\mathrm{i}\,(\boldsymbol{k}\cdot\boldsymbol{r} - \omega_{\boldsymbol{k}n}t')\}\,, \tag{2.76}$$

where we included the time-dependent part. When we return to the original $(\boldsymbol{r},t)$ space, we have

$$\boldsymbol{E}(\boldsymbol{r},t) = [\boldsymbol{u}^*_{\boldsymbol{k}n}(\boldsymbol{r}) \exp\{\mathrm{i}\,(-\boldsymbol{k}\cdot\boldsymbol{r} - \omega_{\boldsymbol{k}n}t)\}]^*\,. \tag{2.77}$$

The actual electric field is given by the real part of this equation. When we compare this equation with (2.18) and (2.22), we obtain

$$\omega_{-\boldsymbol{k}n} = \omega_{\boldsymbol{k}n} \qquad \text{and} \tag{2.78}$$

$$\boldsymbol{u}_{-\boldsymbol{k}n}(\boldsymbol{r}) = \boldsymbol{u}^*_{\boldsymbol{k}n}(\boldsymbol{r}). \tag{2.79}$$

Equation (2.78) implies that the dispersion relation has inversion symmetry. This property is irrespective of whether the structure of the photonic crystal has inversion symmetry.

2.4 Photonic Band Calculation

2.4.1 Fourier Expansion of Dielectric Functions

It is necessary to calculate the expansion coefficients $\{\kappa(\boldsymbol{G})\}$ in (2.10) for the band calculation by the plane-wave expansion method. The inverse Fourier transform gives

$$\kappa(\boldsymbol{G}) = \frac{1}{V_0}\int_{V_0} \mathrm{d}\boldsymbol{r}\,\frac{1}{\varepsilon(\boldsymbol{r})}\exp(-\mathrm{i}\boldsymbol{G}\cdot\boldsymbol{r}), \tag{2.80}$$

where V_0 denotes the volume of the unit cell of the photonic crystal. In general, this integral should be evaluated numerically. However, if the shapes of the dielectric components in the unit cell are simple enough, we can calculate it analytically. In what follows, we treat two such cases. One is the three-dimensional (3D) crystal whose unit cell contains one dielectric sphere, and the other is the 2D crystal whose unit cell contains one circular dielectric rod.

Dielectric Sphere

We denote the radius and the dielectric constant of the sphere by r_a and ε_a, respectively, and the dielectric constant of the background material by ε_b. $1/\varepsilon(\boldsymbol{r})$ is thus given by

$$\frac{1}{\varepsilon(\boldsymbol{r})} = \frac{1}{\varepsilon_\mathrm{b}} + \left(\frac{1}{\varepsilon_\mathrm{a}} - \frac{1}{\varepsilon_\mathrm{b}}\right) S(\boldsymbol{r}), \tag{2.81}$$

where $S(\boldsymbol{r})$ is defined such that

$$S(\boldsymbol{r}) = \begin{cases} 1 & \text{for } |\boldsymbol{r}| \le r_\mathrm{a}, \\ 0 & \text{for } |\boldsymbol{r}| > r_\mathrm{a}. \end{cases} \tag{2.82}$$

Substituting (2.81) and (2.82) into (2.80),

$$\kappa(\boldsymbol{G}) = \frac{1}{\varepsilon_{\mathrm{b}}}\delta_{\boldsymbol{G}0} + \frac{1}{V_0}\left(\frac{1}{\varepsilon_{\mathrm{a}}} - \frac{1}{\varepsilon_{\mathrm{b}}}\right)\int_{V_0} \mathrm{d}\boldsymbol{r}\ S(\boldsymbol{r})\exp(-\mathrm{i}\boldsymbol{G}\cdot\boldsymbol{r}). \tag{2.83}$$

In order to calculate the integral in (2.83), we use spherical coordinates (r, θ, φ). We take the direction with $\theta = 0$ as the direction of vector $\boldsymbol{G}$. For $\boldsymbol{G} \neq 0$, the integral is thus modified to

$$\begin{aligned}
&\int_{V_0} \mathrm{d}\boldsymbol{r}\ S(\boldsymbol{r})\exp(-\mathrm{i}\boldsymbol{G}\cdot\boldsymbol{r}) \\
&= 2\pi\int_0^{r_{\mathrm{a}}} \mathrm{d}r \int_0^{\pi} \mathrm{d}\theta\ r^2 \sin\theta \exp(-\mathrm{i}Gr\cos\theta) \\
&= \frac{4\pi}{G^3}\left(\sin Gr_{\mathrm{a}} - Gr_{\mathrm{a}}\cos Gr_{\mathrm{a}}\right),
\end{aligned} \tag{2.84}$$

where $G = |\boldsymbol{G}|$. For $\boldsymbol{G} = 0$,

$$\int_{V_0} \mathrm{d}\boldsymbol{r}\ S(\boldsymbol{r})\exp(-\mathrm{i}\boldsymbol{G}\cdot\boldsymbol{r}) = \frac{4\pi r_{\mathrm{a}}^3}{3}. \tag{2.85}$$

If we denote the volume fraction of the sphere by f, i.e., if we take

$$f = \frac{4\pi r_{\mathrm{a}}^3}{3V_0}, \tag{2.86}$$

we obtain

$$\kappa(0) = \frac{f}{\varepsilon_{\mathrm{a}}} + \frac{f-1}{\varepsilon_{\mathrm{b}}}. \tag{2.87}$$

For $\boldsymbol{G} \neq 0$, we obtain

$$\kappa(\boldsymbol{G}) = 3f\left(\frac{1}{\varepsilon_{\mathrm{a}}} - \frac{1}{\varepsilon_{\mathrm{b}}}\right)\left\{\frac{\sin Gr_{\mathrm{a}}}{(Gr_{\mathrm{a}})^3} - \frac{\cos Gr_{\mathrm{a}}}{(Gr_{\mathrm{a}})^2}\right\}. \tag{2.88}$$

Circular Dielectric Rod

Because the structure is uniform in the z direction for this case, the integral in (2.80) is equal to zero if $G_z \neq 0$. So, we restrict our discussion to 2D vectors $\{\boldsymbol{G}_{/\!/}\}$. If we denote the 2D unit cell by $V_0^{(2)}$ as before,

$$\kappa(\boldsymbol{G}_{/\!/}) = \frac{1}{V_0^{(2)}}\int_{V_0^{(2)}} \mathrm{d}\boldsymbol{r}_{/\!/}\frac{1}{\varepsilon(\boldsymbol{r}_{/\!/})}\exp(-\mathrm{i}\boldsymbol{G}_{/\!/}\cdot\boldsymbol{r}_{/\!/}). \tag{2.89}$$

If we denote the radius and the dielectric constant of the circular rod by r_{a} and ε_{a}, respectively, and the dielectric constant of the background material by ε_{b}, $1/\varepsilon(\boldsymbol{r}_{/\!/})$ is given by

$$\frac{1}{\varepsilon(\boldsymbol{r}_{/\!/})} = \frac{1}{\varepsilon_{\mathrm{b}}} + \left(\frac{1}{\varepsilon_{\mathrm{a}}} - \frac{1}{\varepsilon_{\mathrm{b}}}\right)S^{(2)}(\boldsymbol{r}_{/\!/}), \tag{2.90}$$

where $S^{(2)}(\boldsymbol{r}_{/\!/})$ is defined such that

$$S^{(2)}(\boldsymbol{r}_{/\!/}) = \begin{cases} 1 & \text{for } \left|\boldsymbol{r}_{/\!/}\right| \leq r_{\mathrm{a}}, \\ 0 & \text{for } \left|\boldsymbol{r}_{/\!/}\right| > r_{\mathrm{a}}. \end{cases} \tag{2.91}$$

Substituting (2.90) and (2.91) into (2.89),

$$\kappa(\boldsymbol{G}_{/\!/}) = \frac{1}{\varepsilon_{\mathrm{b}}}\delta_{\boldsymbol{G}_{/\!/}0} + \frac{1}{V_0}\left(\frac{1}{\varepsilon_{\mathrm{a}}} - \frac{1}{\varepsilon_{\mathrm{b}}}\right)\int_{V_0^{(2)}} \mathrm{d}\boldsymbol{r}_{/\!/} S^{(2)}(\boldsymbol{r}_{/\!/}) \exp(-\mathrm{i}\boldsymbol{G}_{/\!/} \cdot \boldsymbol{r}_{/\!/}). \tag{2.92}$$

In order to calculate the integral in (2.92), we use polar coordinates (r, φ). We take the direction with $\varphi = 0$ as the direction of vector $\boldsymbol{G}_{/\!/}$. For $\boldsymbol{G}_{/\!/} \neq 0$, the integral is modified to

$$\begin{aligned} &\int_{V_0^{(2)}} \mathrm{d}\boldsymbol{r}_{/\!/} S(\boldsymbol{r}_{/\!/}) \exp(-\mathrm{i}\boldsymbol{G}_{/\!/} \cdot \boldsymbol{r}_{/\!/}) \\ &= \int_0^{r_{\mathrm{a}}} \mathrm{d}r \int_0^{2\pi} \mathrm{d}\varphi\; r \exp\left\{\mathrm{i}Gr\sin\left(\varphi - \frac{\pi}{2}\right)\right\} \\ &= \int_0^{r_{\mathrm{a}}} \mathrm{d}r \int_0^{2\pi} \mathrm{d}\varphi\; r \sum_{l=-\infty}^{\infty} J_l(Gr) \exp\left\{\mathrm{i}l\left(\varphi - \frac{\pi}{2}\right)\right\} \\ &= 2\pi \int_0^{r_{\mathrm{a}}} \mathrm{d}r\; rJ_0(Gr), \end{aligned} \tag{2.93}$$

where $G = |\boldsymbol{G}_{/\!/}|$ and J_l is the Bessel function of the lth order. When we derived (2.93), we used the following relation:

$$\exp(\mathrm{i}w\sin\phi) = \sum_{l=-\infty}^{\infty} J_l(w) \exp(\mathrm{i}l\phi). \tag{2.94}$$

If we further use the following relation,

$$\{wJ_1(w)\}' = wJ_0(w), \tag{2.95}$$

we obtain

$$\int_{V_0^{(2)}} \mathrm{d}\boldsymbol{r}_{/\!/} S(\boldsymbol{r}_{/\!/}) \exp(-\mathrm{i}\boldsymbol{G}_{/\!/} \cdot \boldsymbol{r}_{/\!/}) = \frac{2\pi r_{\mathrm{a}}}{G} J_1(Gr_{\mathrm{a}}). \tag{2.96}$$

If we denote the volume fraction of the circular rod by f,

$$f = \frac{\pi r_{\mathrm{a}}^2}{V_0^{(2)}}, \tag{2.97}$$

we finally obtain for $\boldsymbol{G}_{/\!/} \neq 0$

$$\kappa(\boldsymbol{G}_{/\!/}) = 2f\left(\frac{1}{\varepsilon_{\mathrm{a}}} - \frac{1}{\varepsilon_{\mathrm{b}}}\right)\frac{J_1(Gr_{\mathrm{a}})}{Gr_{\mathrm{a}}}. \tag{2.98}$$

For $\boldsymbol{G}_{/\!/} = 0$,

$$\kappa(0) = \frac{f}{\varepsilon_{\rm a}} + \frac{f-1}{\varepsilon_{\rm b}} \tag{2.99}$$

as before.

2.4.2 Some Examples

We consider two examples of the band diagrams calculated by the plane-wave expansion method here. First, we examine the band structure of a simple cubic lattice composed of dielectric spheres. We assume that one dielectric sphere is located at each lattice point. The following values were used for the numerical calculation: the dielectric constant of the spheres and the background are 13.0 and 1.0, and the ratio of the lattice constant to the radius of the sphere is 1:0.3. Figure 2.1 shows the first Brillouin zone of the simple cubic lattice. Highly symmetric points are denoted by the standard notations in this figure. Figure 2.2 shows the photonic band structure obtained by solving (2.32) with 1174 plane waves ($N = 587$). The ordinate is the normalized frequency where a and c denote the lattice constant and the light velocity in free space. The CPU time necessary for the whole calculation was about 23 minutes when we used a supercomputer with a vector processor. The numerical error was estimated to be better than 5% for lower frequencies.

For this structure, there is no complete bandgap that extends throughout the Brillouin zone. However, we can find several important properties peculiar to the photonic crystal in Fig. 2.2. First, there are partial bandgaps for optical waves travelling along the (1, 0, 0) direction, i.e., those waves which have wave vectors on the Δ point. The frequency ranges of the partial gaps are

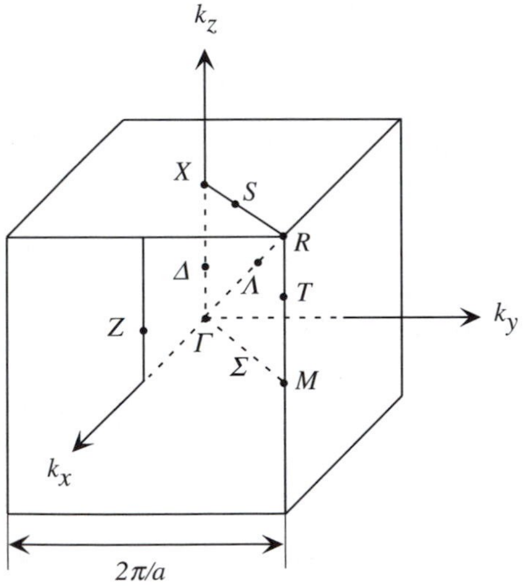

Fig. 2.1. First Brillouin zone of the simple cubic lattice

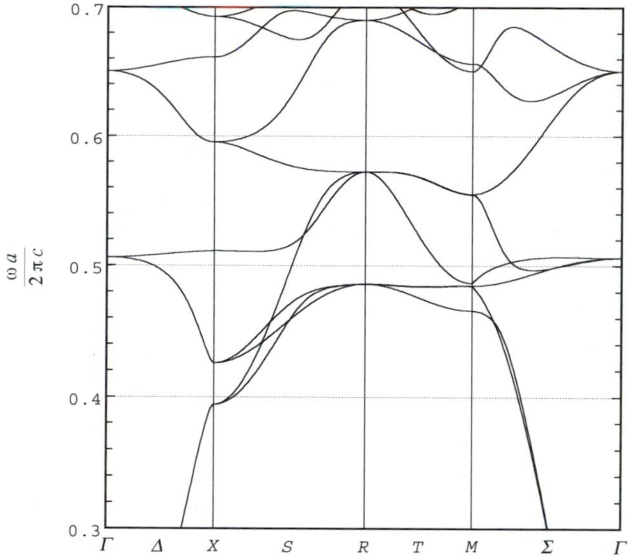

Fig. 2.2. Photonic band structure of a simple cubic lattice with a dielectric sphere at each lattice point. The ordinate is the normalized frequency. The following values were assumed for the numerical calculation: the dielectric constant of the spheres and the background are 13.0 and 1.0, and the ratio of the lattice constant to the radius of the sphere is 1:0.3

$$\frac{\omega a}{2\pi c} = \begin{cases} 0.395\text{–}0.426, \\ 0.512\text{–}0.594, \\ 0.661\text{–}0.691. \end{cases} \tag{2.100}$$

The optical transmittance can be extremely low for those waves. There is also a partial gap on the Σ point around $\omega a/2\pi c = 0.7$. Secondly, there are extraordinary flat bands, such as the third lowest band on the Δ point and the third and fifth lowest bands on the Σ point. Since the group velocity is given by the slope of the dispersion curve, the group velocity of these bands is extremely small. This property is commonly observed for 2D and 3D photonic crystals as was mentioned in Chap. 1. Because various optical phenomena are enhanced when the group velocity of the relevant radiational modes is small, this property may be used for the development of efficient optical devices. The enhancement effect will be described in detail in later chapters. Thirdly, two or three modes sometimes have the same eigenfrequency on the highly symmetric points such as the Γ and the R points. In addition, several bands on the Δ and the T points are doubly degenerate as a whole. This property is closely related to the symmetry of the crystal structure and will be examined in detail by using group theory in Chap. 3.

Next, we consider the band structure of a 2D crystal composed of a regular square array of circular dielectric cylinders. Figure 2.3 shows the structure of the photonic crystal and the 2D first Brillouin zone of the square lattice.

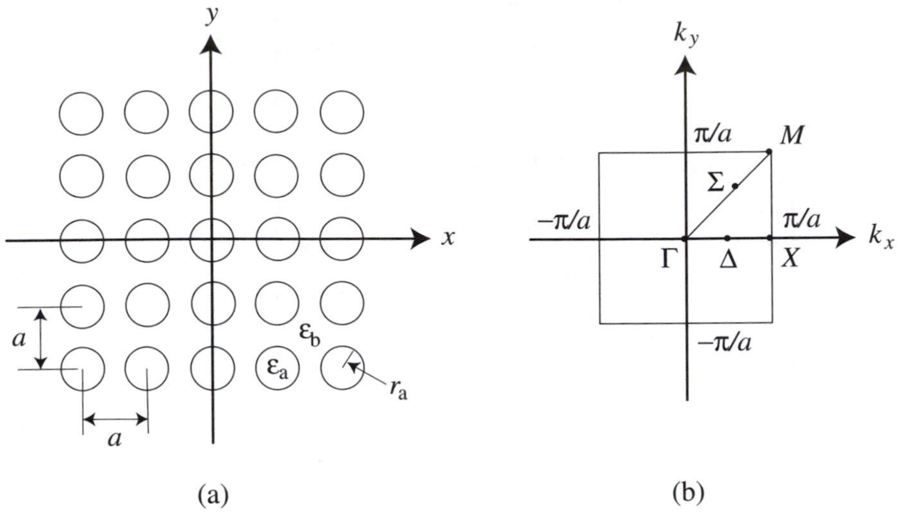

Fig. 2.3. (**a**) Intersection of the 2D square lattice composed of circular cylinders. (**b**) First Brillouin zone of the square lattice

We restrict our discussion to the case for which the wave vectors of the eigenmodes lie in the 2D x-y plane. As was explained in Sect. 2.2, the eigenmodes are classified into two categories according to their polarization for this case, that is, the E polarization for which the electric field is perpendicular to the x-y plane and the H polarization for which the magnetic field is perpendicular to the x-y plane. We treat the E polarization here. The following values were assumed for the numerical calculation: the dielectric constants of the cylinders and the background are 9.0 and 1.0, and the ratio of the lattice constant to the radius of the cylinder is 1:0.38. Figure 2.4 shows the photonic band structure obtained by solving (2.61) with 441 plane waves. The numerical error was estimated to be about 1%. The ordinate is the normalized frequency as in Fig. 2.2. The left-hand side of this figure shows the photonic band structure of the E polarization, whereas the right-hand side shows the density of states of the radiational modes that was obtained by examining the distribution of the eigenfrequencies for 16 000 different wave vectors in the Brillouin zone.

It is clearly seen that there are three bandgaps that extend throughout the 2D Brillouin zone whose frequency ranges are

$$\frac{\omega a}{2\pi c} = \begin{cases} 0.247\text{–}0.277, \\ 0.415\text{–}0.466, \\ 0.623\text{–}0.676. \end{cases} \tag{2.101}$$

These are, of course, not complete gaps since those eigenmodes of the H polarization and those with off-plane wave vectors have eigenfrequencies in

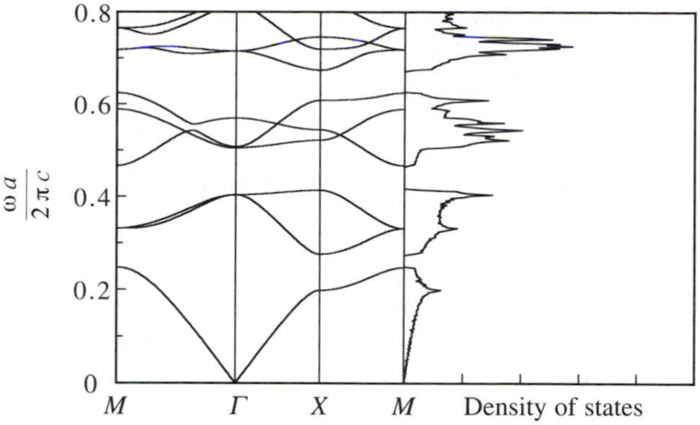

Fig. 2.4. Photonic band structure of a 2D square lattice composed of circular air-cylinders formed in a dielectric material. The ordinate is the normalized frequency as before. The following values were assumed for the numerical calculation: the dielectric constants of the cylinders and the background are 9.0 and 1.0, and the ratio of the lattice constant to the radius of the cylinder is 1:0.38. (After [69])

the above gaps. Therefore, the complete inhibition of spontaneous emission of photons from atoms embbeded in the photonic crystal, for example, cannot be expected. However, it can be partially inhibited. The partial inhibition may be effective when the transition dipole moment of the embedded atoms is aligned perpendicular to the x-y plane. Situations of this kind may be realized with artificial layered structures such as quantum wells made of semiconductors [34].

As in Fig. 2.2, there are several extremely flat bands in Fig. 2.4. These bands can be used to enhance various optical processes such as stimulated emission, second harmonic generation, and quadrature-phase squeezing. For all of these optical processes, the incident electromagnetic field is usually prepared with a laser. Hence, the incident beam has a well-defined propagation direction, and we can propagate it along the x-y plane very easily. Therefore, the 2D band structures, such as Fig. 2.4, have not only mathematical interest but also practical importance. Especially where the laser oscillation is concerned, it takes place at such a frequency and in such a direction that the optical gain due to the stimulated emission is a maximum and the optical loss by transmission at the surface of the crystal is a minimum. It is well recognized that the transmission coefficients for thoses modes with small group velocities are usually small. The lasing thus readily takes place for the flat bands. This point will be examined in detail in Chap. 9.

As for the localized defect modes, the presence of the partial bandgaps shown in Fig. 2.4 is enough for their existence. In fact, McCall et al. [35] observed a defect mode in this 2D crystal by removing a single cylinder. The defect modes usually have quite large quality factors and they can be used as good resonators. Since the interaction between the radiational field and matter depends strongly on the field strength and the interaction time, the large quality factor also enhannces the optical interactions. An efficient numerical method to deal with the defect modes will be given in Chap. 6.

2.5 Phase Velocity, Group Velocity, and Energy Velocity

In the previous three sections, we learned how to calculate the dispersion relations of the radiation modes in the photonic crystals. We can also calculate their wave functions. In addition to the eigenfrequencies and eigenfunctions, there are several parameters that characterize the radiational waves. One of them is the wave velocity. In contrast to the case of particles for which the velocity has a single meaning, waves have three different kinds of velocities, i.e., the phase velocity, the group velocity, and the energy velocity. These velocities are equal to each other in uniform materials with dielectric constants which are real and independent of frequency.

The phase velocity is defined as the velocity of the propagation of an equi-phase surface. This velocity has a definite meaning, for example, for plane waves and spherical waves for which the equi-phase surface can be defined without ambiguity. In the photonic crystal, however, the equi-phase surface cannot be defined rigorously, since its eigenfunction is a superposition of plane waves, as can be seen in (2.26) and (2.27). This means that the phase velocity cannot be defined appropriately in the photonic crystal.

On the other hand, the group velocity, which is the velocity of the propagation of a wave packet, can be defined as usual:

$$\boldsymbol{v}_{\mathrm{g}} = \frac{\partial \omega}{\partial \boldsymbol{k}}. \tag{2.102}$$

The energy velocity is defined as the velocity of the propagation of the electromagnetic energy. The propagation of the electromagnetic energy is described by Poynting's vector. The time-averaged Poynting's vector $\boldsymbol{S}_{\boldsymbol{k}n}(\boldsymbol{r})$ is given by

$$\begin{aligned} \boldsymbol{S}_{\boldsymbol{k}n}(\boldsymbol{r}) &\equiv \overline{\mathrm{Re}\left[\boldsymbol{E}_{\boldsymbol{k}n}(\boldsymbol{r})\mathrm{e}^{-\mathrm{i}\omega_{\boldsymbol{k}n}t}\right] \times \mathrm{Re}\left[\boldsymbol{H}_{\boldsymbol{k}n}(\boldsymbol{r})\mathrm{e}^{-\mathrm{i}\omega_{\boldsymbol{k}n}t}\right]} \\ &= \frac{1}{2}\mathrm{Re}\left[\boldsymbol{E}_{\boldsymbol{k}n}(\boldsymbol{r}) \times \boldsymbol{H}^{*}_{\boldsymbol{k}n}(\boldsymbol{r})\right]. \end{aligned} \tag{2.103}$$

On the other hand, the time-averaged electromagnetic energy density $\mathcal{U}_{\boldsymbol{k}n}(\boldsymbol{r})$ is given by

$$\begin{aligned} \mathcal{U}_{\boldsymbol{k}n}(\boldsymbol{r}) &\equiv \frac{\varepsilon_0\varepsilon(\boldsymbol{r})}{2}\overline{\{\mathrm{Re}\left[\boldsymbol{E}_{\boldsymbol{k}n}(\boldsymbol{r})\mathrm{e}^{-\mathrm{i}\omega_{\boldsymbol{k}n}t}\right]\}^2} + \frac{\mu_0}{2}\overline{\{\mathrm{Re}\left[\boldsymbol{H}_{\boldsymbol{k}n}(\boldsymbol{r})\mathrm{e}^{-\mathrm{i}\omega_{\boldsymbol{k}n}t}\right]\}^2} \\ &= \frac{1}{4}\left\{\varepsilon_0\varepsilon(\boldsymbol{r})\left|\boldsymbol{E}_{\boldsymbol{k}n}(\boldsymbol{r})\right|^2 + \mu_0\left|\boldsymbol{H}_{\boldsymbol{k}n}(\boldsymbol{r})\right|^2\right\}. \end{aligned} \tag{2.104}$$

Thus, the energy velocity $\boldsymbol{v}_{\mathrm{e}}$ is defined as

$$\boldsymbol{v}_{\mathrm{e}} = \frac{\langle \boldsymbol{S}_{\boldsymbol{k}n}(\boldsymbol{r})\rangle}{\langle \mathcal{U}_{\boldsymbol{k}n}(\boldsymbol{r})\rangle}, \tag{2.105}$$

where $\langle \cdots \rangle$ means the spatial average.

Now, the group velocity is equal to the energy velocity even though the dielectric constant is modulated periodically. The proof, which we follow, was given by Yeh [36]. If we substitute (2.18), (2.19), (2.22), and (2.23) into Maxwell's equations (2.13) and (2.14), we have

$$\nabla \times \boldsymbol{u}_{\boldsymbol{k}n}(\boldsymbol{r}) + \mathrm{i}\boldsymbol{k} \times \boldsymbol{u}_{\boldsymbol{k}n}(\boldsymbol{r}) = \mathrm{i}\mu_0\omega_{\boldsymbol{k}n}\boldsymbol{v}_{\boldsymbol{k}n}(\boldsymbol{r}), \tag{2.106}$$

$$\nabla \times \boldsymbol{v}_{\boldsymbol{k}n}(\boldsymbol{r}) + \mathrm{i}\boldsymbol{k} \times \boldsymbol{v}_{\boldsymbol{k}n}(\boldsymbol{r}) = -\mathrm{i}\varepsilon_0\varepsilon(\boldsymbol{r})\omega_{\boldsymbol{k}n}\boldsymbol{u}_{\boldsymbol{k}n}(\boldsymbol{r}). \tag{2.107}$$

Suppose that $\boldsymbol{k}$ is changed by an infinitesimal amount $\delta\boldsymbol{k}$. This results in changes in $\omega_{\boldsymbol{k}n}$, $\boldsymbol{u}_{\boldsymbol{k}n}$, and $\boldsymbol{v}_{\boldsymbol{k}n}$, which are denoted by $\delta\omega_{\boldsymbol{k}n}$, $\delta\boldsymbol{u}_{\boldsymbol{k}n}$, and $\delta\boldsymbol{v}_{\boldsymbol{k}n}$. From (2.106) and (2.107), we have

$$\begin{aligned} &\nabla \times \delta\boldsymbol{u}_{\boldsymbol{k}n} + \mathrm{i}\delta\boldsymbol{k} \times \boldsymbol{u}_{\boldsymbol{k}n} + \mathrm{i}\boldsymbol{k} \times \delta\boldsymbol{u}_{\boldsymbol{k}n} \\ &= \mathrm{i}\mu_0\left(\delta\omega_{\boldsymbol{k}n}\boldsymbol{v}_{\boldsymbol{k}n} + \omega_{\boldsymbol{k}n}\delta\boldsymbol{v}_{\boldsymbol{k}n}\right), \end{aligned} \tag{2.108}$$

$$\begin{aligned} &\nabla \times \delta\boldsymbol{v}_{\boldsymbol{k}n} + \mathrm{i}\delta\boldsymbol{k} \times \boldsymbol{v}_{\boldsymbol{k}n} + \mathrm{i}\boldsymbol{k} \times \delta\boldsymbol{v}_{\boldsymbol{k}n} \\ &= -\mathrm{i}\varepsilon_0\varepsilon\left(\delta\omega_{\boldsymbol{k}n}\boldsymbol{u}_{\boldsymbol{k}n} + \omega_{\boldsymbol{k}n}\delta\boldsymbol{u}_{\boldsymbol{k}n}\right). \end{aligned} \tag{2.109}$$

If we calculate $\boldsymbol{v}^*_{\boldsymbol{k}n} \cdot (2.108) + \boldsymbol{u}_{\boldsymbol{k}n} \cdot (2.109)^*$, we obtain

$$\begin{aligned} &\boldsymbol{v}^*_{\boldsymbol{k}n} \cdot (\nabla \times \delta\boldsymbol{u}_{\boldsymbol{k}n}) + \boldsymbol{u}_{\boldsymbol{k}n} \cdot (\nabla \times \delta\boldsymbol{v}^*_{\boldsymbol{k}n}) + 2\mathrm{i}\delta\boldsymbol{k} \cdot (\boldsymbol{u}_{\boldsymbol{k}n} \times \boldsymbol{v}^*_{\boldsymbol{k}n}) \\ &\qquad + \mathrm{i}\boldsymbol{k} \cdot (\delta\boldsymbol{u}_{\boldsymbol{k}n} \times \boldsymbol{v}^*_{\boldsymbol{k}n} + \boldsymbol{u}_{\boldsymbol{k}n} \times \delta\boldsymbol{v}^*_{\boldsymbol{k}n}) \\ &= \mathrm{i}\omega_{\boldsymbol{k}n}\left(\mu_0\boldsymbol{v}^*_{\boldsymbol{k}n} \cdot \delta\boldsymbol{v}_{\boldsymbol{k}n} + \varepsilon_0\varepsilon\boldsymbol{u}_{\boldsymbol{k}n} \cdot \delta\boldsymbol{u}^*_{\boldsymbol{k}n}\right) \\ &\qquad + \mathrm{i}\delta\omega_{\boldsymbol{k}n}\left(\mu_0\left|\boldsymbol{v}_{\boldsymbol{k}n}\right|^2 + \varepsilon_0\varepsilon\left|\boldsymbol{u}_{\boldsymbol{k}n}\right|^2\right). \end{aligned} \tag{2.110}$$

If we calculate $\delta\boldsymbol{u}^*_{\boldsymbol{k}n} \cdot (2.107)$ and $\delta\boldsymbol{v}_{\boldsymbol{k}n} \cdot (2.106)^*$,

$$\mathrm{i}\varepsilon_0\varepsilon\omega_{\boldsymbol{k}n}\delta\boldsymbol{u}^*_{\boldsymbol{k}n} \cdot \boldsymbol{u}_{\boldsymbol{k}n} = -\delta\boldsymbol{u}^*_{\boldsymbol{k}n} \cdot (\nabla \times \boldsymbol{v}_{\boldsymbol{k}n}) - \mathrm{i}\boldsymbol{k} \cdot (\boldsymbol{v}_{\boldsymbol{k}n} \times \delta\boldsymbol{u}^*_{\boldsymbol{k}n}), \tag{2.111}$$

$$\mathrm{i}\mu_0\omega_{\boldsymbol{k}n}\delta\boldsymbol{v}_{\boldsymbol{k}n} \cdot \boldsymbol{v}^*_{\boldsymbol{k}n} = -\delta\boldsymbol{v}_{\boldsymbol{k}n} \cdot (\nabla \times \boldsymbol{u}^*_{\boldsymbol{k}n}) + \mathrm{i}\boldsymbol{k} \cdot (\boldsymbol{u}^*_{\boldsymbol{k}n} \times \delta\boldsymbol{v}_{\boldsymbol{k}n}). \tag{2.112}$$

Using (2.111) and (2.112), (2.110) is written as

$$\begin{aligned} &\boldsymbol{v}^*_{\boldsymbol{k}n} \cdot (\nabla \times \delta\boldsymbol{u}_{\boldsymbol{k}n}) + \boldsymbol{u}_{\boldsymbol{k}n} \cdot (\nabla \times \delta\boldsymbol{v}^*_{\boldsymbol{k}n}) + 2\mathrm{i}\delta\boldsymbol{k} \cdot (\boldsymbol{u}_{\boldsymbol{k}n} \times \boldsymbol{v}^*_{\boldsymbol{k}n}) \\ &\qquad + \delta\boldsymbol{u}^*_{\boldsymbol{k}n} \cdot (\nabla \times \boldsymbol{v}_{\boldsymbol{k}n}) + \delta\boldsymbol{v}_{\boldsymbol{k}n} \cdot (\nabla \times \boldsymbol{u}^*_{\boldsymbol{k}n}) \\ &\qquad + \mathrm{i}\boldsymbol{k} \cdot (\delta\boldsymbol{u}_{\boldsymbol{k}n} \times \boldsymbol{v}^*_{\boldsymbol{k}n} - \delta\boldsymbol{v}^*_{\boldsymbol{k}n} \times \boldsymbol{u}_{\boldsymbol{k}n} + \boldsymbol{v}_{\boldsymbol{k}n} \times \delta\boldsymbol{u}^*_{\boldsymbol{k}n} - \boldsymbol{u}^*_{\boldsymbol{k}n} \times \delta\boldsymbol{v}_{\boldsymbol{k}n}) \\ &= \mathrm{i}\delta\omega_{\boldsymbol{k}n}\left(\mu_0\left|\boldsymbol{v}_{\boldsymbol{k}n}\right|^2 + \varepsilon_0\varepsilon\left|\boldsymbol{u}_{\boldsymbol{k}n}\right|^2\right). \end{aligned} \tag{2.113}$$

The right-hand side of (2.113) is purely imaginary. Hence, we take the imaginary part of (2.113) by calculating (2.113) – (2.113)*:

$$\begin{aligned} &\boldsymbol{v}^*_{\boldsymbol{k}n} \cdot (\nabla \times \delta\boldsymbol{u}_{\boldsymbol{k}n}) - \delta\boldsymbol{u}_{\boldsymbol{k}n} \cdot (\nabla \times \boldsymbol{v}^*_{\boldsymbol{k}n}) \\ &\qquad + \boldsymbol{u}_{\boldsymbol{k}n} \cdot (\nabla \times \delta\boldsymbol{v}^*_{\boldsymbol{k}n}) - \delta\boldsymbol{v}^*_{\boldsymbol{k}n} \cdot (\nabla \times \boldsymbol{u}_{\boldsymbol{k}n}) \\ &\qquad + \delta\boldsymbol{u}^*_{\boldsymbol{k}n} \cdot (\nabla \times \boldsymbol{v}_{\boldsymbol{k}n}) - \boldsymbol{v}_{\boldsymbol{k}n} \cdot (\nabla \times \delta\boldsymbol{u}^*_{\boldsymbol{k}n}) \\ &\qquad + \delta\boldsymbol{v}_{\boldsymbol{k}n} \cdot (\nabla \times \boldsymbol{u}^*_{\boldsymbol{k}n}) - \boldsymbol{u}^*_{\boldsymbol{k}n} \cdot (\nabla \times \delta\boldsymbol{v}_{\boldsymbol{k}n}) \\ &\qquad + 4\mathrm{i}\delta\boldsymbol{k} \cdot \mathrm{Re}[\boldsymbol{u}_{\boldsymbol{k}n} \times \boldsymbol{v}^*_{\boldsymbol{k}n}] \\ &= 2\mathrm{i}\delta\omega_{\boldsymbol{k}n}\left(\mu_0\left|\boldsymbol{v}_{\boldsymbol{k}n}\right|^2 + \varepsilon_0\varepsilon\left|\boldsymbol{u}_{\boldsymbol{k}n}\right|^2\right). \end{aligned} \tag{2.114}$$

If we define a periodic function $\boldsymbol{F}$ by

$$\boldsymbol{F} = \delta \boldsymbol{u}_{\boldsymbol{k}n} \times \boldsymbol{v}^*_{\boldsymbol{k}n} + \delta \boldsymbol{v}^*_{\boldsymbol{k}n} \times \boldsymbol{u}_{\boldsymbol{k}n} + \boldsymbol{v}_{\boldsymbol{k}n} \times \delta \boldsymbol{u}^*_{\boldsymbol{k}n} + \boldsymbol{u}^*_{\boldsymbol{k}n} \times \delta \boldsymbol{v}_{\boldsymbol{k}n}, \quad (2.115)$$

(2.114) is modified to

$$\boldsymbol{\nabla} \cdot \boldsymbol{F} + 4\mathrm{i}\delta\boldsymbol{k} \cdot \mathrm{Re}\left[\boldsymbol{u}_{\boldsymbol{k}n} \times \boldsymbol{v}^*_{\boldsymbol{k}n}\right] = 2\mathrm{i}\delta\omega_{\boldsymbol{k}n}\left(\mu_0 \left|\boldsymbol{v}_{\boldsymbol{k}n}\right|^2 + \varepsilon_0 \varepsilon \left|\boldsymbol{u}_{\boldsymbol{k}n}\right|^2\right), \quad (2.116)$$

where we have used the following vector identity:

$$\boldsymbol{\nabla} \cdot (\boldsymbol{A} \times \boldsymbol{B}) = \boldsymbol{B} \cdot (\boldsymbol{\nabla} \times \boldsymbol{A}) - \boldsymbol{A} \cdot (\boldsymbol{\nabla} \times \boldsymbol{B}). \quad (2.117)$$

If we perform an integration over a unit cell, we have

$$\begin{aligned} &\langle \boldsymbol{\nabla} \cdot \boldsymbol{F} \rangle + 4\mathrm{i}\delta\boldsymbol{k} \cdot \langle \mathrm{Re}\left[\boldsymbol{u}_{\boldsymbol{k}n} \times \boldsymbol{v}^*_{\boldsymbol{k}n}\right] \rangle \\ &\quad = 2\mathrm{i}\delta\omega_{\boldsymbol{k}n}\left\langle \mu_0 \left|\boldsymbol{v}_{\boldsymbol{k}n}\right|^2 + \varepsilon_0 \varepsilon \left|\boldsymbol{u}_{\boldsymbol{k}n}\right|^2 \right\rangle. \end{aligned} \quad (2.118)$$

Because of the periodic nature of $\boldsymbol{F}$,

$$\langle \boldsymbol{\nabla} \cdot \boldsymbol{F} \rangle = \frac{1}{V_0}\int_{V_0} \mathrm{d}\boldsymbol{r} \boldsymbol{\nabla} \cdot \boldsymbol{F} = \frac{1}{V_0}\int_{S_0} \mathrm{d}S F_n = 0. \quad (2.119)$$

In this equation, V_0 is the volume of the unit cell and S_0 is its surface. F_n denotes the outward normal component of $\boldsymbol{F}$ on S_0. We used Gauss's theorem to derive the second equality in (2.119). Using the definition of $\boldsymbol{v}_\mathrm{e}$, (2.105), we finally have

$$\delta\omega_{\boldsymbol{k}n} = \boldsymbol{v}_\mathrm{e} \cdot \delta\boldsymbol{k}. \quad (2.120)$$

From the definition of the group velocity, (2.102), we have

$$\boldsymbol{v}_\mathrm{e} = \boldsymbol{v}_\mathrm{g}. \quad (2.121)$$

2.6 Calculation of Group Velocity

As we learned in the last section, the group velocity of the radiation modes has very important role in light propagation and optical response in the photonic crystals. Hence, the calculation of the group velocity is an essencial task for the understanding of their optical properties. Since the group velocity is defined as the derivative of the angular frequency with respect to the wave vector (see (2.102)), we may calculate it by numerical differentiation. That is, we may actually evaluate the following limit:

$$\boldsymbol{v}_\mathrm{g} = \lim_{\Delta\boldsymbol{k} \to 0} \frac{\omega_{\boldsymbol{k}+\Delta\boldsymbol{k},n} - \omega_{\boldsymbol{k},n}}{\Delta\boldsymbol{k}}. \quad (2.122)$$

This numerical differentiation needs a limiting procedure for which we have to know a series of eigenfrequencies as a function of the wave vector.

There is quite a convenient method to avoid this procedure and give an accurate evaluation of the group velocity. We use the Hellmann–Feynman theorem for this purpose, with which the readers may be familiar as it relates to quantum mechanical calculations. In the case of quantum mechanics, the Hellmann–Feynman theorem is stated as follows. First, we assume an Hermitian operator $\widehat{H}$ that depends on an external variable α, and denote it by $\widehat{H}_\alpha$. We also assume that we know the orthonormal set composed of the eigenfunctions of $\widehat{H}_\alpha$, which we denote by $\{|\alpha n\rangle\,; n = 1, 2, \cdots\}$:

$$\widehat{H}_\alpha \left|\alpha n\right\rangle = \lambda_{\alpha n} \left|\alpha n\right\rangle, \tag{2.123}$$

$$\langle \alpha n | \alpha n \rangle = 1. \tag{2.124}$$

In the above two equations, we used the standard notation of "bra" and "ket" vectors. $\lambda_{\alpha n}$ is the eigenvalue of $\widehat{H}_\alpha$ for state $|\alpha n\rangle$ and $\langle \cdots | \cdots \rangle$ denotes the inner product. Now, our problem is to calculate the derivative of $\lambda_{\alpha n}$ with respect to α. Because the state vector (or the eigenfunction) $|\alpha n\rangle$ is normalized to unity (2.124), the following holds.

$$\begin{aligned}
\frac{\partial}{\partial \alpha}\lambda_{\alpha n} &= \frac{\partial}{\partial \alpha}\left\langle \alpha n | \widehat{H}_\alpha | \alpha n \right\rangle \\
&= \frac{\partial \left\langle \alpha n\right|}{\partial \alpha}\widehat{H}_\alpha \left|\alpha n\right\rangle + \left\langle \alpha n \left| \frac{\partial \widehat{H}_\alpha}{\partial \alpha} \right| \alpha n \right\rangle + \left\langle \alpha n\right| \widehat{H}_\alpha \frac{\partial \left|\alpha n\right\rangle}{\partial \alpha} \\
&= \lambda_{\alpha n}\frac{\partial}{\partial \alpha}\left\langle \alpha n | \alpha n \right\rangle + \left\langle \alpha n \left| \frac{\partial \widehat{H}_\alpha}{\partial \alpha} \right| \alpha n \right\rangle \\
&= \left\langle \alpha n \left| \frac{\partial \widehat{H}_\alpha}{\partial \alpha} \right| \alpha n \right\rangle .
\end{aligned} \tag{2.125}$$

Once we know the analytical expression of $\partial \widehat{H}_\alpha / \partial \alpha$, $\partial \lambda_{\alpha n} / \partial \alpha$ can readily be obtained using this equation without the limiting procedure in (2.122).

Now, we will see how to use the Hellmann–Feynman theorem for the calculation of the group velocity. Here, we show the method for the E polarization. The dispersion relation is obtained by solving the eigenvalue equation (2.61). When we define a column vector $\boldsymbol{A}_{\boldsymbol{k}_{/\!/} n}$ by

$$A_{\boldsymbol{k}_{/\!/} n}(\boldsymbol{G}_{/\!/}) = \left|\boldsymbol{k}_{/\!/} + \boldsymbol{G}_{/\!/}\right| E_{z,\boldsymbol{k}_{/\!/} n}(\boldsymbol{G}_{/\!/}), \tag{2.126}$$

(2.61) is transformed to

$$\mathrm{M}_{\boldsymbol{k}_{/\!/}} \boldsymbol{A}_{\boldsymbol{k}_{/\!/} n} = \frac{\omega_{\boldsymbol{k}n}^{(E)2}}{c^2} \boldsymbol{A}_{\boldsymbol{k}_{/\!/} n}, \tag{2.127}$$

where $\mathrm{M}_{\boldsymbol{k}_{/\!/}}$ is a $\boldsymbol{k}_{/\!/}$-dependent matrix whose $(\boldsymbol{G}_{/\!/}, \boldsymbol{G}'_{/\!/})$ component is given by the following equation:

$$\mathrm{M}_{\boldsymbol{k}_{/\!/}}(\boldsymbol{G}_{/\!/}, \boldsymbol{G}'_{/\!/}) = \left|\boldsymbol{k}_{/\!/} + \boldsymbol{G}_{/\!/}\right| \left|\boldsymbol{k}_{/\!/} + \boldsymbol{G}'_{/\!/}\right| \kappa(\boldsymbol{G}_{/\!/} - \boldsymbol{G}'_{/\!/}). \tag{2.128}$$

The $\boldsymbol{k}_{/\!/}$-dependent vector $\boldsymbol{A}_{\boldsymbol{k}_{/\!/} n}$ gives the eigenfunction $E_{z,\boldsymbol{k}_{/\!/} n}(\boldsymbol{r}_{/\!/})$ as

$$E_{z,\boldsymbol{k}_{/\!/}n}(\boldsymbol{r}_{/\!/}) = h\sum_{\boldsymbol{G}_{/\!/}} \frac{A_{\boldsymbol{k}_{/\!/}n}(\boldsymbol{G}_{/\!/})}{|\boldsymbol{k}_{/\!/}+\boldsymbol{G}_{/\!/}|} \exp\left\{\mathrm{i}\left(\boldsymbol{k}_{/\!/}+\boldsymbol{G}_{/\!/}\right)\cdot \boldsymbol{r}_{/\!/}\right\}, \tag{2.129}$$

and it is normalized to unity, i.e., $|\boldsymbol{A}_{\boldsymbol{k}_{/\!/}n}| = 1$. In (2.129), h is the normalization constant.

Here, we assume that $\varepsilon(\boldsymbol{r}_{/\!/})$ is real. Then, $\kappa(-\boldsymbol{G}_{/\!/}) = \kappa^*(\boldsymbol{G}_{/\!/})$ and $\mathrm{M}_{\boldsymbol{k}_{/\!/}}$ is an Hermitian matrix. Therefore, we can apply the Hellmann–Feynman theorem to the present problem, and we obtain

$$\boldsymbol{A}^{t*}_{\boldsymbol{k}_{/\!/}n}\frac{\partial \mathrm{M}_{\boldsymbol{k}_{/\!/}}}{\partial \boldsymbol{k}_{/\!/}}\boldsymbol{A}_{\boldsymbol{k}_{/\!/}n} = \frac{\partial}{\partial \boldsymbol{k}_{/\!/}}\left(\frac{\omega^{(E)2}_{\boldsymbol{k}_{/\!/}n}}{c^2}\right) = \frac{2\omega^{(E)}_{\boldsymbol{k}_{/\!/}n}}{c^2}\frac{\partial \omega^{(E)}_{\boldsymbol{k}_{/\!/}n}}{\partial \boldsymbol{k}_{/\!/}}, \tag{2.130}$$

where t denotes the transposed matrix and

$$\frac{\partial \mathrm{M}_{\boldsymbol{k}_{/\!/}}(\boldsymbol{G}_{/\!/},\boldsymbol{G}'_{/\!/})}{\partial \boldsymbol{k}_{/\!/}} = \left\{\frac{\left|\boldsymbol{k}_{/\!/}+\boldsymbol{G}'_{/\!/}\right|}{\left|\boldsymbol{k}_{/\!/}+\boldsymbol{G}_{/\!/}\right|}\left(\boldsymbol{k}_{/\!/}+\boldsymbol{G}_{/\!/}\right) + \frac{\left|\boldsymbol{k}_{/\!/}+\boldsymbol{G}_{/\!/}\right|}{\left|\boldsymbol{k}_{/\!/}+\boldsymbol{G}'_{/\!/}\right|}\left(\boldsymbol{k}_{/\!/}+\boldsymbol{G}'_{/\!/}\right)\right\}\kappa(\boldsymbol{G}_{/\!/}-\boldsymbol{G}'_{/\!/}). \tag{2.131}$$

Therefore, the group velocity $v_{\mathrm{g}}(\omega, n)$ can be readily evaluated once the eigenvector $\boldsymbol{A}_{\boldsymbol{k}_{/\!/}n}$ and the eigenvalue $\omega^{(E)2}_{\boldsymbol{k}_{/\!/}n}/c^2$ are obtained by the band calculation based on the plane-wave expansion method.

2.7 Complete Set of Eigenfunctions

When we derive the Green's function related to the electric field and discuss the optical response of the photonic crystals, we need a complete set of eigenfunctions concerned with the electric field. However, $\mathcal{L}_{\boldsymbol{E}}$ defined by (2.20) is not an Hermitian operator, hence its eigenfunctions $\{\boldsymbol{E}_{\boldsymbol{k}n}(\boldsymbol{r})\}$ are not complete or orthogonal. Here, we define a differential operator that is Hermitian and examine the properties of its eigenfunctions [37–39]. We introduce a complex vectorial function $\boldsymbol{Q}(\boldsymbol{r}, \mathrm{t})$ and a differential operator $\mathcal{H}$:

$$\boldsymbol{Q}(\boldsymbol{r},t) \equiv \sqrt{\varepsilon(\boldsymbol{r})}\boldsymbol{E}(\boldsymbol{r},t), \tag{2.132}$$

$$\mathcal{H}\boldsymbol{Q}(\boldsymbol{r},t) \equiv \frac{1}{\sqrt{\varepsilon(\boldsymbol{r})}}\boldsymbol{\nabla}\times\left\{\boldsymbol{\nabla}\times\frac{1}{\sqrt{\varepsilon(\boldsymbol{r})}}\boldsymbol{Q}(\boldsymbol{r},t)\right\}. \tag{2.133}$$

Then, (2.20) leads to

$$\left(\frac{1}{c^2}\frac{\partial^2}{\partial t^2}+\mathcal{H}\right)\boldsymbol{Q}(\boldsymbol{r},t) = 0. \tag{2.134}$$

We assume the periodic boundary condition for $\boldsymbol{Q}(\boldsymbol{r},t)$ as usual. We will verify that $\mathcal{H}$ is an Hermitian operator. Then, its eigenfunctions form an orthogonal complete set.

Now, the inner product of two complex vectorial functions $\boldsymbol{Q}_1(\boldsymbol{r})$ and $\boldsymbol{Q}_2(\boldsymbol{r})$ is defined by

$$\langle \boldsymbol{Q}_1, \boldsymbol{Q}_2 \rangle \equiv \int_V \mathrm{d}\boldsymbol{r} \boldsymbol{Q}_1^*(\boldsymbol{r}) \cdot \boldsymbol{Q}_2(\boldsymbol{r}), \tag{2.135}$$

where V is the volume on which the periodic boundary condition is imposed. As a vector identity, the next equation holds.

$$\nabla \cdot (\boldsymbol{A} \times \boldsymbol{B}) \equiv (\nabla \times \boldsymbol{A}) \cdot \boldsymbol{B} - \boldsymbol{A} \cdot (\nabla \times \boldsymbol{B}). \tag{2.136}$$

Using this identity, we obtain

$$\begin{aligned}
\langle \mathcal{H}\boldsymbol{Q}_1, \boldsymbol{Q}_2 \rangle &\equiv \int_V \mathrm{d}\boldsymbol{r} \left[\nabla \times \left\{ \nabla \times \frac{\boldsymbol{Q}_1^*(\boldsymbol{r})}{\sqrt{\varepsilon(\boldsymbol{r})}} \right\} \right] \cdot \frac{\boldsymbol{Q}_2(\boldsymbol{r})}{\sqrt{\varepsilon(\boldsymbol{r})}} \\
&= \int_S \mathrm{d}S \left[\left\{ \nabla \times \frac{\boldsymbol{Q}_1^*(\boldsymbol{r})}{\sqrt{\varepsilon(\boldsymbol{r})}} \right\} \times \frac{\boldsymbol{Q}_2(\boldsymbol{r})}{\sqrt{\varepsilon(\boldsymbol{r})}} \right]_n \\
&\quad + \int_V \mathrm{d}\boldsymbol{r} \left\{ \nabla \times \frac{\boldsymbol{Q}_1^*(\boldsymbol{r})}{\sqrt{\varepsilon(\boldsymbol{r})}} \right\} \cdot \left\{ \nabla \times \frac{\boldsymbol{Q}_2(\boldsymbol{r})}{\sqrt{\varepsilon(\boldsymbol{r})}} \right\},
\end{aligned} \tag{2.137}$$

where S denotes the surface of V and the first integral on the right-hand side is the surface integral of the normal component of the integrand. This surface integral is equal to zero because of the periodic boundary condition. Then, applying the identity (2.136) again, we obtain

$$\begin{aligned}
\langle \mathcal{H}\boldsymbol{Q}_1, \boldsymbol{Q}_2 \rangle &= \int_S \mathrm{d}S \left[\frac{\boldsymbol{Q}_1^*(\boldsymbol{r})}{\sqrt{\varepsilon(\boldsymbol{r})}} \times \left\{ \nabla \times \frac{\boldsymbol{Q}_2(\boldsymbol{r})}{\sqrt{\varepsilon(\boldsymbol{r})}} \right\} \right]_n \\
&\quad + \int_V \mathrm{d}\boldsymbol{r} \frac{\boldsymbol{Q}_1^*(\boldsymbol{r})}{\sqrt{\varepsilon(\boldsymbol{r})}} \cdot \left[\nabla \times \left\{ \nabla \times \frac{\boldsymbol{Q}_2(\boldsymbol{r})}{\sqrt{\varepsilon(\boldsymbol{r})}} \right\} \right] \\
&= \langle \boldsymbol{Q}_1, \mathcal{H}\boldsymbol{Q}_2 \rangle,
\end{aligned} \tag{2.138}$$

where we have again used the fact that the surface integral is equal to zero. This equation implies that $\mathcal{H}$ is an Hermitian operator.

Next, we can easily show that $\boldsymbol{Q}_{\boldsymbol{k}n}^{(\mathrm{L})}(\boldsymbol{r})$, given by

$$\boldsymbol{Q}_{\boldsymbol{k}n}^{(\mathrm{L})}(\boldsymbol{r}) = C\sqrt{\varepsilon(\boldsymbol{r})} \frac{\boldsymbol{k} + \boldsymbol{G}_n}{|\boldsymbol{k} + \boldsymbol{G}_n|} \exp\left\{ \mathrm{i}(\boldsymbol{k} + \boldsymbol{G}_n) \cdot \boldsymbol{r} \right\}, \tag{2.139}$$

where $\boldsymbol{G}_n$ is a reciprocal lattice vector and C is a normalization constant, satisfies

$$\nabla \times \left\{ \frac{1}{\sqrt{\varepsilon(\boldsymbol{r})}} \boldsymbol{Q}_{\boldsymbol{k}n}^{(\mathrm{L})}(\boldsymbol{r}) \right\} = 0. \tag{2.140}$$

Then, from the definition of $\mathcal{H}$,

$$\mathcal{H}\boldsymbol{Q}_{\boldsymbol{k}n}^{(\mathrm{L})}(\boldsymbol{r}) = 0. \tag{2.141}$$

This equation implies that $\boldsymbol{Q}_{\boldsymbol{k}n}^{(\mathrm{L})}(\boldsymbol{r})$ is an eigenfunction of $\mathcal{H}$ and its eigen-angular frequency is zero. There exists one such solution for each $\boldsymbol{k}$ and n. In general,

$$\begin{aligned}\boldsymbol{\nabla}\cdot\{\varepsilon(\boldsymbol{r})\boldsymbol{E}_{\boldsymbol{k}n}^{(\mathrm{L})}(\boldsymbol{r})\} &\equiv \boldsymbol{\nabla}\cdot\left\{\sqrt{\varepsilon(\boldsymbol{r})}\boldsymbol{Q}_{\boldsymbol{k}n}^{(\mathrm{L})}(\boldsymbol{r})\right\} \\ &\neq 0.\end{aligned} \tag{2.142}$$

We call $\boldsymbol{Q}_{\boldsymbol{k}n}^{(\mathrm{L})}(\boldsymbol{r})$ a quasi-longitudinal solution. Strictly speaking, the $\boldsymbol{Q}_{\boldsymbol{k}n}^{(\mathrm{L})}(\boldsymbol{r})$ defined by (2.139) are not orthogonal to each other, but their orthogonalization can be readily accomplished by Schmidt's method. That is, we can take linear combinations of the $\boldsymbol{Q}_{\boldsymbol{k}n}^{(\mathrm{L})}(\boldsymbol{r})$ to form an orthonormal set. We denote the orthogonalized quasi-longitudinal solutions by the same symbols. Because the $\boldsymbol{Q}_{\boldsymbol{k}n}^{(\mathrm{L})}(\boldsymbol{r})$ do not satisfy (2.11), they are unphysical solutions for the present problem. However, they are important mathematically, since we cannot obtain a complete set without them.

On the other hand, there exist quasi-transverse solutions that correspond to the transverse plane waves in a uniform crystal. They satisfy the following equation:

$$\mathcal{H}\boldsymbol{Q}_{\boldsymbol{k}n}^{(\mathrm{T})}(\boldsymbol{r}) = \frac{\omega_{\boldsymbol{k}n}^{(\mathrm{T})2}}{c^2}\boldsymbol{Q}_{\boldsymbol{k}n}^{(\mathrm{T})}(\boldsymbol{r}). \tag{2.143}$$

The eigen-angular frequency $\omega_{\boldsymbol{k}n}^{(\mathrm{T})}$ is generally non-zero. From (2.143) and (2.133),

$$\begin{aligned}\boldsymbol{\nabla}\cdot\left\{\sqrt{\varepsilon(\boldsymbol{r})}\boldsymbol{Q}_{\boldsymbol{k}n}^{(\mathrm{T})}(\boldsymbol{r})\right\} &= \frac{c^2}{\omega_{\boldsymbol{k}n}^{(T)2}}\boldsymbol{\nabla}\cdot\left[\boldsymbol{\nabla}\times\left\{\boldsymbol{\nabla}\times\frac{\boldsymbol{Q}_{\boldsymbol{k}n}^{(\mathrm{T})}(\boldsymbol{r})}{\sqrt{\varepsilon(\boldsymbol{r})}}\right\}\right] \\ &\equiv 0.\end{aligned} \tag{2.144}$$

Note that $\boldsymbol{Q}_{\boldsymbol{k}n}^{(\mathrm{T})}$ or $\boldsymbol{Q}_{\boldsymbol{k}n}^{(\mathrm{L})}$ are *not* purely transverse nor longitudinal because of the spatial variation of $\varepsilon(\boldsymbol{r})$. The terms "quasi-transverse" and "quasi-longitudinal" are used here to emphasize that (2.144) and (2.140) reduce, when $\varepsilon(\boldsymbol{r})$ is a constant, to the usual relations concerned with transverse and longitudinal waves, respectively.

Now, we normalize these eigenfunctions as follows:

$$\int_V \mathrm{d}\boldsymbol{r}\boldsymbol{Q}_{\boldsymbol{k}n}^{(\alpha)*}(\boldsymbol{r})\cdot\boldsymbol{Q}_{\boldsymbol{k}'n'}^{(\beta)}(\boldsymbol{r}) = V\delta_{\alpha\beta}\delta_{\boldsymbol{k}\boldsymbol{k}'}\delta_{nn'}, \tag{2.145}$$

where α, β = T or L. Note that $\boldsymbol{Q}_{\boldsymbol{k}n}^{(\alpha)}(\boldsymbol{r})$ and $\boldsymbol{E}_{\boldsymbol{k}n}^{(\alpha)}(\boldsymbol{r})$ ($\equiv \boldsymbol{Q}_{\boldsymbol{k}n}^{(\alpha)}(\boldsymbol{r})/\sqrt{\varepsilon(\boldsymbol{r})}$) are dimensionless by this definition. The completeness of the eigenfunctions leads to

$$\sum_{kn} \boldsymbol{Q}_{kn}^{(\mathrm{T})}(\boldsymbol{r}) \otimes \boldsymbol{Q}_{kn}^{(T)*}(\boldsymbol{r}') + \sum_{kn} \boldsymbol{Q}_{kn}^{(\mathrm{L})}(\boldsymbol{r}) \otimes \boldsymbol{Q}_{kn}^{(L)*}(\boldsymbol{r}')$$
$$= V \overleftrightarrow{\mathrm{I}} \delta(\boldsymbol{r} - \boldsymbol{r}'), \tag{2.146}$$

where $\otimes$ denotes a tensor whose elements are given by the product of the elements of two vectors, i.e., $(\boldsymbol{A} \otimes \boldsymbol{B})_{ij} = A_i B_j$, and $\overleftrightarrow{\mathrm{I}}$ is the unit tensor. $\delta(\boldsymbol{r})$ is Dirac's delta function.

We make one remark on (2.146) here. Readers who are familiar with non-relativistic quantum mechanics may wonder why the tensors introduced in the above equation are necessary to express the completeness. In non-relativistic quantum mechanics, the wave function that describes an electron is a complex scalar function. We denote the eigenfunctions of a Hamiltonian operator $\widehat{H}$ by $\{\phi_n; n = 1, 2, \cdots\}$. We normalize them as

$$\int_V \mathrm{d}\boldsymbol{r} \phi_n^*(\boldsymbol{r}) \phi_{n'}(\boldsymbol{r}) = V \delta_{nn'}. \tag{2.147}$$

The completeness is thus expressed as

$$\sum_n \phi_n(\boldsymbol{r}) \phi_n^*(\boldsymbol{r}') = V \delta(\boldsymbol{r} - \boldsymbol{r}'). \tag{2.148}$$

From this equation, we obtain for any complex function $f(\boldsymbol{r})$

$$f(\boldsymbol{r}) = \frac{1}{V} \sum_n \phi_n(\boldsymbol{r}) \int_V \mathrm{d}\boldsymbol{r} \phi_n^*(\boldsymbol{r}') f(\boldsymbol{r}'). \tag{2.149}$$

This equation means that any complex function can be expanded with $\{\phi_n(\boldsymbol{r})\}$. This is what the completeness implies for a set of scalar functions. As for a set of vectorial functions, they should also be able to expand any vectors in 3D Euclidean space. As an example, consider the unit vectors $\boldsymbol{e}_1$, $\boldsymbol{e}_2$, and $\boldsymbol{e}_3$ which are parallel to the x, y, and z axes, respectively. They are a complete set in 3D Euclidean space, since any vector $\boldsymbol{v}$ can be expanded with them:

$$\boldsymbol{v} = \sum_n \boldsymbol{e}_n (\boldsymbol{e}_n \cdot \boldsymbol{v}). \tag{2.150}$$

We can write this equation in a slightly different manner. We define three tensors, $\overleftrightarrow{\mathrm{T}}_1$, $\overleftrightarrow{\mathrm{T}}_2$, and $\overleftrightarrow{\mathrm{T}}_3$ as

$$\overleftrightarrow{\mathrm{T}}_1 \equiv \boldsymbol{e}_1 \otimes \boldsymbol{e}_1 = \begin{pmatrix} 1 & 0 & 0 \\ 0 & 0 & 0 \\ 0 & 0 & 0 \end{pmatrix}, \tag{2.151}$$

$$\overleftrightarrow{\mathrm{T}}_2 \equiv \boldsymbol{e}_2 \otimes \boldsymbol{e}_2 = \begin{pmatrix} 0 & 0 & 0 \\ 0 & 1 & 0 \\ 0 & 0 & 0 \end{pmatrix}, \tag{2.152}$$

$$\overleftrightarrow{\mathrm{T}}_3 \equiv \boldsymbol{e}_3 \otimes \boldsymbol{e}_3 = \begin{pmatrix} 0 & 0 & 0 \\ 0 & 0 & 0 \\ 0 & 0 & 1 \end{pmatrix}. \tag{2.153}$$

Since

$$\sum_{n=1}^{3} \overleftrightarrow{\mathrm{T}}_n = \overleftrightarrow{\mathrm{I}}, \tag{2.154}$$

(2.150) is rewritten as

$$\boldsymbol{v} = \sum_{n=1}^{3} \overleftrightarrow{\mathrm{T}}_n \boldsymbol{v} = \sum_{n=1}^{3} \boldsymbol{e}_n \otimes \boldsymbol{e}_n \boldsymbol{v}. \tag{2.155}$$

From this equation, we can see that

$$\sum_{n=1}^{3} \boldsymbol{e}_n \otimes \boldsymbol{e}_n = \overleftrightarrow{\mathrm{I}} \tag{2.156}$$

is the condition for the completeness in the 3D Euclidean space. As a conclusion, (2.146) is the condition for the completeness both in functional space and in Euclidean space.

As for the 2D crystal, we examine the case of the E polarization here. The H polarization can be treated in a similar manner. First, we define a function $Q_z(\boldsymbol{r}_{/\!/}, t)$ and a differential operator $\mathcal{H}^{(2)}$ such that

$$Q_z(\boldsymbol{r}_{/\!/}, t) = \sqrt{\varepsilon(\boldsymbol{r}_{/\!/})} E_z(\boldsymbol{r}_{/\!/}), \tag{2.157}$$

$$\mathcal{H}^{(2)} \equiv -\frac{1}{\sqrt{\varepsilon(\boldsymbol{r}_{/\!/})}} \left(\frac{\partial^2}{\partial x^2} + \frac{\partial^2}{\partial y^2} \right) \frac{1}{\sqrt{\varepsilon(\boldsymbol{r}_{/\!/})}}. \tag{2.158}$$

Then (2.53) leads to

$$\left(\frac{1}{c^2} \frac{\partial^2}{\partial t^2} + \mathcal{H}^{(2)} \right) Q_z(\boldsymbol{r}_{/\!/}, t) = 0. \tag{2.159}$$

We can verify as before that $\mathcal{H}^{(2)}$ defined by (2.159) is an Hermitian operator. Its eigenfunctions $\{Q_{z,\boldsymbol{k}_{/\!/}n}(\boldsymbol{r}_{/\!/})\}$ thus form an orthogonal complete set. We normalize these eigenfunctions as follows:

$$\int_{V^{(2)}} \mathrm{d}\boldsymbol{r}_{/\!/} Q^*_{z,\boldsymbol{k}_{/\!/}n}(\boldsymbol{r}_{/\!/}) Q_{z,\boldsymbol{k}'_{/\!/}n'}(\boldsymbol{r}_{/\!/}) = V^{(2)} \delta_{\boldsymbol{k}_{/\!/}\boldsymbol{k}'_{/\!/}} \delta_{nn'}, \tag{2.160}$$

where $V^{(2)}$ is the 2D volume on which the periodic boundary condition is imposed. The completeness of the eigenfunctions leads to

$$\sum_{\boldsymbol{k}_{/\!/}n} Q_{z,\boldsymbol{k}_{/\!/}n}(\boldsymbol{r}_{/\!/}) Q^*_{z,\boldsymbol{k}_{/\!/}n}(\boldsymbol{r}'_{/\!/}) = V^{(2)} \delta(\boldsymbol{r}_{/\!/} - \boldsymbol{r}'_{/\!/}) \tag{2.161}$$

in this case, since $\{Q_{z,\boldsymbol{k}_{/\!/}n}(\boldsymbol{r}_{/\!/})\}$ are scalar functions.

2.8 Retarded Green's Function

In later chapters, we will deal with various optical processes in photonic crystals. For that purpose it is essential to calculate the electromagnetic field radiated from oscillating polarization fields. This task can generally be performed by means of a retarded Green's (tensor) function $\overleftrightarrow{\mathrm{G}}(\boldsymbol{r}, \boldsymbol{r}', t)$ [37–39] that satisfies the following two equations.

$$-\left(\frac{1}{c^2}\frac{\partial^2}{\partial t^2} + \mathcal{H}\right)\overleftrightarrow{\mathrm{G}}(\boldsymbol{r}, \boldsymbol{r}', t - t') = \overleftrightarrow{\mathrm{I}}\delta(\boldsymbol{r} - \boldsymbol{r}')\delta(t - t'), \tag{2.162}$$

$$\overleftrightarrow{\mathrm{G}}(\boldsymbol{r}, \boldsymbol{r}', t) = 0 \quad \text{for} \quad t < 0. \tag{2.163}$$

Green's function in frequency space $\overleftrightarrow{\mathcal{G}}(\boldsymbol{r}, \boldsymbol{r}', \omega)$ is defined by the Fourier transform of $\overleftrightarrow{\mathrm{G}}(\boldsymbol{r}, \boldsymbol{r}', t)$:

$$\overleftrightarrow{\mathrm{G}}(\boldsymbol{r}, \boldsymbol{r}', t) \equiv \frac{1}{2\pi}\int_{-\infty}^{\infty} \mathrm{d}\omega \overleftrightarrow{\mathcal{G}}(\boldsymbol{r}, \boldsymbol{r}', \omega)\mathrm{e}^{-\mathrm{i}\omega t}. \tag{2.164}$$

Then,

$$\left(\frac{\omega^2}{c^2} - \mathcal{H}\right)\overleftrightarrow{\mathcal{G}}(\boldsymbol{r}, \boldsymbol{r}', \omega) = \overleftrightarrow{\mathrm{I}}\delta(\boldsymbol{r} - \boldsymbol{r}'). \tag{2.165}$$

Then, from (2.141), (2.143), and (2.146), we can obtain its explicit expression:

$$\begin{aligned}\overleftrightarrow{\mathcal{G}}(\boldsymbol{r}, \boldsymbol{r}', \omega) &\equiv \int_{-\infty}^{\infty} \mathrm{d}t\, \overleftrightarrow{\mathrm{G}}(\boldsymbol{r}, \boldsymbol{r}', t)\mathrm{e}^{\mathrm{i}\omega t} \\ &= \frac{c^2}{V}\sum_{\boldsymbol{k}n}\left[\frac{\boldsymbol{Q}_{\boldsymbol{k}n}^{(\mathrm{T})}(\boldsymbol{r}) \otimes \boldsymbol{Q}_{\boldsymbol{k}n}^{(\mathrm{T})*}(\boldsymbol{r}')}{\left(\omega - \omega_{\boldsymbol{k}n}^{(\mathrm{T})} + \mathrm{i}\delta\right)\left(\omega + \omega_{\boldsymbol{k}n}^{(\mathrm{T})} + \mathrm{i}\delta\right)}\right. \\ &\qquad \left. + \frac{\boldsymbol{Q}_{\boldsymbol{k}n}^{(\mathrm{L})}(\boldsymbol{r}) \otimes \boldsymbol{Q}_{\boldsymbol{k}n}^{(\mathrm{L})*}(\boldsymbol{r}')}{(\omega + \mathrm{i}\delta)^2}\right].\end{aligned} \tag{2.166}$$

Here, δ is a positive infinitesimal that assures the causality, i.e., (2.163). For $t \geq 0$, the inverse transform of (2.166) gives

$$\begin{aligned}\overleftrightarrow{\mathrm{G}}(\boldsymbol{r}, \boldsymbol{r}', t) &= \frac{c^2}{2\pi V}\sum_{\boldsymbol{k}n}\frac{\boldsymbol{Q}_{\boldsymbol{k}n}^{(\mathrm{T})}(\boldsymbol{r}) \otimes \boldsymbol{Q}_{\boldsymbol{k}n}^{(\mathrm{T})*}(\boldsymbol{r}')}{2\omega_{\boldsymbol{k}n}^{(\mathrm{T})}} \\ &\quad \times \int_C \left(\frac{1}{\omega - \omega_{\boldsymbol{k}n}^{(\mathrm{T})} + \mathrm{i}\delta} - \frac{1}{\omega + \omega_{\boldsymbol{k}n}^{(\mathrm{T})} + \mathrm{i}\delta}\right)\mathrm{e}^{-\mathrm{i}\omega t}\mathrm{d}\omega \\ &\quad + \frac{c^2}{2\pi V}\sum_{\boldsymbol{k}n}\boldsymbol{Q}_{\boldsymbol{k}n}^{(\mathrm{L})}(\boldsymbol{r}) \otimes \boldsymbol{Q}_{\boldsymbol{k}n}^{(\mathrm{L})*}(\boldsymbol{r}')\int_C \frac{\mathrm{e}^{-\mathrm{i}\omega t}}{(\omega + \mathrm{i}\delta)^2}\mathrm{d}\omega \\ &= -\frac{c^2}{V}\sum_{\boldsymbol{k}n}\left\{\frac{\sin\omega_{\boldsymbol{k}n}^{(\mathrm{T})}t}{\omega_{\boldsymbol{k}n}^{(\mathrm{T})}}\boldsymbol{Q}_{\boldsymbol{k}n}^{(\mathrm{T})}(\boldsymbol{r}) \otimes \boldsymbol{Q}_{\boldsymbol{k}n}^{(\mathrm{T})*}(\boldsymbol{r}')\right.\end{aligned}$$

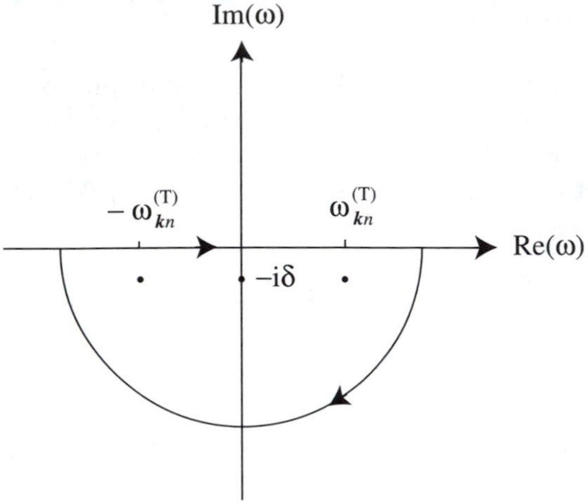

Fig. 2.5. Contour of the integration in (2.167) for $t \geq 0$. For $t < 0$, the contour should enclose the upper half plane

$$+t\boldsymbol{Q}_{\boldsymbol{k}n}^{(\mathrm{L})}(\boldsymbol{r}) \otimes \boldsymbol{Q}_{\boldsymbol{k}n}^{(\mathrm{L})*}(\boldsymbol{r}')\Bigg\}, \tag{2.167}$$

where the contour C is shown in Fig. 2.5. For $t \geq 0$, we close the path of integration in the lower half of the complex ω plane. For $t < 0$, we close it in the upper half plane, and we obtain (2.163).

As for the E polarization in the 2D crystal, we define the retarded Green's function of (2.159) by

$$-\left(\frac{1}{c^2}\frac{\partial^2}{\partial t^2} + \mathcal{H}^{(2)}\right) G^{(2)}(\boldsymbol{r}_{/\!/}, \boldsymbol{r}'_{/\!/}, t-t') = \delta(\boldsymbol{r}_{/\!/} - \boldsymbol{r}'_{/\!/})\delta(t-t'), \tag{2.168}$$

$$G^{(2)}(\boldsymbol{r}_{/\!/}, \boldsymbol{r}'_{/\!/}, t) = 0 \quad \text{for} \quad t < 0. \tag{2.169}$$

In this case, the Fourier transform of the Green's function, $\mathcal{G}^{(2)}(\boldsymbol{r}_{/\!/}, \boldsymbol{r}'_{/\!/}, \omega)$, is given by

$$\begin{aligned}\mathcal{G}^{(2)}(\boldsymbol{r}_{/\!/}, \boldsymbol{r}'_{/\!/}, \omega) &\equiv \int_{-\infty}^{\infty} \mathrm{d}t G^{(2)}(\boldsymbol{r}_{/\!/}, \boldsymbol{r}'_{/\!/}, t)\mathrm{e}^{\mathrm{i}\omega t} \\ &= \frac{c^2}{V^{(2)}} \sum_{\boldsymbol{k}_{/\!/}n} \frac{Q_{z,\boldsymbol{k}_{/\!/}n}(\boldsymbol{r}_{/\!/}) Q^*_{z,\boldsymbol{k}_{/\!/}n}(\boldsymbol{r}'_{/\!/})}{\left(\omega - \omega^{(E)}_{\boldsymbol{k}_{/\!/}n} + \mathrm{i}\delta\right)\left(\omega + \omega^{(E)}_{\boldsymbol{k}_{/\!/}n} + \mathrm{i}\delta\right)}.\end{aligned} \tag{2.170}$$

This satisfies the following equation:

$$\left(\frac{\omega^2}{c^2} - \mathcal{H}^{(2)}\right) \mathcal{G}^{(2)}(\boldsymbol{r}_{/\!/}, \boldsymbol{r}'_{/\!/}, \omega) = \delta(\boldsymbol{r}_{/\!/} - \boldsymbol{r}'_{/\!/}). \tag{2.171}$$

The inverse transform of (2.170) leads to

$$G^{(2)}(\boldsymbol{r}_{/\!/}, \boldsymbol{r}'_{/\!/}, t) = -\frac{c^2}{V^{(2)}} \sum_{\boldsymbol{k}_{/\!/} n} \frac{\sin \omega^{(E)}_{\boldsymbol{k}_{/\!/} n} t}{\omega^{(E)}_{\boldsymbol{k}_{/\!/} n}} Q_{z,\boldsymbol{k}_{/\!/} n}(\boldsymbol{r}_{/\!/}) Q^*_{z,\boldsymbol{k}_{/\!/} n}(\boldsymbol{r}'_{/\!/}) \qquad (2.172)$$

for $t \geq 0$. These Green's functions will be used in later chapters to analyze the optical response and the defect modes in photonic crystals.

3. Symmetry of Eigenmodes

The symmetry of the eigenmodes of photonic crystals plays an important role in their optical response. In this chapter, we will formulate the group theory of the radiation field in the photonic crystals in order to classify the spatial symmetry of their eigenfunctions. An efficient method for the symmetry assignment that excludes unphysical quasi-longitudinal modes, which was discussed in the last chapter, will be presented. The existence of such eigenmodes that cannot be excited by external plane waves because of the mismatching of the symmetry will be shown. This method will be applied to some examples.

3.1 Group Theory for Two-Dimensional Crystals

The symmetry of the eigenfunctions of the photonic crystals plays an important role in their optical response. In fact, Robertson et al. [40, 41] showed clear evidence of the presence of such eigenmodes in two-dimensional (2D) photonic crystals that cannot be excited by an external plane wave due to the mismatching of their spatial symmetry. We refer to these modes as uncoupled modes. On the other hand, the symmetry of the eigenmodes leads to an additional selection rule for nonlinear optical processes and absence of diffraction loss in the photonic crystal (see Chaps. 5 and 8). The classification of the eigenmodes according to their spatial symmetry by means of group theory is thus necessary and often very powerful to understand the optical properties of the photonic crystals. In particular, when we compare the transmission spectra observed experimentally with the calculated photonic band structure, the knowledge of the uncoupled modes is indispensable.

Since it is easier to deal with scalar waves, we begin with the group theory for 2D photonic crystals [42]. As was shown in Sect. 2.2, the vectorial wave equations derived from Maxwell's equations are reduced to two independent scalar equations as long as the wave vector lies in the 2D x-y plane. These two waves are called the E polarization for which the electric field is parallel to the z axis and the H polarization for which the magnetic field is parallel to the z axis, respectively. The eigenfunctions satisfy the following eigenvalue equations:

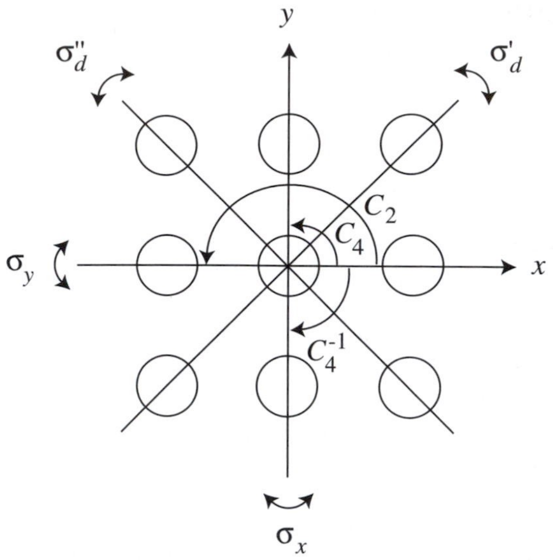

Fig. 3.1. Symmetry operations for the square array of dielectric cylinders

$$\mathcal{L}_E^{(2)} E_z(\boldsymbol{r}_{/\!/}) = \frac{\omega^2}{c^2} E_z(\boldsymbol{r}_{/\!/}), \tag{3.1}$$

$$\mathcal{L}_H^{(2)} H_z(\boldsymbol{r}_{/\!/}) = \frac{\omega^2}{c^2} H_z(\boldsymbol{r}_{/\!/}), \tag{3.2}$$

where operators $\mathcal{L}_E^{(2)}$ and $\mathcal{L}_H^{(2)}$ are defined by (2.57) and (2.58).

Now, all 2D photonic crystals have translational symmetry with respect to the elementary lattice vectors, $\boldsymbol{a}_{1/\!/}$ and $\boldsymbol{a}_{2/\!/}$:

$$\varepsilon(\boldsymbol{r} + l\boldsymbol{a}_{1/\!/} + m\boldsymbol{a}_{2/\!/}) = \varepsilon(\boldsymbol{r}), \tag{3.3}$$

where l and m are integers. In addition to this translational symmetry, the 2D photonic crystal often has other spatial symmetries. As an example, consider the square array of dielectric cylinders discussed in Sect. 2.2. We take the x and y axes as in Fig. 3.1. It is apparent that the structure is invariant when we change x to $-x$:

$$\varepsilon(-x, y) = \varepsilon(x, y). \tag{3.4}$$

If we denote the symmetry operation that changes x to $-x$ by σ_x, we say that the structure is invariant under the mirror reflection σ_x. Equation (3.4) may be rewritten as

$$[\sigma_x \varepsilon](\boldsymbol{r}_{/\!/}) \equiv \varepsilon(\sigma_x^{-1} \boldsymbol{r}_{/\!/}) = \varepsilon(\boldsymbol{r}_{/\!/}). \tag{3.5}$$

The convention of group theory is to use not σ_x but σ_x^{-1} in the middle term of this equation. The reason will become clear in the following. The structure

Table 3.1. Products of the operations in the C_{4v} point group. This table shows the product R_1R_2 where R_1 denotes an operation in the first column and R_2 denotes an operation in the first row

C_{4v}	E	C_4	C_4^{-1}	C_4^2	σ_x	σ_y	σ_d'	σ_d''
E	E	C_4	C_4^{-1}	C_4^2	σ_x	σ_y	σ_d'	σ_d''
C_4	C_4	C_2	E	C_4^{-1}	σ_d''	σ_d'	σ_x	σ_y
C_4^{-1}	C_4^{-1}	E	C_2	C_4	σ_d'	σ_d''	σ_y	σ_x
C_4^2	C_4^2	C_4^{-1}	C_4	E	σ_y	σ_x	σ_d''	σ_d'
σ_x	σ_x	σ_d'	σ_d''	σ_y	E	C_2	C_4	C_4^{-1}
σ_y	σ_y	σ_d''	σ_d'	σ_x	C_2	E	C_4^{-1}	C_4
σ_d'	σ_d'	σ_y	σ_x	σ_d''	C_4^{-1}	C_4	E	C_2
σ_d''	σ_d''	σ_x	σ_y	σ_d'	C_4	C_4^{-1}	C_2	E

is also invariant under the mirror reflection σ_y that changes y to $-y$. There is another set of mirror reflections under which the structure is invariant. They are σ_d', which changes (x, y) to (y, x), and σ_d'', which changes (x, y) to $(-y, -x)$. The structure is also invariant when it is rotated by 90, 180, or 270 degrees counterclockwise about the origin. These symmetry operations are denoted by C_4, $C_2(\equiv C_4^2)$, and $C_4^{-1}(\equiv C_4^3)$. C_n generally means rotation by $2\pi/n$ radian.

Together with the identity operation E that keeps the structure as it is, all these symmetry operations constitute the C_{4v} point group:

$$C_{4v} = \{E, C_4, C_4^{-1}, C_2, \sigma_x, \sigma_y, \sigma_d', \sigma_d''\}. \tag{3.6}$$

Namely, (1) any two subsequent symmetry operations in C_{4v} are equivalent to a single operation in it, (2) any three symmetry operations $\{R_1, R_2, R_3\}$ have the property that $R_1(R_2R_3) = (R_1R_2)R_3$, and (3) the inverse of any symmetry operation in C_{4v} is also an operation in it. These conditions may easily be confirmed. As for item (1), Table 3.1 shows the products of two operations in C_{4v}. If the following relation holds, we say that R_1 is conjugate with R_2:

$$R_1 = RR_2R^{-1} \qquad \text{for some } R \in C_{4v}. \tag{3.7}$$

(σ_x, σ_y), (σ_d', σ_d''), and (C_4, C_4^{-1}) are conjugate with each other since we have the following relations:

$$\sigma_y = C_4\sigma_x C_4^{-1}, \tag{3.8}$$

$$\sigma_d'' = C_4\sigma_d' C_4^{-1}, \tag{3.9}$$

$$C_4^{-1} = \sigma_x C_4 \sigma_x^{-1}. \tag{3.10}$$

We may regard the two conjugate operations in each set to be substantially the same since they may interchange with each other by changing the coordinate system. For example, when the x axis is interchanged with the y axis, then σ_x becomes σ_y and σ_y becomes σ_x. σ_x and σ_y are sometimes denoted by σ_v, and σ'_d and σ''_d are denoted by σ_d.

Now, any rotation or mirror reflection in two dimensions can be represented by a 2×2 matrix. As for the C_{4v} point group, we have

$$\begin{aligned}
\overleftrightarrow{\mathrm{R}}_E &= \begin{pmatrix} 1 & 0 \\ 0 & 1 \end{pmatrix}, & \overleftrightarrow{\mathrm{R}}_{C_2} &= \begin{pmatrix} -1 & 0 \\ 0 & -1 \end{pmatrix}, \\
\overleftrightarrow{\mathrm{R}}_{C_4} &= \begin{pmatrix} 0 & -1 \\ 1 & 0 \end{pmatrix}, & \overleftrightarrow{\mathrm{R}}_{C_4^{-1}} &= \begin{pmatrix} 0 & 1 \\ -1 & 0 \end{pmatrix}, \\
\overleftrightarrow{\mathrm{R}}_{\sigma_x} &= \begin{pmatrix} -1 & 0 \\ 0 & 1 \end{pmatrix}, & \overleftrightarrow{\mathrm{R}}_{\sigma_y} &= \begin{pmatrix} 1 & 0 \\ 0 & -1 \end{pmatrix}, \\
\overleftrightarrow{\mathrm{R}}_{\sigma'_d} &= \begin{pmatrix} 0 & 1 \\ 1 & 0 \end{pmatrix}, & \overleftrightarrow{\mathrm{R}}_{\sigma''_d} &= \begin{pmatrix} 0 & -1 \\ -1 & 0 \end{pmatrix}.
\end{aligned} \tag{3.11}$$

We denote the matrix representation of a 2D symmetry operation R of an arbitrary point group $\mathcal{M}$ by $\overleftrightarrow{\mathrm{R}}$ in this section. Since R does not change the length of vectors, $\overleftrightarrow{\mathrm{R}}$ is an orthogonal matrix:

$$\overleftrightarrow{\mathrm{R}}\,\overleftrightarrow{\mathrm{R}}^t = \overleftrightarrow{\mathrm{I}}, \tag{3.12}$$

where $\overleftrightarrow{\mathrm{R}}^t$ is the transposed matrix of $\overleftrightarrow{\mathrm{R}}$, and $\overleftrightarrow{\mathrm{I}}$ is the unit matrix. For an arbitrary function f of $\boldsymbol{r}_{/\!/}$, the action of R on f is defined by

$$[Rf](\boldsymbol{r}_{/\!/}) \equiv f(\overleftrightarrow{\mathrm{R}}^{-1}\boldsymbol{r}_{/\!/}). \tag{3.13}$$

The subsequent action of operations R_1 and R_2 is thus given by

$$\begin{aligned}
[R_1R_2f](\boldsymbol{r}_{/\!/}) &\equiv [R_1[R_2f]](\boldsymbol{r}_{/\!/}) = [R_2f](\overleftrightarrow{\mathrm{R}}_1^{-1}\boldsymbol{r}_{/\!/}) \\
&= f(\overleftrightarrow{\mathrm{R}}_2^{-1}\overleftrightarrow{\mathrm{R}}_1^{-1}\boldsymbol{r}_{/\!/}) = f((\overleftrightarrow{\mathrm{R}}_1\overleftrightarrow{\mathrm{R}}_2)^{-1}\boldsymbol{r}_{/\!/}).
\end{aligned} \tag{3.14}$$

Hence, the order of R_1 and R_2 when they operate on f is consistent with that of $\overleftrightarrow{\mathrm{R}}_1$ and $\overleftrightarrow{\mathrm{R}}_2$ when they operate on $\boldsymbol{r}_{/\!/}$. This is a consequence of the definition (3.13), where $\overleftrightarrow{\mathrm{R}}^{-1}$ is operated on $\boldsymbol{r}_{/\!/}$.

In general, we assume that the 2D photonic crystal is invariant under any symmetry operation R that belongs to the point group $\mathcal{M}$:

$$[R\varepsilon](\boldsymbol{r}_{/\!/}) \equiv \varepsilon(R^{-1}\boldsymbol{r}_{/\!/}) = \varepsilon(\boldsymbol{r}_{/\!/}) \qquad (\forall R \in \mathcal{M}). \tag{3.15}$$

Using this equation, we can prove that R commutes with $\mathcal{L}_E^{(2)}$ and $\mathcal{L}_H^{(2)}$:

$$R\mathcal{L}_E^{(2)}R^{-1} = \mathcal{L}_E^{(2)}, \tag{3.16}$$

$$R\mathcal{L}_H^{(2)}R^{-1} = \mathcal{L}_H^{(2)}. \tag{3.17}$$

This implies that the conventional classification of the eigenmodes according to the irreducible representations of the $\boldsymbol{k}$-group $\mathcal{M}_{\boldsymbol{k}}$ can be applicable to the present problem.[1]

[1] See [43] and [44] for details of the application of group theory to solid-state physics.

Proof of (3.16) and (3.17)

We denote $\overleftrightarrow{\mathrm{R}}$ by

$$\overleftrightarrow{\mathrm{R}} = \begin{pmatrix} R_{11}, & R_{12} \\ R_{21}, & R_{22} \end{pmatrix}. \tag{3.18}$$

From (3.12), we have

$$\begin{pmatrix} R_{11}^2 + R_{12}^2, & R_{11}R_{21} + R_{12}R_{22} \\ R_{21}R_{11} + R_{22}R_{12}, & R_{21}^2 + R_{22}^2 \end{pmatrix} = \begin{pmatrix} 1, & 0 \\ 0, & 1 \end{pmatrix}. \tag{3.19}$$

For an arbitrary function $f(\boldsymbol{r}_{/\!/})$,

$$\begin{aligned} R\frac{\partial}{\partial x}R^{-1}f(\boldsymbol{r}_{/\!/}) &= R\frac{\partial}{\partial x}f(\overleftrightarrow{\mathrm{R}}\boldsymbol{r}_{/\!/}) \\ &= R\left\{ R_{11}\left.\frac{\partial f}{\partial x}\right|_{\overleftrightarrow{\mathrm{R}}\boldsymbol{r}_{/\!/}} + R_{21}\left.\frac{\partial f}{\partial y}\right|_{\overleftrightarrow{\mathrm{R}}\boldsymbol{r}_{/\!/}} \right\} \\ &= R_{11}\left.\frac{\partial f}{\partial x}\right|_{\boldsymbol{r}_{/\!/}} + R_{21}\left.\frac{\partial f}{\partial y}\right|_{\boldsymbol{r}_{/\!/}}. \end{aligned} \tag{3.20}$$

Thus,

$$R\frac{\partial}{\partial x}R^{-1} = R_{11}\frac{\partial}{\partial x} + R_{21}\frac{\partial}{\partial y}. \tag{3.21}$$

Similarly, we have

$$R\frac{\partial}{\partial y}R^{-1} = R_{12}\frac{\partial}{\partial x} + R_{22}\frac{\partial}{\partial y}. \tag{3.22}$$

Hence,

$$\begin{aligned} R\mathcal{L}_E^{(2)}R^{-1} &= -R\frac{1}{\varepsilon(\boldsymbol{r}_{/\!/})}R^{-1}\left\{ \left(R\frac{\partial}{\partial x}R^{-1}\right)^2 + \left(R\frac{\partial}{\partial y}R^{-1}\right)^2 \right\} \\ &= -\frac{1}{\varepsilon(\boldsymbol{r}_{/\!/})}\left\{ \left(R_{11}^2 + R_{12}^2\right)\frac{\partial^2}{\partial x^2} \right. \\ &\quad \left. +2\left(R_{11}R_{21} + R_{12}R_{22}\right)\frac{\partial^2}{\partial x \partial y} + \left(R_{21}^2 + R_{22}^2\right)\frac{\partial^2}{\partial y^2} \right\} \\ &= \mathcal{L}_E^{(2)}. \end{aligned} \tag{3.23}$$

Here, we used (3.19). Equation (3.17) can be verified similarly. ■

Now, the $\boldsymbol{k}$-group $\mathcal{M}_{\boldsymbol{k}}$ is defined as the subgroup of $\mathcal{M}$ that keeps the wave vector $\boldsymbol{k}$ invariant.[2] Consider this point with the square lattice as an

[2] Rigorously speaking, we should deal with the spatial group instead of the point group to treat our problem in a mathematically correct manner. However, as long as we treat symmorphic spatial groups, it makes no difference to the results.

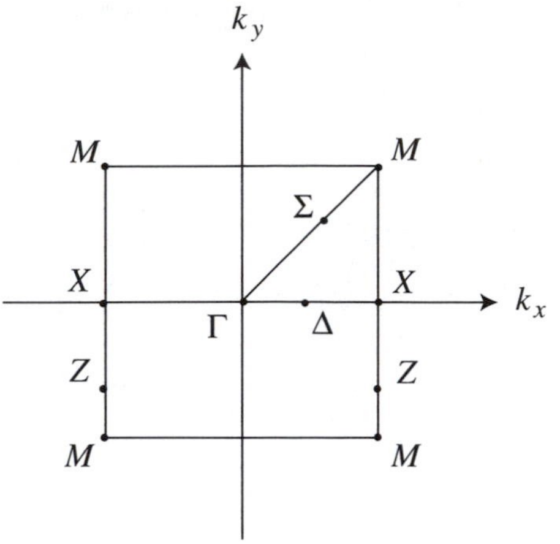

Fig. 3.2. First Brillouin zone of the 2D square lattice

example. Figure 3.2 shows its first Brillouin zone. The elementary lattice vectors $\{\boldsymbol{a}_{1/\!/}, \boldsymbol{a}_{2/\!/}\}$ and the elementary reciprocal lattice vectors $\{\boldsymbol{b}_{1/\!/}, \boldsymbol{b}_{2/\!/}\}$ are

$$\boldsymbol{a}_{1/\!/} = \begin{pmatrix} a \\ 0 \end{pmatrix}, \qquad \boldsymbol{a}_{2/\!/} = \begin{pmatrix} 0 \\ a \end{pmatrix}, \tag{3.24}$$

$$\boldsymbol{b}_{1/\!/} = \begin{pmatrix} 2\pi/a \\ 0 \end{pmatrix}, \qquad \boldsymbol{b}_{2/\!/} = \begin{pmatrix} 0 \\ 2\pi/a \end{pmatrix}, \tag{3.25}$$

where a is the lattice constant of the square photonic crystal. The Γ, X, and M points in Fig. 3.2 denote $(0, 0)$, $(\pm\pi/a, 0)$, and $(\pm\pi/a, \pm\pi/a)$, respectively. These two X points and four M points are equivalent to each other, since the difference between them is just a linear combination of the elementary reciprocal lattice vectors. First, consider the Γ point. We can readily see that all symmetry operations in the C_{4v} point group keep the Γ point invariant. Hence, by definition,

$$\mathcal{M}_\Gamma = C_{4v}. \tag{3.26}$$

In the case of the X point,

$$\mathcal{M}_X = \{E, C_2, \sigma_x, \sigma_y\} \equiv C_{2v}. \tag{3.27}$$

Here, we should note that operations C_2 and σ_x keep the X point invariant, since

$$\overleftrightarrow{\mathrm{R}}_{C_2} \begin{pmatrix} \pi/a \\ 0 \end{pmatrix} = \overleftrightarrow{\mathrm{R}}_{\sigma_x} \begin{pmatrix} \pi/a \\ 0 \end{pmatrix} = \begin{pmatrix} -\pi/a \\ 0 \end{pmatrix} \equiv \begin{pmatrix} \pi/a \\ 0 \end{pmatrix}. \tag{3.28}$$

As for the others,

$$\mathcal{M}_M = C_{4v}, \tag{3.29}$$
$$\mathcal{M}_\Delta = \{E, \sigma_y\} \equiv C_{1h}, \tag{3.30}$$
$$\mathcal{M}_\Sigma = \{E, \sigma_d\} \equiv C_{1h}, \tag{3.31}$$
$$\mathcal{M}_Z = \{E, \sigma_x\} \equiv C_{1h}. \tag{3.32}$$

Group theory tells us that any eigenfunction, $E_{z,\boldsymbol{k}n}(\boldsymbol{r}_{/\!/})$ or $H_{z,\boldsymbol{k}n}(\boldsymbol{r}_{/\!/})$, is an irreducible representation of $\mathcal{M}_{\boldsymbol{k}}$. For example, the C_{4v} point group has four one-dimensional irreducible representations A_1, A_2, B_1, B_2 and one two-dimensional irreducible representation E. Here,"one-dimensional" implies that the eigenmode is not degenerate and "two-dimensional" implies that the eigenmode is doubly degenerate. Each irreducible representation has its own spatial symmetry which is expressed by its *character*, χ. Consider this point with a B_1 mode as an example. We denote the character of the B_1 representation by χ_{B_1} and assume that $f_{B_1}(\boldsymbol{r}_{/\!/})$ is attributed to the B_1 representation. We refer to $f_{B_1}(\boldsymbol{r}_{/\!/})$ as a basis of the B_1 representation. Group theory tells us that for any symmetry operation R of the C_{4v} point group,

$$Rf_{B_1}(\boldsymbol{r}_{/\!/}) = \chi_{B_1}(R) f_{B_1}(\boldsymbol{r}_{/\!/}), \tag{3.33}$$

and

$$\chi_{B_1}(R) = \begin{cases} 1 & \text{for } R = E,\ C_2,\ \sigma_x,\ \sigma_y, \\ -1 & \text{for } R = C_4,\ C_4^{-1},\ \sigma_d',\ \sigma_d''. \end{cases} \tag{3.34}$$

For other one-dimensional irreducible representations, the same kind of equation holds. For the two-dimensional representation E, there are two eigenfunctions that have the same eigenfrequency. We denote them by $f_E^{(1)}(\boldsymbol{r}_{/\!/})$ and $f_E^{(2)}(\boldsymbol{r}_{/\!/})$. We refer to them as a basis set of the E representation. In general, the operation of R on these functions transforms them into the following linear combinations.

$$Rf_E^{(1)}(\boldsymbol{r}_{/\!/}) = A_{11} f_E^{(1)}(\boldsymbol{r}_{/\!/}) + A_{12} f_E^{(2)}(\boldsymbol{r}_{/\!/}), \tag{3.35}$$
$$Rf_E^{(2)}(\boldsymbol{r}_{/\!/}) = A_{21} f_E^{(1)}(\boldsymbol{r}_{/\!/}) + A_{22} f_E^{(2)}(\boldsymbol{r}_{/\!/}). \tag{3.36}$$

Group theory tells us that

$$\mathrm{Tr(A)} \equiv A_{11} + A_{22} = \chi_E(R). \tag{3.37}$$

The characters of these irreducible representations are listed in Table 3.2. Generally speaking, the conjugate operations such as σ_x and σ_y, or, C_4 and C_4^{-1} have the same character. Hence, in Table 3.2, σ_x and σ_y are denoted by σ_v, σ_d' and σ_d'' are denoted by σ_d, and C_4 and C_4^{-1} are represented by C_4. The symbol "2" before C_4, σ_v, and σ_d in the first row of this table implies that there are two such elements that are conjugate with each other and have the same character. As for the C_{2v} and C_{1h} point groups, the irreducble representations and their characters are listed in Tables 3.3 and

Table 3.2. Character table for the C_{4v} point group

C_{4v}	E	$2C_4$	C_2	$2\sigma_v$	$2\sigma_d$
A_1	1	1	1	1	1
A_2	1	1	1	−1	−1
B_1	1	−1	1	1	−1
B_2	1	−1	1	−1	1
E	2	0	−2	0	0

Table 3.3. The character table for the C_{2v} point group

C_{2v}	E	C_2	σ_y	σ_x
A_1	1	1	1	1
A_2	1	1	−1	−1
B_1	1	−1	1	−1
B_2	1	−1	−1	1

Table 3.4. Character table for the C_{1h} point group

C_{1h}	E	σ
A	1	1
B	1	−1

3.4, respectively. All irreducible representations are one-dimensional for these two groups.

For the assignment of the irreducible representation to each mode, it is very convenient to use so-called *compatibility relations* that tell us the mutual relations between the irreducible representations for adjacent $\boldsymbol{k}$ vectors. We examine this for the Γ and Δ points as an example. As mentioned previously, the Γ point has the C_{4v} symmetry and the Δ point has the C_{1h} symmetry. Since the Γ point can be regarded as a special case of the Δ point, the irreducible representations for the Γ point can also be regarded as those for the Δ point. The $\sigma_y(\sigma_v)$ mirror reflection for the former is identical to the σ mirror reflection for the latter. When we compare the character tables of these two point groups, we readily find that the A_1 and B_1 irreducible representations of C_{4v} have the same characters as the A irreducible representation of C_{1h} for symmetry operations E and $\sigma_v(\sigma)$. Hence, we can conclude that the A_1 and B_1 modes on the Γ point should connect to the A mode on the Δ point. Similarly, the A_2 and B_2 modes on the former should connect to the B mode on the latter. On the other hand, a so-called *reduction* procedure is necessary to relate the doubly degenerate E mode on the Γ point to one-dimensional irreducible representations on the Δ point, since the E mode is a reducible representation for the C_{1h} point group. The rule for the reduction is simple.

Table 3.5. Compatibility relations for the square lattice

		Σ	Δ
Γ :	A_1	A	A
	A_2	B	B
	B_1	B	A
	B_2	A	B
	E	$A+B$	$A+B$
M :	$A_1,\ B_2$	A	-
	$A_2,\ B_1$	B	-
	E	$A+B$	-
X :	$A_1,\ B_1$	-	A
	$A_2,\ B_2$	-	B

For example,

$$\begin{aligned}&\text{the number of the } A \text{ representation on the } \Delta \text{ point}\\ &= \frac{\sum\limits_{R\in C_{1h}} \chi_E(R)\chi_A(R)}{\text{the number of elements in } C_{1h}}. \end{aligned} \tag{3.38}$$

Hence, it is calculated as below, and we obtain one A representation.

		E	σ
Γ :	E	2	0
Δ :	A	1	1
		$(2\times 1+0\times 1)/2=1 \longrightarrow 1\times A$	

Similarly, we see that the number of the B representation is also one. As a result, we can conclude that the doubly degenerate E mode on the Γ point connects to one A mode and one B mode on the Δ point. Similar relations can be obtained for the rest of the highly symmetric points. The results are summarized in Table 3.5.

The assignment of the spatial symmetry to each mode can, of course, be accomplished by examining its eigenfunctions by numerical calculation. However, there is another simple but powerful method to do that [42]. This method also provides the information in a very intuitive manner on how many eigenmodes with a particular symmetry should appear in the band diagram. The method is similar to that used for the assignment of the symmetry of molecular orbitals composed of atomic orbitals. In order to show the detail of this method, consider the extended zone scheme for the reciprocal lattice space, which is shown in Fig. 3.3 for the square lattice.

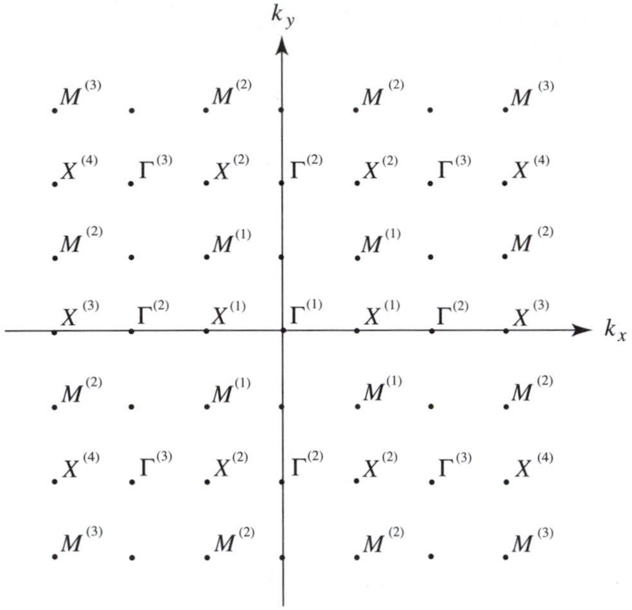

Fig. 3.3. Reciprocal lattice space of the 2D square lattice in the extended zone scheme. Highly symmetric points Γ, X, and M are denoted by dots. The number in each pair of parentheses is given in ascending order of the angular frequency of the plane wave in free space. For example, the angular frequencies that correspond to $\Gamma^{(1)}$, $\Gamma^{(2)}$, and $\Gamma^{(3)}$ are 0, $2\pi c/a$, and $2\sqrt{2}\pi c/a$, respectively. Note that, for example, there are four equivalent $\Gamma^{(2)}$ and $\Gamma^{(3)}$ points that are connected to each other with reciprocal lattice vectors

Highly symmetric points Γ, X, and M are denoted by dots, and they are numbered in ascending order of the angular frequency of the corresponding plane wave in free space. First, consider the Γ points, which have the symmetry of C_{4v}. We begin with $\Gamma^{(2)}$ that has the second lowest angular frequency among the Γ points. There are four equivalent $\Gamma^{(2)}$ points that are connected to each other with the reciprocal lattice vectors. They have the same eigenfrequency when the spatial variation of the dielectric constant is absent. When an infinitesimally small periodic perturbation is introduced to the dielectric constant, these four plane waves are mixed and their particular combinations will be the eigenmodes of the radiation field. This implies, in the terminology of group theory, that those four plane waves are the reducible four-dimensional representation of the radiation field. In order to determine what kind of symmetries the eigenmodes can have, we need to perform the *reduction* procedure. For this purpose, it is enough to see how many $\Gamma^{(2)}$ points will be invariant when the symmetry operations of the C_{4v} point group are ap-

Table 3.6. Characters of the lowest three representations at the Γ point of the square lattice

	C_{4v}	E	$2C_4$	C_4^2	$2\sigma_v$	$2\sigma_d$
$\Gamma^{(1)}$	N_R	1	1	1	1	1
	$\chi^{(1)}(R)$	1	1	1	1	1
	$\chi(R)$	1	1	1	1	1
$\Gamma^{(2)}$	N_R	4	0	0	2	0
	$\chi^{(1)}(R)$	1	-	-	1	-
	$\chi(R)$	4	0	0	2	0
$\Gamma^{(3)}$	N_R	4	0	0	0	2
	$\chi^{(1)}(R)$	1	-	-	-	1
	$\chi(R)$	4	0	0	0	2

plied.[3] Table 3.6 lists the number of the invariant Γ points, N_R, the character for one Γ point, $\chi^{(1)}(R)$, and the character as a whole, $\chi(R) = N_R \times \chi^{(1)}(R)$. $\chi^{(1)}(R)$ is the quantity that characterizes the *overlapping* of the initial and the transformed *atomic orbitals*, or, for the present problem, the overlapping of the initial and the transformed polarization vectors. This quantity plays an important role when we make the symmetry assignment for the 3D crystals. However, for the 2D crystals, it is always equal to unity as long as we treat the electric field E_z for the E polarization and the magnetic field H_z for the H polarization, since no 2D symmetry operation changes E_z or H_z.

The reduction procedure is similar to that used to obtain the compatibility relations. For example, in the case of the B_1 representation,

$$\text{the number of the } B_1 \text{ representation on the } \Gamma^{(2)} \text{ point} \\ = \frac{\sum_{R \in C_{4v}} \chi(R)\chi_{B_1}(R)}{\text{the number of elements in } C_{4v}} . \tag{3.39}$$

Hence, it is calculated as below, and we obtain one B_1 representation.

	E	$2C_4$	C_4^2	$2\sigma_v$	$2\sigma_d$
$\chi(R)$	4	0	0	2	0
$\chi_{B_1}(R)$	1	-1	1	1	-1
	$\{4 \times 1\ +$	$0\ +$	$0\ +$	$2 \times 1 \times 2\ +$	$0\}/8 = 1 \longrightarrow 1 \times B_1$

[3] For this case, the word "invariant" should be considered exactly. N_R does not include the number of the wave vectors that are transformed to their equivalent points, but is the number of those wave vectors that R keeps at their original positions.

Table 3.7. Irreducible representations given by the superposition of electromagnetic plane waves in free space whose wave vectors are regarded as those in the reciprocal lattice space of the square lattice. The number in the square brackets denotes the number of the equivalent $\boldsymbol{k}$ vectors

Symmetry	Point	Representative $\boldsymbol{k}$ vector	$\omega a/2\pi c$ in free space	Irreducible representations
C_{4v}	$\Gamma^{(1)}$	$\pi/a(0,0)$ [1]	0	A_1
	$\Gamma^{(2)}$	$\pi/a(2,0)$ [4]	1	A_1+B_1+E
	$\Gamma^{(3)}$	$\pi/a(2,2)$ [4]	$\sqrt{2}$	A_1+B_2+E
	$\Gamma^{(4)}$	$\pi/a(4,0)$ [4]	2	A_1+B_1+E
C_{4v}	$M^{(1)}$	$\pi/a(1,1)$ [4]	$1/\sqrt{2}$	A_1+B_2+E
	$M^{(2)}$	$\pi/a(3,1)$ [8]	$\sqrt{10}/2$	$A_1+A_2+B_1+B_2+2E$
C_{2v}	$X^{(1)}$	$\pi/a(1,0)$ [2]	$1/2$	A_1+B_1
	$X^{(2)}$	$\pi/a(1,2)$ [4]	$\sqrt{5}/2$	$A_1+A_2+B_1+B_2$
	$X^{(3)}$	$\pi/a(3,0)$ [2]	$3/2$	A_1+B_1
	$X^{(4)}$	$\pi/a(3,2)$ [4]	$\sqrt{13}/2$	$A_1+A_2+B_1+B_2$

We also obtain one A_1 mode and one E mode. The irreducible representations can be obtained for other Γ points in a similar manner. The results are summarized in Table 3.7. The reduction procedure is the same for the M points, because they also have the C_{4v} symmetry. On the other hand, the X points have C_{2v} symmetry. Hence, their irreducible representations can be obtained by consulting the character table of the C_{2v} point group shown in Table 3.3. The results for the M and X points are also listed in Table 3.7.

We have so far treated the case of the dielectric function with an infinitesimally small periodic modulation. When the spatial variation of the dielectric function becomes large, in such a manner that its symmetry is maintained, then the angular frequency of the eigenmode will change, but the symmetry of the eigenfunctions will remain unchanged. This is because the rigorous eigenfunction of the photonic crystal with a particular type of symmetry should be formed as a linear combination of unperturbed wave functions with the same symmetry. Otherwise, the rigorous eigenfunction would have no symmetry, and this contradicts the general rule that any eigenfunction of a photonic crystal should be attributed to one of its irreducible representations. Therefore, the mode assignment of the actual photonic bands can be accomplished by means of the comparison with those in free space. This task is easy, especially for the low frequency region where $\omega a/2\pi c \leq 1$, and examples will be given in the following sections.

3.2 Classification of Eigenmodes in the Square Lattice

Figure 3.4 shows the photonic band structure of the 2D crystal composed of a square array of circular air cylinders made in a uniform dielectric material with a dielectric constant of 2.1. The filling factor, or the volume fraction of air cylinders, is 0.25. The dielectric constant of the air rod was taken to be unity. 2D photonic crystals of this type were fabricated and studied experimentally [45]. The dispersion curves were obtained by the plane-wave expansion method described in Sect. 2.2. In order to obtain the symmetry assignment shown in Fig. 3.4, it was sufficient to know the symmetry of a few modes by examining their eigenfunctions by numerical calculation. The rest of the assignment was conducted by consulting Tables 3.5 and 3.7. If we compare the symmetries in the band diagram and those in Table 3.7, their correspondence is apparent. For example, the lowest mode on the Γ point with an eigenfrequency of zero comes from the $\Gamma^{(1)}$ point in Table 3.7. The second to the fifth modes on the Γ point come from the $\Gamma^{(2)}$ point. Group theory tells us that two of them are non-degenerate A_1 and B_1 modes and one of them is doubly degenerate E mode, which is actually observed in Fig. 3.4. Similar correspondence is found for the sixth to the ninth modes on the Γ point and for the rest of this figure. Hence, we may conclude that group theory is quite powerful for the symmetry assignment for the photonic band structure.

For the present problem, the contrast of the dielectric constant, 2.1:1.0, is not very high, and thus the dispersion curves are not so far from those in a uniform material with an averaged dielectric constant, which was shown in Fig. 1.9. This is why the extremely good correspondence shown above was obtained. However, even if the contrast of the dielectric constant is high, such as 10:1, we usually also obtain a good correspondence for $\omega a/2\pi c \leq 1$.

There is a very interesting phenomenon related to the symmetry of the eigenmodes that was first reported by Robertson et al. [40, 41]. This is concerned with the B modes on the Δ and the Σ points. These modes are antisymmetric about the mirror plane spanned by the x and z axes (the symmetry operation is σ_y) or by the vector (1, 1) and the z axis (the symmetry operation is σ_d). We examine the case of the E polarization first. We assume that an external plane wave polarized in the z direction is incident on the photonic crystal and propagated in the x direction. Hence, its electric field has only the z component of the following form:

$$E_z = E_{z0} \exp\{\mathrm{i}(kx - \omega t)\}. \tag{3.40}$$

Because this field does not depend on the y coordinate, it is symmetric under the mirror reflection σ_y. On the other hand, the B modes on the Δ point are antisymmetric. Therefore, their effective coupling on the surface of the photonic crystal is equal to zero, and the incident plane wave cannot excite the B modes. Hence, we refer to those B modes as uncoupled modes or

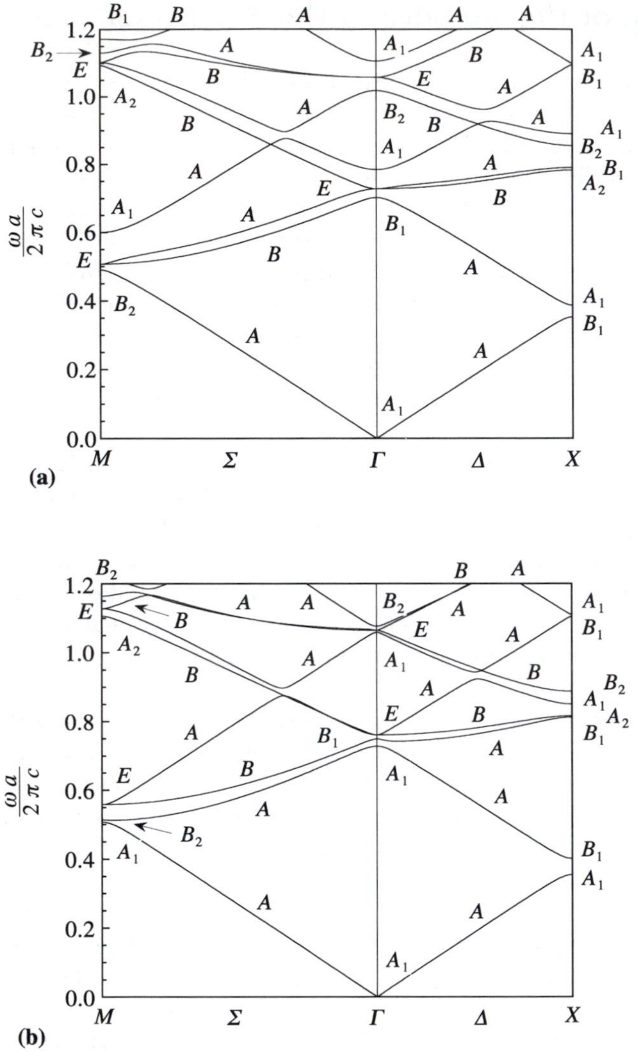

Fig. 3.4. Dispersion relation of the 2D square lattice composed of circular air-rods formed in the dielectric material with a dielectric constant of 2.1 for (**a**) the E polarization and (**b**) the H polarization. The filling factor is 0.25. (After [46])

inactive modes. The same phenomenon takes place when the incident plane wave is pointed to the (1, 1) direction or the Γ to M direction. It cannot excite the B modes on the Σ point. Hence, the B modes on the Σ point are also uncoupled modes. As for the incident plane waves polarized in the x-y plane, the same arguments can be made by considering the H polarization. For example, the incident plane wave that is propagated in the x direction now has the z component of the magnetic field of the following form:

$$H_z = H_{z0} \exp\{\mathrm{i}(kx - \omega t)\}, \tag{3.41}$$

which is symmetric under the mirror reflection σ_y. By a similar discussion to that given above for the E polarization, we can conclude that all B modes of the H polarization on the Δ and the Σ points are uncoupled modes.

Robertson et al. [40, 41] found by experimental observation that these B modes do not contribute to the light propagation. This was also confirmed by numerical calculation of the transmission spectra [42, 47]. The same feature was predicted by the vector KKR method for an fcc lattice [30]. The presence of the uncoupled modes is peculiar to 2D and 3D photonic crystals. For 1D crystals, all eigenmodes are symmetric when it is propagated in the direction perpendicular to the dielectric layers. In Chap. 4, some examples of opaque frequency regions in the transmission spectra due to the uncoupled modes will be presented.

3.3 Classification of Eigenmodes in the Hexagonal Lattice

Figure 3.5 is another example that shows the symmetry operations for the hexagonal array of circular dielectric cylinders. The structure is invariant when it is rotated by multiples of 60 degrees. Those rotations are denoted by C_6, $C_3(=C_6^2)$, $C_2(=C_6^3)$, $C_3^{-1}(=C_3^2)$, and $C_6^{-1}(=C_6^5)$. The structure is also invariant when the σ_x or σ_y mirror reflection is operated. It is readily seen that there are two other mirror reflections denoted by σ_x' and σ_x'' that are equivalent to σ_x. These three mirror planes intersect each other at angles of 60 degrees. σ_y' and σ_y'' are, on the other hand, equivalent to σ_y. Together with the identity operation E, these symmetry operations constitute the C_{6v} point group:

$$C_{6v} = \{E, C_6, C_6^{-1}, C_3, C_3^{-1}, C_2, \sigma_x, \sigma_x', \sigma_x'', \sigma_y, \sigma_y', \sigma_y''\}. \tag{3.42}$$

We can easily prove that the elements in the following sets of operations are conjugate with each other:

$$(\sigma_x,\ \sigma_x',\ \sigma_x''), \quad (\sigma_y,\ \sigma_y',\ \sigma_y''), \quad (C_6,\ C_6^{-1}), \quad (C_3,\ C_3^{-1}). \tag{3.43}$$

Hence, we may write

$$C_{6v} = \{E, 2C_6, 2C_3, C_2, 3\sigma_x, 3\sigma_y\}. \tag{3.44}$$

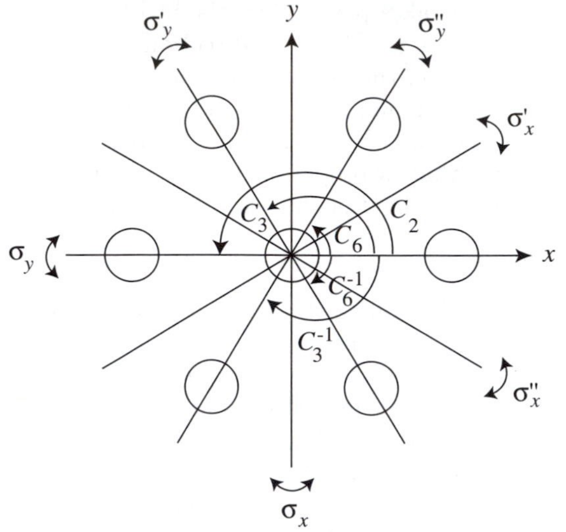

Fig. 3.5. Symmetry operations for the 2D hexagonal array of dielectric cylinders

Here, we represented C_6 and C_6^{-1} by $2C_6$, and so on.

Figure 3.6 shows the first Brillouin zone of the hexagonal lattice. The elementary lattice vectors $\{\boldsymbol{a}_{1/\!/}, \boldsymbol{a}_{2/\!/}\}$ and the elementary reciprocal lattice vectors $\{\boldsymbol{b}_{1/\!/}, \boldsymbol{b}_{2/\!/}\}$ are

$$\boldsymbol{a}_{1/\!/} = \begin{pmatrix} a \\ 0 \end{pmatrix}, \qquad \boldsymbol{a}_{2/\!/} = \begin{pmatrix} a/2 \\ \sqrt{3}a/2 \end{pmatrix}, \tag{3.45}$$

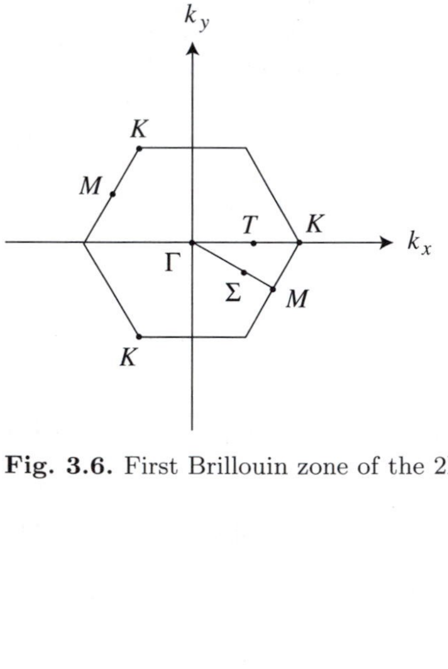

Fig. 3.6. First Brillouin zone of the 2D hexagonal lattice

$$\boldsymbol{b}_{1/\!/} = \begin{pmatrix} 2\pi/a \\ -2\pi/\sqrt{3}a \end{pmatrix}, \qquad \boldsymbol{b}_{2/\!/} = \begin{pmatrix} 0 \\ 4\pi/\sqrt{3}a \end{pmatrix}, \tag{3.46}$$

where a is the lattice constant. There are three highly symmetric points: The Γ point, (0, 0), the K point, $(4\pi/3a, 0)$, and the M point, $(\pi/a, -\pi/\sqrt{3}a)$. Those points between the Γ and the K points are denoted by T, and those between the Γ and the M points are denoted by Σ. Examining the symmetry of these points, it is apparent that

$$\mathcal{M}_\Gamma = C_{6v}. \tag{3.47}$$

As for the K point,

$$\mathcal{M}_K = \{E, 2C_3, 3\sigma_y\} = C_{3v}, \tag{3.48}$$

since we have

$$C_3 \begin{pmatrix} 4\pi/3a \\ 0 \end{pmatrix} = \begin{pmatrix} -2\pi/3a \\ 2\pi/\sqrt{3}a \end{pmatrix} = \begin{pmatrix} 4\pi/3a \\ 0 \end{pmatrix} + \boldsymbol{b}_{1/\!/}$$
$$\equiv \begin{pmatrix} 4\pi/3a \\ 0 \end{pmatrix}, \qquad \text{etc.} \tag{3.49}$$

On the other hand, for the M point,

$$\mathcal{M}_M = \{E, C_2, \sigma_y'', \sigma_x''\} = C_{2v}, \tag{3.50}$$

since we have

$$C_2 \begin{pmatrix} \pi/a \\ -\pi/\sqrt{3}a \end{pmatrix} = \begin{pmatrix} -\pi/a \\ \pi/\sqrt{3}a \end{pmatrix} = \begin{pmatrix} \pi/a \\ -\pi/\sqrt{3}a \end{pmatrix} - \boldsymbol{b}_{1/\!/}$$
$$\equiv \begin{pmatrix} \pi/a \\ 0 \end{pmatrix}, \qquad \text{etc.} \tag{3.51}$$

Table 3.8. Character table for the C_{6v} point group

C_{6v}	E	$2C_6$	$2C_3$	C_2	$3\sigma_y$	$3\sigma_x$
A_1	1	1	1	1	1	1
A_2	1	1	1	1	−1	−1
B_1	1	−1	1	−1	1	−1
B_2	1	−1	1	−1	−1	1
E_1	2	1	−1	−2	0	0
E_2	2	−1	−1	2	0	0

Table 3.9. Character table for the C_{3v} point group

C_{3v}	E	$2C_3$	$3\sigma_v$
A_1	1	1	1
A_2	1	1	-1
E	2	-1	0

Table 3.10. Compatibility relations in the hexagonal lattice

		T	Σ
Γ:	A_1	A	A
	A_2	B	B
	B_1	A	B
	B_2	B	A
	$E_1,\ E_2$	$A+B$	$A+B$
K:	A_1	A	-
	A_2	B	-
	E	$A+B$	-
M:	$A_1,\ B_1$	-	A
	$A_2,\ B_2$	-	B

Table 3.11. Irreducible representations for the electromagnetic waves in free space, whose wave vectors are reduced in the first Brillouin zone of the hexagonal lattice

Symmetry	Point	Representative $\boldsymbol{k}$ vector	$\omega a/2\pi c$ in free space	Irreducible representations
C_{6v}	$\Gamma^{(1)}$	$2\pi/a(0,0)$	0	A_1
	$\Gamma^{(2)}$	$2\pi/a(1,-1/\sqrt{3})$	$2/\sqrt{3}$	$A_1+B_2+E_1+E_2$
	$\Gamma^{(3)}$	$2\pi/a(2,0)$	2	$A_1+B_1+E_1+E_2$
C_{3v}	$K^{(1)}$	$2\pi/a(2/3,0)$	$2/3$	A_1+E
	$K^{(2)}$	$2\pi/a(2/3,2/\sqrt{3})$	$4/3$	A_1+E
	$K^{(3)}$	$2\pi/a(5/3,-1/\sqrt{3})$	$2\sqrt{7}/3$	A_1+A_2+2E
C_{2v}	$M^{(1)}$	$\pi/a(1,-1/\sqrt{3})$	$1/\sqrt{3}$	A_1+B_1
	$M^{(2)}$	$\pi/a(1,\sqrt{3})$	1	A_1+B_2
	$M^{(3)}$	$\pi/a(3,1/\sqrt{3})$	$\sqrt{7/3}$	$A_1+A_2+B_1+B_2$

As for the others,

$$\mathcal{M}_\Sigma = \{E, \sigma_x''\} = C_{1h}, \tag{3.52}$$

$$\mathcal{M}_T = \{E, \sigma_y\} = C_{1h}. \tag{3.53}$$

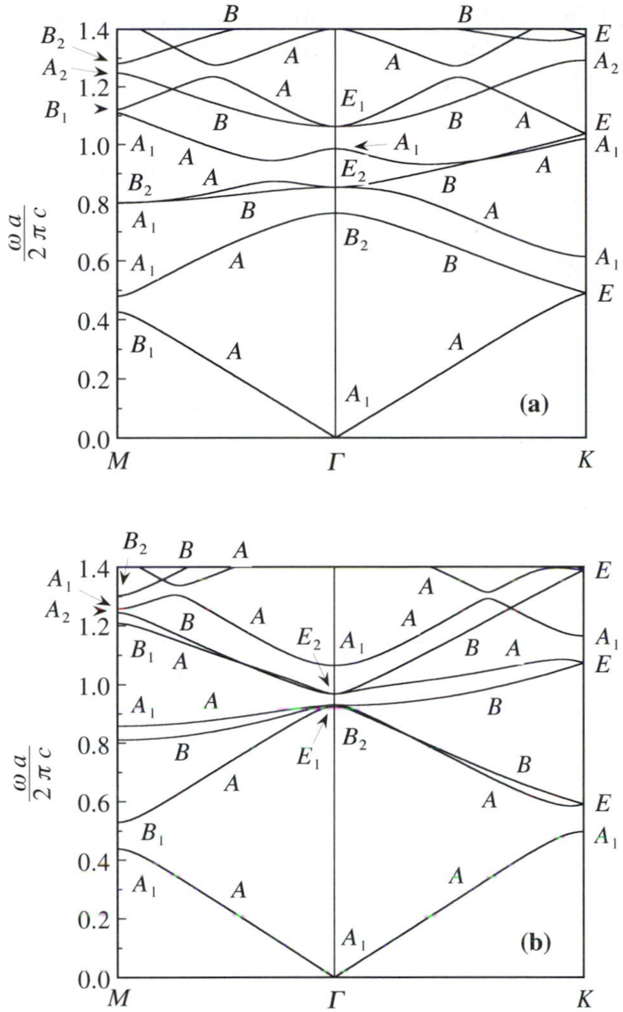

Fig. 3.7. Photonic band structure of the 2D hexagonal lattice for (**a**) the E polarization and (**b**) the H polarization

The character tables for the C_{6v} and the C_{3v} point groups are shown in Tables 3.8 and 3.9. The compatibility relations can be obtained as before. The result is summarized in Table 3.10.

As we did for the square lattice, we can obtain the irreducible representations that are expected to appear in the band diagram of the hexagonal lattice by examining the set of plane waves in free space with the same unperturbed eigenfrequencies. The reduction procedure to obtain the irreducible representations is completely similar to that used for the square lattice. The results are shown in Table 3.11. The assignment of the symmetry to each mode is easily performed by consulting Tables 3.10 and 3.11 if we know the symmetry of a few modes by solving the eigenvalue problem by numerical calculation. Figure 3.7 shows an example of the band diagram and the symmetry assignment that was obtained for the hexagonal lattice composed of circular air-cylinders formed in a uniform dielectric material with a dielectric constant of 2.72. The filling factor is 0.65. Photonic crystals of this type were fabricated and studied in [48].

3.4 Group Theory for Three-Dimensional Crystals

Now, we proceed to the case of 3D photonic crystals [49, 50]. The eigenfunctions of the electric field $\boldsymbol{E}(\boldsymbol{r})$ and the magnetic field $\boldsymbol{H}(\boldsymbol{r})$ satisfy the following eigenvalue equations:

$$\mathcal{L}_E \boldsymbol{E}(\boldsymbol{r}) \equiv \frac{1}{\varepsilon(\boldsymbol{r})} \nabla \times \{\nabla \times \boldsymbol{E}(\boldsymbol{r})\} = \frac{\omega^2}{c^2} \boldsymbol{E}(\boldsymbol{r}), \tag{3.54}$$

$$\mathcal{L}_H \boldsymbol{H}(\boldsymbol{r}) \equiv \nabla \times \left\{ \frac{1}{\varepsilon(\boldsymbol{r})} \nabla \times \boldsymbol{H}(\boldsymbol{r}) \right\} = \frac{\omega^2}{c^2} \boldsymbol{H}(\boldsymbol{r}). \tag{3.55}$$

We denote a symmetry operation that belongs to the point group $\mathcal{M}$ of the crystal by R as before. R acts on any scalar function $f(\boldsymbol{r})$ such that

$$R f(\boldsymbol{r}) \equiv f(\overleftrightarrow{\mathrm{R}}^{-1} \boldsymbol{r}), \tag{3.56}$$

where $\overleftrightarrow{\mathrm{R}}$ is now the 3×3 matrix representation of the operator R. Because R is a rotation, inversion, or their product, $\overleftrightarrow{\mathrm{R}}$ is an orthogonal matrix:

$$\overleftrightarrow{\mathrm{R}}\, \overleftrightarrow{\mathrm{R}}^t = \overleftrightarrow{\mathrm{I}}, \tag{3.57}$$

where t denotes the transposed matrix and $\overleftrightarrow{\mathrm{I}}$ is the 3×3 unit matrix. This implies that

$$\overleftrightarrow{\mathrm{R}}^t = \overleftrightarrow{\mathrm{R}}^{-1}. \tag{3.58}$$

We note that

$$\det \overleftrightarrow{\mathrm{R}} = \pm 1, \tag{3.59}$$

since from (3.57) and

$$\det \overleftrightarrow{\mathrm{R}} = \det \overleftrightarrow{\mathrm{R}}^t, \tag{3.60}$$

we have

$$(\det \overleftrightarrow{\mathrm{R}})^2 = \det\, (\overleftrightarrow{\mathrm{R}}\, \overleftrightarrow{\mathrm{R}}^t) = 1. \tag{3.61}$$

When R acts on a vector field, it changes both the position vector and the field vector:

$$R \boldsymbol{F}(\boldsymbol{r}) \equiv \overleftrightarrow{\mathrm{R}} \boldsymbol{F}(\overleftrightarrow{\mathrm{R}}^{-1} \boldsymbol{r}). \tag{3.62}$$

$\varepsilon(\boldsymbol{r})$ is invariant when it is operated by R:

$$R\varepsilon(\boldsymbol{r}) \equiv \varepsilon(\overleftrightarrow{\mathrm{R}}^{-1}\boldsymbol{r}) = \varepsilon(\boldsymbol{r}) \qquad \text{for} \quad \forall R \in \mathcal{M}, \tag{3.63}$$

Then, we can verify by direct calculation that R commutes with $\mathcal{L}_E$ and $\mathcal{L}_H$:

$$R\mathcal{L}_E R^{-1} = \mathcal{L}_E, \tag{3.64}$$

$$R\mathcal{L}_H R^{-1} = \mathcal{L}_H. \tag{3.65}$$

Therefore, the usual classification of the eigenmodes according to the irreducible representations of the $\boldsymbol{k}$-group $\mathcal{M}_{\boldsymbol{k}}$ can be applied to $\boldsymbol{E}_{\boldsymbol{k}n}(\boldsymbol{r})$ and $\boldsymbol{H}_{\boldsymbol{k}n}(\boldsymbol{r})$.

Proof of (3.64) and (3.65)

We denote $\overleftrightarrow{\mathrm{R}}$ by

$$\overleftrightarrow{\mathrm{R}} = \begin{pmatrix} R_{11}, & R_{12}, & R_{13} \\ R_{21}, & R_{22}, & R_{23} \\ R_{31}, & R_{32}, & R_{33} \end{pmatrix}. \tag{3.66}$$

For an arbitrary vector field $\boldsymbol{F}(\boldsymbol{r})$,

$$\begin{aligned} & R(\boldsymbol{\nabla}\times)R^{-1}\boldsymbol{F}(\boldsymbol{r}) \\ &= R(\boldsymbol{\nabla}\times)\overleftrightarrow{\mathrm{R}}^{-1}\boldsymbol{F}(\overleftrightarrow{\mathrm{R}}\boldsymbol{r}) \\ &= \overleftrightarrow{\mathrm{R}} \begin{pmatrix} (R_{22}R_{33} - R_{23}R_{32})(\partial_y F_3 - \partial_z F_2) \\ \quad +(R_{13}R_{32} - R_{12}R_{33})(\partial_z F_1 - \partial_x F_3) \\ \quad +(R_{12}R_{23} - R_{13}R_{22})(\partial_x F_2 - \partial_y F_1) \\ (R_{23}R_{31} - R_{21}R_{33})(\partial_y F_3 - \partial_z F_2) \\ \quad +(R_{11}R_{33} - R_{13}R_{31})(\partial_z F_1 - \partial_x F_3) \\ \quad +(R_{13}R_{21} - R_{11}R_{23})(\partial_x F_2 - \partial_y F_1) \\ (R_{21}R_{32} - R_{22}R_{31})(\partial_y F_3 - \partial_z F_2) \\ \quad +(R_{12}R_{31} - R_{11}R_{32})(\partial_z F_1 - \partial_x F_3) \\ \quad +(R_{11}R_{22} - R_{12}R_{21})(\partial_x F_2 - \partial_y F_1) \end{pmatrix}, \end{aligned} \tag{3.67}$$

where $\partial_y F_3$ denotes $\partial F_3/\partial y$, etc. When we denote the determinant of the (i,j)th cofactor of $\overleftrightarrow{\mathrm{R}}$ by Δ_{ij}, and define a matrix $\overleftrightarrow{\mathrm{A}}$ by

$$\overleftrightarrow{\mathrm{A}} = \overleftrightarrow{\mathrm{R}} \begin{pmatrix} \Delta_{11}, & \Delta_{21}, & \Delta_{31} \\ \Delta_{12}, & \Delta_{22}, & \Delta_{32} \\ \Delta_{13}, & \Delta_{23}, & \Delta_{33} \end{pmatrix}, \tag{3.68}$$

(3.67) can be rewritten as

$$R(\boldsymbol{\nabla}\times)R^{-1}\boldsymbol{F}(\boldsymbol{r}) = \overleftrightarrow{\mathrm{A}}\boldsymbol{\nabla} \times \boldsymbol{F}(\boldsymbol{r}). \tag{3.69}$$

It is a fundamental theorem of linear algebra that

$$A_{ij} = \sum_{l=1}^{3} R_{il}\Delta_{jl} = (\det \overleftrightarrow{\mathrm{R}})\delta_{ij}. \tag{3.70}$$

We thus have

$$R(\boldsymbol{\nabla}\times)R^{-1}\boldsymbol{F}(\boldsymbol{r}) = (\det \overleftrightarrow{\mathrm{R}})\boldsymbol{\nabla} \times \boldsymbol{F}(\boldsymbol{r}). \tag{3.71}$$

Since this equation holds for arbitrary vector fields, we obtain

$$R(\boldsymbol{\nabla}\times)R^{-1} = (\det \overleftrightarrow{\mathrm{R}})\boldsymbol{\nabla} \times . \tag{3.72}$$

Finally, we obtain

$$\begin{aligned} R\mathcal{L}_E R^{-1} &= R\frac{1}{\varepsilon(\boldsymbol{r})}R^{-1}\left\{R(\boldsymbol{\nabla}\times)R^{-1}\right\}^2 \\ &= \mathcal{L}_E. \end{aligned} \tag{3.73}$$

Here, we used (3.61). Equation (3.65) can be proved in a similar manner. ∎

We denote the characters of the irreducible representations of the electric and the magnetic fields by $\chi^{(E)}(R, \boldsymbol{k}n)$ and $\chi^{(H)}(R, \boldsymbol{k}n)$, respectively:

$$R\boldsymbol{E}_{\boldsymbol{k}n}(\boldsymbol{r}) = \chi^{(E)}(R, \boldsymbol{k}n)\boldsymbol{E}_{\boldsymbol{k}n}(\boldsymbol{r}), \tag{3.74}$$

$$R\boldsymbol{H}_{\boldsymbol{k}n}(\boldsymbol{r}) = \chi^{(H)}(R, \boldsymbol{k}n)\boldsymbol{H}_{\boldsymbol{k}n}(\boldsymbol{r}). \tag{3.75}$$

From one of Maxwell's equations, the following relation holds:

$$\boldsymbol{\nabla} \times \boldsymbol{E}_{\boldsymbol{k}n}(\boldsymbol{r}) = \mathrm{i}\omega_{\boldsymbol{k}n}\mu_0\boldsymbol{H}_{\boldsymbol{k}n}(\boldsymbol{r}). \tag{3.76}$$

Thus, we have,

$$R(\boldsymbol{\nabla}\times)R^{-1}R\boldsymbol{E}_{\boldsymbol{k}n}(\boldsymbol{r}) = \mathrm{i}\omega_{\boldsymbol{k}n}\mu_0 R\boldsymbol{H}_{\boldsymbol{k}n}(\boldsymbol{r}). \tag{3.77}$$

Substituting (3.72), (3.74), and (3.75), we obtain

$$(\det\ \overleftrightarrow{\mathrm{R}})\chi^{(E)}(R, \boldsymbol{k}n)\nabla \times \boldsymbol{E}_{\boldsymbol{k}n}(\boldsymbol{r}) = \mathrm{i}\omega_{\boldsymbol{k}n}\mu_0\chi^{(H)}(R, \boldsymbol{k}n)\boldsymbol{H}_{\boldsymbol{k}n}(\boldsymbol{r}). \tag{3.78}$$

Comparing (3.76) and (3.78), we obtain

$$\chi^{(H)}(R, \boldsymbol{k}n) = (\det\ \overleftrightarrow{\mathrm{R}})\chi^{(E)}(R, \boldsymbol{k}n), \tag{3.79}$$

where

$$\det\ \overleftrightarrow{\mathrm{R}} = \begin{cases} +1 & \text{(proper transformation)}, \\ -1 & \text{(improper transformation)}. \end{cases} \tag{3.80}$$

This difference between $\chi^{(E)}$ and $\chi^{(H)}$ originates from the fact that the electric field is a true vector whereas the magnetic field is an axial vector. Equation (3.79) implies that the irreducible representation for the magnetic field is generally different from that for the electric field. On the other hand, the character for the electric displacement $\boldsymbol{D}_{\boldsymbol{k}n}(\boldsymbol{r})$ is the same as that for the electric field because the following relation holds:

$$\begin{aligned}(R\boldsymbol{D}_{\boldsymbol{k}n})(\boldsymbol{r}) &= (R\varepsilon)(\boldsymbol{r})(R\boldsymbol{E}_{\boldsymbol{k}n})(\boldsymbol{r}) \\ &= \chi^{(E)}(R,\boldsymbol{k}n)\boldsymbol{D}_{\boldsymbol{k}n}(\boldsymbol{r}). \end{aligned} \tag{3.81}$$

Therefore, $\boldsymbol{D}_{\boldsymbol{k}n}(\boldsymbol{r})$ and $\boldsymbol{E}_{\boldsymbol{k}n}(\boldsymbol{r})$ are attributed to the same irreducible representation. Because $\boldsymbol{D}_{\boldsymbol{k}n}(\boldsymbol{r})$ and $\boldsymbol{H}_{\boldsymbol{k}n}(\boldsymbol{r})$ are purely transverse waves even with the spatial variation of the dielectric function, we can constitute their eigenfunctions with transverse plane waves alone. We can use this fact to obtain the irreducible representations of the eigenfunctions.

3.5 Classification of Eigenmodes in the Simple Cubic Lattice

We consider the details of the reduction procedure for the transverse vector fields with a simple cubic lattice composed of dielectric spheres as an example. Figure 3.8 shows its symmetry operations. First, the structure is invariant under the rotations about the x, y, and z axes by 90, 180, and 270 degrees. These rotations are denoted by C_4, C_2, and C_4^{-1} as before. C_4 and C_4^{-1} are conjugate with each other. We should also note that the C_4 rotation about the x axis is conjugate with that about the y and the z axes. Hence, these three C_4 and three C_4^{-1} rotations have the same characters and may be denoted by $6C_4$. Similarly, three C_2 rotations are conjugate with each other and denoted by $3C_2$. There are two other sets of rotation axes that are denoted by C_2' and C_3. Six C_2' operations are conjugate with each other. As for the C_3 rotation, eight operations are conjugate. The structure is also invariant under the spatial inversion I. The rest of the symmetry operations are given by the product of I and the operations described above: $6IC_4$, $3IC_2$, $6IC_2'$, and $8IC_3$. Among these, IC_2 and IC_2' are usually denoted by σ_h and σ_d, respectively. They are mirror reflections about the horizontal plane and the diagonal plane. Together with the identity operation E, these symmetry operations constitute the O_h point group:

$$O_h = \{E, 6C_4, 3C_2, 3C_2', 8C_3, I, 6IC_4, 3\sigma_h, 6\sigma_d, 8IC_3\}. \tag{3.82}$$

The character table of the O_h point group is shown in Table 3.12.

The first Brillouin zone of the simple cubic lattice is depicted in Fig. 2.1. We examine the symmetry of its highly symmetric points. It is apparent that the Γ point has the O_h symmetry:

$$\mathcal{M}_\Gamma = O_h. \tag{3.83}$$

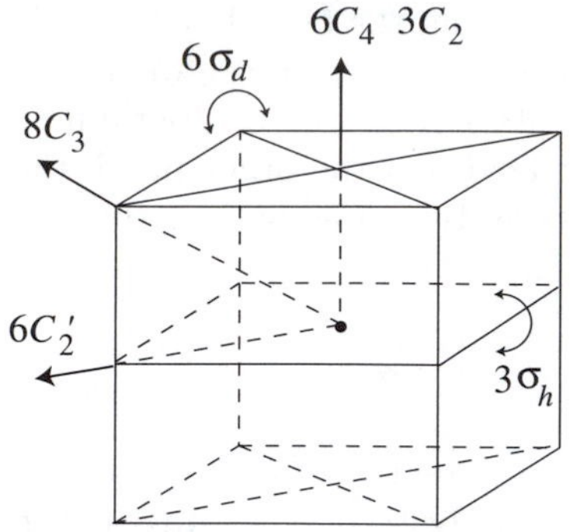

Fig. 3.8. Symmetry operations for the O_h point group

Table 3.12. Character table for the O_h point group

O_h	E	$6C_4$	$3C_2$	$6C_2'$	$8C_3$	I	$6IC_4$	$3\sigma_h$	$6\sigma_d$	$8IC_3$
A_{1g}	1	1	1	1	1	1	1	1	1	1
A_{2g}	1	−1	1	−1	1	1	−1	1	−1	1
E_g	2	0	2	0	−1	2	0	2	0	−1
T_{1g}	3	1	−1	−1	0	3	1	−1	−1	0
T_{2g}	3	−1	−1	1	0	3	−1	−1	1	0
A_{1u}	1	1	1	1	1	−1	−1	−1	−1	−1
A_{2u}	1	−1	1	−1	1	−1	1	−1	1	−1
E_u	2	0	2	0	−1	−2	0	−2	0	1
T_{1u}	3	1	−1	−1	0	−3	−1	1	1	0
T_{2u}	3	−1	−1	1	0	−3	1	1	−1	0

The symmetries of other points are

$$\mathcal{M}_R = O_h, \tag{3.84}$$

$$\mathcal{M}_X = \{E, 2C_4, C_2, 2C_2', 2C_2'', I, 2IC_4, \sigma_h, 2\sigma_v, 2\sigma_d\} = D_{4h}, \tag{3.85}$$

$$\mathcal{M}_M = D_{4h}, \tag{3.86}$$

$$\mathcal{M}_\Delta = \{E, 2C_4, C_2, 2\sigma_v, 2\sigma_d\} = C_{4v}, \tag{3.87}$$

$$\mathcal{M}_T = C_{4v}, \tag{3.88}$$

$$\mathcal{M}_\Lambda = \{E, 2C_3, 3\sigma_d\} = C_{3v}, \tag{3.89}$$

$$\mathcal{M}_\Sigma = \{E, C_2', \sigma_h, \sigma_d\} = C_{2v}, \tag{3.90}$$

$$\mathcal{M}_Z = \{E, C_2, 2\sigma_h\} = C_{2v}, \tag{3.91}$$

$$\mathcal{M}_S = \{E, C_2', \sigma_h, \sigma_d\} = C_{2v}. \tag{3.92}$$

The symmetry operations for these points are illustrated in Fig. 3.9. The character table for the D_{4h} point group is given in Table 3.13. The character

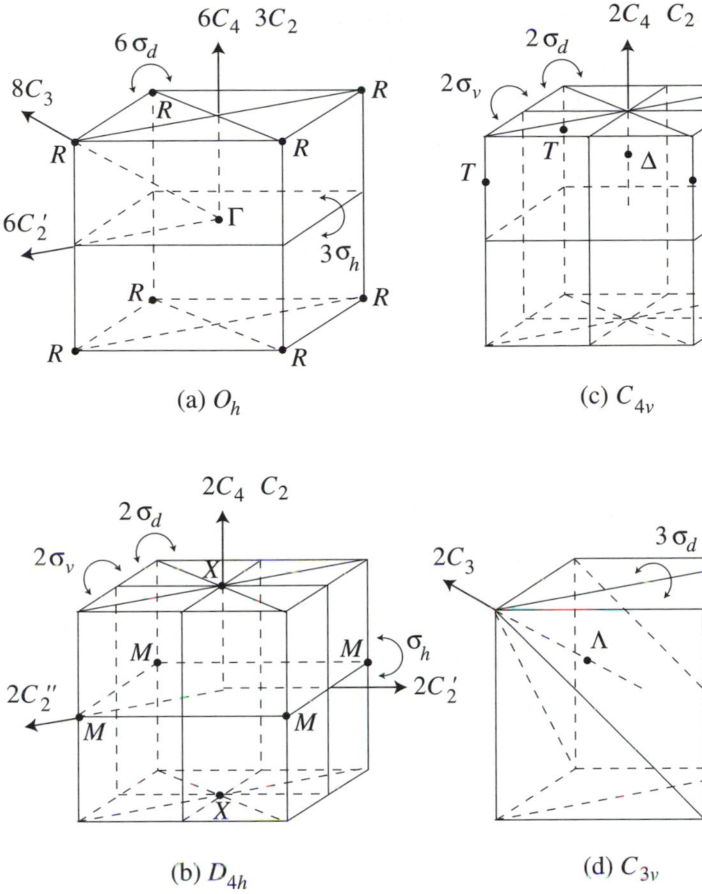

Fig. 3.9. Symmetry operations for the highly symmetric points in the first Brillouin zone of the simple cubic lattice

tables for the C_{4v}, C_{3v}, and C_{2v} point groups are given in Tables 3.2, 3.9, and 3.3, respectively. Our next task is to obtain the compatibility relations. The procedure is exactly the same as that for the 2D crystals described in the previous sections. The final results are summarized in Tables 3.14 and 3.15.

In Sects. 3.1–3.3, we obtained the possible combinations of irreducible representations that appear in the band diagrams of the 2D crystals by performing the reduction procedure for the reducible representations composed of plane waves. We would like to do the same thing for the electromagnetic eigenmodes in the 3D crystals. However, there is an apparent difference between the present and the previous problems. That is, we now have to deal with vector fields instead of scalar fields. Hence, we should first examine how

Table 3.13. Character table for the D_{4h} point group

D_{4h}	E	$2C_4$	C_2	$2C_2'$	$2C_2''$	I	$2IC_4$	σ_h	$2\sigma_v$	$2\sigma_d$
A_{1g}	1	1	1	1	1	1	1	1	1	1
A_{2g}	1	1	1	−1	−1	1	1	1	−1	−1
B_{1g}	1	−1	1	1	−1	1	−1	1	1	−1
B_{2g}	1	−1	1	−1	1	1	−1	1	−1	1
E_g	2	0	−2	0	0	2	0	−2	0	0
A_{1u}	1	1	1	1	1	−1	−1	−1	−1	−1
A_{2u}	1	1	1	−1	−1	−1	−1	−1	1	1
B_{1u}	1	−1	1	1	−1	−1	1	−1	−1	1
B_{2u}	1	−1	1	−1	1	−1	1	−1	1	−1
E_u	2	0	−2	0	0	−2	0	2	0	0

we can constitute the reducible representations with related plane waves and how we can reduce them to obtain the irreducible representations.

We begin with the case of the magnetic field. Because we assumed throughout this book that the magnetic permeability is constant, the magnetic field is purely transverse. Hence, its eigenfunction can be expressed as a linear combination of transverse plane waves. For the infinitesimally small spatial modulation of the dielectric constant, the eigenfunction is a particular combination of the plane waves with the same eigenfrequency in free space that satisfies Bloch's theorem. In other words, the eigenfunction is composed of the plane waves with equivalent $\boldsymbol{k}$ vectors in the extended Brillouin zone. We have to take into account the presence of two independent polarizations for each $\boldsymbol{k}$ vector. We can thus constitute the reducible representation of the eigenfunction of the magnetic field. Our final task is to reduce it to obtain the irreducible representations. The reduction procedure for the 3D crystals is slightly different from that for the 2D crystals because of the presence of the two polarizations.

We examine the reduction procedure with the $X^{(1)}$ point as an example. It has the D_{4h} symmetry as was shown before. There are two equivalent $X^{(1)}$ points in the first Brillouin zone whose coordinates are $\pi/a(0,0,1)$ and $\pi/a(0,0,-1)$. First, we have to count the number of exactly invariant $X^{(1)}$ points, N_R, when operation R in D_{4h} is applied. This is listed in Table 3.16. It can readily be seen that non-zero N_R appears for the rotations about the z axis and the mirror reflections whose mirror plane contains the z axis. In other words, the wave vector should be on the rotation axis or on the mirror plane for N_R to be non-zero.

The next task is to find the character for one $X^{(1)}$ point, $\chi^{(1)}(R)$. In general, we denote the plane waves with the two independent polarization vectors by $\boldsymbol{f}_1(\boldsymbol{r})$ and $\boldsymbol{f}_2(\boldsymbol{r})$:

Table 3.14. Compatibility relations for the simple cubic lattice

		Δ	Σ	Λ
Γ :	A_{1g}	A_1	A_1	A_1
	A_{2g}	B_1	B_1	A_2
	E_g	$A_1 + B_1$	$A_1 + B_1$	E
	T_{1g}	$A_2 + E$	$A_2 + B_1 + B_2$	$A_2 + E$
	T_{2g}	$B_2 + E$	$A_1 + A_2 + B_2$	$A_1 + E$
	A_{1u}	A_2	A_2	A_2
	A_{2u}	B_2	B_2	A_1
	E_u	$A_2 + B_2$	$A_2 + B_2$	E
	T_{1u}	$A_1 + E$	$A_1 + B_1 + B_2$	$A_1 + E$
	T_{2u}	$B_1 + E$	$A_1 + A_2 + B_1$	$A_2 + E$
		Δ	Z	S
X :	A_{1g}	A_1	A_1	A_1
	A_{2g}	A_2	B_2	B_2
	B_{1g}	B_1	A_1	B_2
	B_{2g}	B_2	B_2	A_1
	E_g	E	$A_2 + B_1$	$A_2 + B_1$
	A_{1u}	A_2	A_2	A_2
	A_{2u}	A_1	B_1	B_1
	B_{1u}	B_2	A_2	B_1
	B_{2u}	B_1	B_1	A_2
	E_u	E	$A_1 + B_2$	$A_1 + B_2$

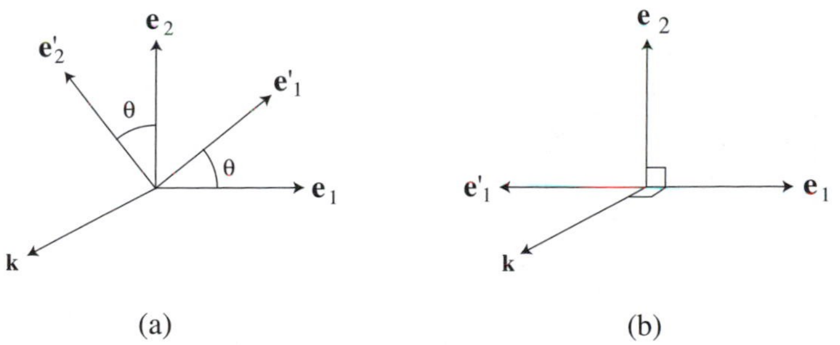

Fig. 3.10. Variation of the polarization vectors by (**a**) rotation and (**b**) mirror reflection

Table 3.15. Compatibility relations for the simple cubic lattice (continued from Table 3.14)

		Σ	Z	T
$M:$	A_{1g}	A_1	A_1	A_1
	A_{2g}	B_1	B_1	A_2
	B_{1g}	B_1	A_1	B_1
	B_{2g}	A_1	B_1	B_2
	E_g	A_2+B_2	A_2+B_2	E
	A_{1u}	A_2	A_2	A_2
	A_{2u}	B_2	B_2	A_1
	B_{1u}	B_2	A_2	B_2
	B_{2u}	A_2	B_2	B_1
	E_u	A_1+B_1	A_1+B_1	E
		Λ	S	T
$R:$	A_{1g}	A_1	A_1	A_1
	A_{2g}	A_2	B_2	B_1
	E_g	E	A_1+B_2	A_1+B_1
	T_{1g}	A_2+E	$A_2+B_1+B_2$	A_2+E
	T_{2g}	A_1+E	$A_1+A_2+B_1$	B_2+E
	A_{1u}	A_2	A_2	A_2
	A_{2u}	A_1	B_1	B_2
	E_u	E	A_2+B_1	A_2+B_2
	T_{1u}	A_1+E	$A_1+B_1+B_2$	A_1+E
	T_{2u}	A_2+E	$A_1+A_2+B_2$	B_1+E

Table 3.16. Characters of the lowest two representations at the X point of the simple cubic lattice. (After [50])

	D_{4h}	E	$2C_4$	C_4^2	$2C_2'$	$2C_2''$	I	$2IC_4$	σ_h	$2\sigma_v$	$2\sigma_d$
$X^{(1)}$	N_R	2	2	2	0	0	0	0	0	2	2
	$\chi^{(1)}(R)$	2	0	-2	–	–	–	–	–	0	0
	$\chi(R)$	4	0	-4	0	0	0	0	0	0	0
$X^{(2)}$	N_R	8	0	0	0	0	0	0	0	4	0
	$\chi^{(1)}(R)$	2	–	–	–	–	–	–	–	0	–
	$\chi(R)$	16	0	0	0	0	0	0	0	0	0

Table 3.17. Character for the rotation and the mirror reflection

R	C_6	C_4	C_3	C_2	σ
$\chi^{(1)}(R)$	1	0	-1	-2	0

$$f_1(r) = e_1 e^{ik\cdot r}, \tag{3.93}$$
$$f_2(r) = e_2 e^{ik\cdot r}. \tag{3.94}$$

Here, we do not restrict our argument to the $X^{(1)}$ point. The following discussion is common to all wave vectors. When the rotation about the axis that contains the wave vector $\boldsymbol{k}$ is operated, according to the definition of the operation (3.62), these vector functions are transformed as follows (see Fig. 3.10(a)):

$$R_\theta \begin{pmatrix} \boldsymbol{f}_1(\boldsymbol{r}) \\ \boldsymbol{f}_2(\boldsymbol{r}) \end{pmatrix} = \begin{pmatrix} \cos\theta, & -\sin\theta \\ \sin\theta, & \cos\theta \end{pmatrix} \begin{pmatrix} \boldsymbol{f}_1(\boldsymbol{r}) \\ \boldsymbol{f}_2(\boldsymbol{r}) \end{pmatrix}. \tag{3.95}$$

Note that $\boldsymbol{k}\cdot\boldsymbol{r}$ is invariant even when the rotation is operated. By definition, $\chi^{(1)}(R)$ is given by the trace of the 2×2 matrix in the right-hand side of (3.95), i.e.,

$$\chi^{(1)}(R_\theta) = 2\cos\theta. \tag{3.96}$$

As for the mirror reflection σ, whose mirror plane contains the $\boldsymbol{k}$ vector, the two plane waves are transformed such that (see Fig. 3.10(b))

$$\sigma \begin{pmatrix} \boldsymbol{f}_1(\boldsymbol{r}) \\ \boldsymbol{f}_2(\boldsymbol{r}) \end{pmatrix} = \begin{pmatrix} -1, & 0 \\ 0, & 1 \end{pmatrix} \begin{pmatrix} \boldsymbol{f}_1(\boldsymbol{r}) \\ \boldsymbol{f}_2(\boldsymbol{r}) \end{pmatrix}. \tag{3.97}$$

Note that $\boldsymbol{k}\cdot\boldsymbol{r}$ is also invariant when the mirror reflection is operated. We thus obtain

$$\chi^{(1)}(\sigma) = 0. \tag{3.98}$$

The results are summarized in Table 3.17. The character for the two equivalent $X^{(1)}$ points as a whole, $\chi(R)$, is then given by

$$\chi(R) = N_R \times \chi^{(1)}(R), \tag{3.99}$$

and it is listed in Table 3.16. The character of the reducible representation composed of the plane waves corresponding to the eight equivalent $X^{(2)}$ points with the second lowest unperturbed angular frequency, whose representative wave vector is given by $\pi/a(0,2,1)$, can be obtained by following the same procedure. The results are also summarized in Table 3.16. Our final task is the reduction of the reducible representation to find the irreducible representations contained in it. This can be accomplished by consulting the character table of D_{4h} (Table 3.13). In the case of the E_g representation, for example,

$$\text{the number of the } E_g \text{ representation on the } X^{(1)} \text{ point}$$
$$= \frac{\sum_{R \in D_{4h}} \chi(R)\chi_{E_g}(R)}{\text{the number of elements in } D_{4h}}. \qquad (3.100)$$

By this calculation, we find that we have one E_g and one E_u irreducible representations for the $X^{(1)}$ point. We can obtain the irreducible represen-taions for other symmetric points in a similar manner. The final results are summarized in Table 3.18.

From the relation given in (3.79), we can obtain the irreducible representations for the electric field. In order to distinguish the representations for the magnetic field and the electric field, we denote the former by superscript (H) and the latter by superscript (E). So, for example, we have $E_g^{(H)}$ and $E_u^{(H)}$ representations for the $X^{(1)}$ point. Now, the character for the electric field accompanied by the magnetic field of the E_g representation is given by

$$\chi^{(E)}(R) = (\det \overleftrightarrow{\mathrm{R}})\chi_{E_g}^{(H)}(R). \qquad (3.101)$$

When we compare $\chi^{(E)}(R)$ given by this equation with the characters for the D_{4h} point group listed in Table 3.13, we can readily find that $\chi^{(E)}(R)$ agrees with the character for the E_u representation. Similarly, we obtain the E_g representation for the electric field accompanied with the magnetic field of the E_u representation:

$$X^{(1)}: \quad E_g^{(H)} \Longleftrightarrow E_u^{(E)}, \qquad (3.102)$$
$$E_u^{(H)} \Longleftrightarrow E_g^{(E)}. \qquad (3.103)$$

We see that the same sets of irreducible representations appear for both magnetic and electric fields. This is not a coincidence, but a necessity. In order to prove this general property it is enough to think of the procedure to obtain the irreducible representations for the electric displacement that is transverse even with the spatial modulation of the dielectric constant like the magnetic field. We can readily understand that the procedure is exactly the same as for the magnetic field. We thus obtain the same set of irreducible representations for $\boldsymbol{D}_{\boldsymbol{k}n}(\boldsymbol{r})$, and hence, for $\boldsymbol{E}_{\boldsymbol{k}n}(\boldsymbol{r})$ as for $\boldsymbol{H}_{\boldsymbol{k}n}(\boldsymbol{r})$. Therefore, the superscripts (E) and (H) are dropped in Table 3.18 because the irreducible representations as a whole are common to both fields. $\Gamma^{(1)}$ should be dealt with separately because its wave vector is equal to zero and there is no distinction between the transverse and longitudinal plane waves. Actually, these are constant polarizations to all 3D directions with zero eigenfrequency, and we have to treat them equally. Then, the degree of freedom, i.e., the number of independent polarizations is three for this particular case. When we follow the argument given above for this case in a similar manner, we see that $\chi(R_\theta) = 2\cos\theta + 1$, $\chi(\sigma) = 1$, etc., which leads to the T_{1u} representation.

As an example, Fig. 3.11 shows the symmetry assignment of the magnetic field of the simple cubic lattice whose band structure was shown in Fig.

Table 3.18. Irreducible representations of the unperturbed wave functions for a simple cubic lattice. (After [50])

Symmetry	Point	Representative $\boldsymbol{k}$ vector	$\omega a/2\pi c$ in free space	Number of equivalent points	Irreducible representations
O_h	$\Gamma^{(1)}$	$\pi/a(0, 0, 0)$	0	1	T_{1u}
	$\Gamma^{(2)}$	$\pi/a(0, 0, 2)$	1	6	$T_{1g} + T_{2g} + T_{1u} + T_{2u}$
	$R^{(1)}$	$\pi/a(1, 1, 1)$	$\sqrt{3}/2$	8	$E_g + T_{1g} + T_{2g} + E_u + T_{1u} + T_{2u}$
D_{4h}	$X^{(1)}$	$\pi/a(0, 0, 1)$	$1/2$	2	$E_g + E_u$
	$X^{(2)}$	$\pi/a(0, 2, 1)$	$\sqrt{5}/2$	8	$A_{1g} + A_{2g} + B_{1g} + B_{2g} + 2E_g$ $+A_{1u} + A_{2u} + B_{1u} + B_{2u} + 2E_u$
	$M^{(1)}$	$\pi/a(1, 1, 0)$	$\sqrt{2}/2$	4	$A_{2g} + B_{1g} + E_g + A_{2u} + B_{1u} + E_u$
	$M^{(2)}$	$\pi/a(1, 1, 2)$	$\sqrt{6}/2$	8	$A_{1g} + A_{2g} + B_{1g} + B_{2g} + 2E_g$ $+A_{1u} + A_{2u} + B_{1u} + B_{2u} + 2E_u$
C_{4v}	$\Delta^{(1)}$	$\pi/a(0, 0, \alpha)$	$\alpha/2$	1	E
	$T^{(1)}$	$\pi/a(1, 1, \alpha)$	$\sqrt{\alpha^2 + 2}/2$	4	$A_1 + A_2 + B_1 + B_2 + 2E$
C_{3v}	$\Lambda^{(1)}$	$\pi/a(\alpha, \alpha, \alpha)$	$\sqrt{3}\alpha/2$	1	E
C_{2v}	$\Sigma^{(1)}$	$\pi/a(\alpha, \alpha, 0)$	$\sqrt{2}\alpha/2$	1	$B_1 + B_2$
	$Z^{(1)}$	$\pi/a(1, 0, \alpha)$	$\sqrt{\alpha^2 + 1}/2$	2	$A_1 + A_2 + B_1 + B_2$
	$S^{(1)}$	$\pi/a(\alpha, \alpha, 1)$	$\sqrt{2\alpha^2 + 1}/2$	2	$A_1 + A_2 + B_1 + B_2$

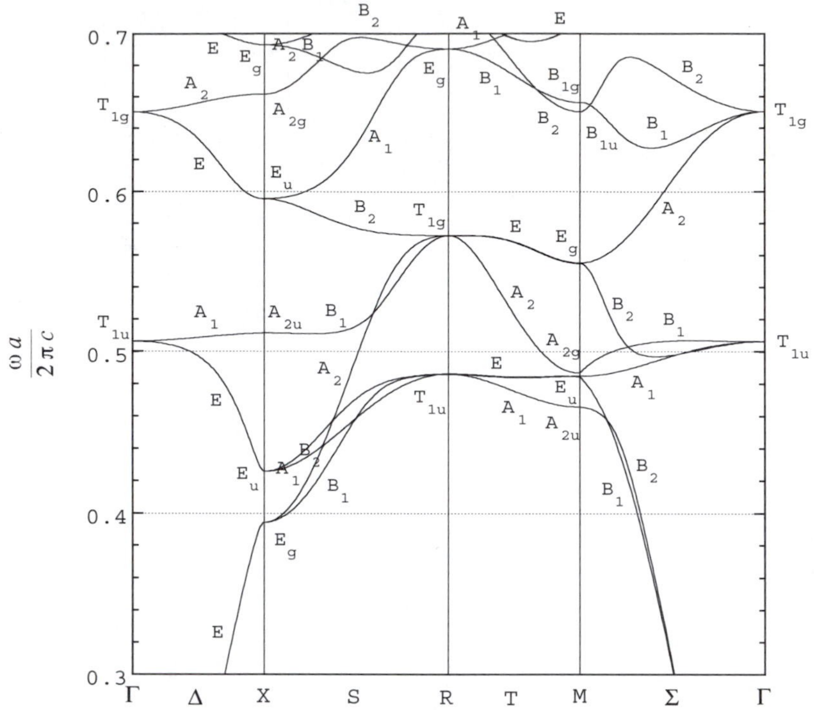

Fig. 3.11. Photonic band structure and the irreducible representations for the magnetic field of the eigenmodes of the simple cubic lattice with a dielectric sphere at each lattice point. The ordinate is the normalized frequncy. The following values were assumed for the numerical calculation; the dielectric constants of the spheres and the background are 13.0 and 1.0, and the ratio of the lattice constant a and the radius of the sphere is 1:0.3. All one-dimensional eigenmodes at the Δ point, i.e., A_1, A_2, B_1, and B_2 modes, and A_1 and A_2 modes at the Σ point are uncoupled modes. (After [50])

2.2. For the actual procedure of the symmetry assignment, we had to know the irreducible representations of only several eigenmodes by studying their wave functions numerically, and then the compatibility relations (Table 3.14) that connect the irreducible representations for adjacent wave vectors with each other and Table 3.18 were utilized in order to accomplish the rest of the assignments. The irreducible representations of the electric field can be derived using (3.79).

Now, we examine the uncoupled modes on the Δ and Σ points. For the Δ point, its symmetry is C_{4v}, and each of its one-dimensional representations has a particular value of the character for the two σ_v mirror reflections; σ_y on the x-z plane and σ_x on the y-z plane. This implies that the eigenmodes on the Δ point should be symmetric or antisymmetric for both σ_x and σ_y. On the other hand, the magnetic field of a plane wave that is linearly polarized to the x or y direction must be symmetric for one mirror reflection and antisymmetric for the other. Therefore, there is a symmetry mismatch between those one-dimensional eigenmodes and the incident plane wave. Then, we can identify them as uncoupled modes. As for the Σ point, its symmetry is C_{2v}, which has the following two mirror reflections; σ_z on the x-y plane and σ_{xy} on the plane spanned by the z axis and the $(1,1,0)$ vector. Among the four one-dimensional representations, A_1 is symmetric and A_2 is antisymmetric for both mirror reflections. Hence, they are uncoupled modes. On the other hand, B_1 and B_2 modes are symmetric for one mirror reflection and antisymmetric for the other. Then they have the same symmetry as the linearly polarized plane wave, and they can be excited by the incident plane wave, although they are coupled to only one polarization. Then, for example, we expect total reflection for both polarizations at $\omega a/2\pi c \approx 0.6$ when the incident plane wave is pointed to the Γ-M direction, i.e., the $(1,1,0)$ direction in spite of the non-zero density of states. Since the frequency range where only the A_2 mode exists is quite large, it will not be difficult to confirm this total reflection experimentally. In addition, the assumed specimen is transparent for one polarization and opaque for the other polarization at $\omega a/2\pi c \approx 0.47$, 0.53, and 0.66. This feature is also an important consequence of the symmetry of the lattice and its eigenfunctions.

3.6 Classification of Eigenmodes in the fcc Lattice

Photonic crystals with the face-centered-cubic (fcc) structure have attracted great interest, since specimens of good quality with a small lattice constant, e.g., less than 1 mm, can be obtained. The representative specimens are artificial opal crystals composed of the fcc lattice of dielectric spheres made of silicon dioxide [2, 51–53]. Fcc-like structures are also realized by *Yablonovite* [54, 55]. In this section, we will examine the symmetry properties of the fcc photonic crystals, and present the compatibility relations and the irreducible representations of the electromagnetic field that are expected to appear in the photonic band diagrams from the group-theoretical argument.

The first Brillouin zone of the fcc lattice is shown in Fig. 3.12. The elementary lattice vectors $\{\boldsymbol{a}_1, \boldsymbol{a}_2, \boldsymbol{a}_3\}$ and the elementary reciprocal lattice vectors $\{\boldsymbol{b}_1, \boldsymbol{b}_2, \boldsymbol{b}_3\}$ of the fcc lattice are given by

$$\boldsymbol{a}_1 = \frac{a}{2}\begin{pmatrix} 0 \\ 1 \\ 1 \end{pmatrix}, \quad \boldsymbol{a}_2 = \frac{a}{2}\begin{pmatrix} 1 \\ 0 \\ 1 \end{pmatrix}, \quad \boldsymbol{a}_3 = \frac{a}{2}\begin{pmatrix} 1 \\ 1 \\ 0 \end{pmatrix}, \tag{3.104}$$

$$\boldsymbol{b}_1 = \frac{2\pi}{a}\begin{pmatrix} -1 \\ 1 \\ 1 \end{pmatrix}, \boldsymbol{b}_2 = \frac{2\pi}{a}\begin{pmatrix} 1 \\ -1 \\ 1 \end{pmatrix}, \boldsymbol{b}_3 = \frac{2\pi}{a}\begin{pmatrix} 1 \\ 1 \\ -1 \end{pmatrix}. \tag{3.105}$$

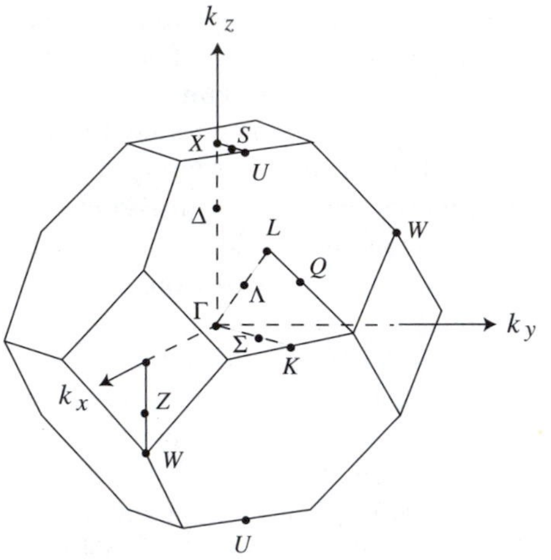

Fig. 3.12. First Brillouin zone of the fcc lattice

Table 3.19. Character table for the D_{3d} point group

D_{3d}	E	$2C_3$	$3C_2'$	I	$2IC_3$	$3\sigma_v$
A_{1g}	1	1	1	1	1	1
A_{2g}	1	1	−1	1	1	−1
E_g	2	−1	0	2	−1	0
A_{1u}	1	1	1	−1	−1	−1
A_{2u}	1	1	−1	−1	−1	1
E_u	2	−1	0	−2	1	0

It can be easily found that the symmetries of the highly symmetric points are as follows.

$$\mathcal{M}_\Gamma = O_h, \tag{3.106}$$

$$\mathcal{M}_L = D_{3d}, \tag{3.107}$$

$$\mathcal{M}_W = D_{2d}, \tag{3.108}$$

$$\mathcal{M}_\Delta = C_{4v}, \tag{3.109}$$

$$\mathcal{M}_\Lambda = C_{3v}, \tag{3.110}$$

$$\mathcal{M}_\Sigma = \mathcal{M}_Z = \mathcal{M}_S = \mathcal{M}_U = \mathcal{M}_K = C_{2v}. \tag{3.111}$$

The character tables of the O_h, C_{4v}, C_{3v}, C_{2v} point groups were already presented. The characters for the D_{3d} and D_{2d} point groups are shown in Tables 3.19 and 3.20. The compatibility relations can be obtained in a similar manner as was done for the simple cubic lattice. The results are listed in Tables 3.21 and 3.22.

Table 3.20. Character table for the D_{2d} point group

D_{2d}	E	$2IC_4$	C_2	$2C_2'$	$2\sigma_d$
A_1	1	1	1	1	1
A_2	1	1	1	−1	−1
B_1	1	−1	1	1	−1
B_2	1	−1	1	−1	1
E	2	0	−2	0	0

Table 3.21. Compatibility relations for the fcc lattice

Γ	Δ	Σ	Λ	$\overline{\Gamma U}$	$\overline{\Gamma W}$
A_{1g}	A_1	A_1	A_1	A	A
A_{2g}	B_1	B_2	A_2	B	A
E_g	A_1+B_1	A_1+B_2	E	$A+B$	$2A$
T_{1g}	A_2+E	$A_2+B_1+B_2$	A_2+E	$A+2B$	$A+2B$
T_{2g}	B_2+E	$A_1+A_2+B_1$	A_1+E	$2A+B$	$A+2B$
A_{1u}	A_2	A_2	A_2	B	B
A_{2u}	B_2	B_1	A_1	A	B
E_u	A_2+B_2	A_2+B_1	E	$A+B$	$2B$
T_{1u}	A_1+E	$A_1+B_1+B_2$	A_1+E	$2A+B$	$2A+B$
T_{2u}	B_1+E	$A_1+A_2+B_2$	A_2+E	$A+2B$	$2A+B$
X	Δ	Z	S		
A_{1g}	A_1	A_1	A_1		
A_{2g}	A_2	B_2	B_2		
B_{1g}	B_1	A_1	B_2		
B_{2g}	B_2	B_2	A_1		
E_g	E	A_2+B_1	A_2+B_1		
A_{1u}	A_2	A_2	A_2		
A_{2u}	A_1	B_1	B_1		
B_{1u}	B_2	A_2	B_1		
B_{2u}	B_1	B_1	A_2		
E_u	E	A_1+B_2	A_1+B_2		

Table 3.22. Compatibility relations for the fcc lattice (continued from Table 3.21)

L	Λ	Q	$\overline{UL}$	$\overline{KL}$
A_{1g}	A_1	A	A	A
A_{2g}	A_2	B	B	B
E_g	E	$A+B$	$A+B$	$A+B$
A_{1u}	A_2	A	B	B
A_{2u}	A_1	B	A	A
E_u	E	$A+B$	$A+B$	$A+B$
W	Z	Q	$\overline{\Gamma W}$	$\overline{WU}$
A_1	A_1	A	A	A
A_2	A_2	B	B	B
B_1	A_2	A	B	B
B_2	A_1	B	A	A
E	B_1+B_2	$A+B$	$A+B$	$A+B$
U	S	$\overline{\Gamma U}$	$\overline{UW}$	$\overline{UL}$
A_1	A_1	A	A	A
A_2	A_2	B	B	B
B_1	B_1	A	B	A
B_2	B_2	B	A	B
K	Σ	$\overline{KL}$	$\overline{KW}$	
A_1	A_1	A	A	
A_2	A_2	B	B	
B_1	B_1	A	B	
B_2	B_2	B	A	

The irreducible representations of the electromagnetic eigenmodes that are expected to appear in the photonic band diagram from the group-theoretical argument can be obtained by simply following the same procedure as we did for the simple cubic lattice. The only difference is the choice of the wave vectors with which we make the basis plane waves. The results are listed in Table 3.23.

Table 3.23. Irreducible representations of the unperturbed wave functions for a fcc lattice

Symmetry	Point	Representative $\boldsymbol{k}$ vector	$\omega a/2\pi c$ in free space	Number of equivalent points	Irreducible representations
O_h	$\Gamma^{(1)}$	$\pi/a(0, 0, 0)$	0	1	T_{1u}
	$\Gamma^{(2)}$	$\pi/a(1, 1, 1)$	$\sqrt{3}$	8	$E_g + T_{1g} + T_{2g} + E_u + T_{1u} + T_{2u}$
D_{3d}	$L^{(1)}$	$\pi/a(1/2, 1/2, 1/2)$	$\sqrt{3/4}$	2	$E_g + E_u$
	$L^{(2)}$	$\pi/a(-3/2, 1/2, 1/2)$	$\sqrt{11/4}$	6	$A_{1g} + A_{2g} + 2E_g + A_{1u} + A_{2u} + 2E_u$
D_{2d}	$W^{(1)}$	$\pi/a(0, 1/2, 1)$	$\sqrt{5/4}$	4	$A_1 + A_2 + B_1 + B_2 + 2E$
	$W^{(2)}$	$\pi/a(1, 3/2, 0)$	$\sqrt{13/4}$	4	$A_1 + A_2 + B_1 + B_2 + 2E$
C_{4v}	$\Delta^{(1)}$	$\pi/a(0, 0, \alpha)$	α	1	E
	$\Delta^{(2)}$	$\pi/a(1, 1, \alpha\text{-}1)$	$\sqrt{3 + \alpha^2 - 2\alpha}$	4	$A_1 + A_2 + B_1 + B_2 + 2E$
	$\Delta^{(3)}$	$\pi/a(0, 0, \alpha - 2)$	$2 - \alpha$	1	E
C_{3v}	$\Lambda^{(1)}$	$\pi/a(\alpha, \alpha, \alpha)$	$\sqrt{3}\alpha$	1	E
	$\Lambda^{(2)}$	$\pi/a(\alpha - 1, \alpha - 1, \alpha - 1)$	$\sqrt{3}(1 - \alpha)$	1	E
	$\Lambda^{(3)}$	$\pi/a(\alpha + 1, \alpha - 1, \alpha - 1)$	$\sqrt{3\alpha^2 - 2\alpha + 3}$	3	$A_1 + A_2 + 2E$
	$\Lambda^{(4)}$	$\pi/a(\alpha, \alpha, \alpha - 2)$	$\sqrt{3\alpha^2 - 4\alpha + 4}$	3	$A_1 + A_2 + 2E$

Table 3.23. (Continued)

Symmetry	Point	Representative $\boldsymbol{k}$ vector	$\omega a/2\pi c$ in free space	Number of equivalent points	Irreducible representations
C_{2v}	$\Sigma^{(1)}$	$\pi/a(\alpha,\ \alpha,\ 0)$	$\sqrt{2}\alpha$	1	$B_1 + B_2$
	$\Sigma^{(2)}$	$\pi/a(\alpha-1,\ \alpha-1,\ 1)$	$\sqrt{2\alpha^2 - 4\alpha + 3}$	2	$A_1 + A_2 + B_1 + B_2$
	$Z^{(1)}$	$\pi/a(0,\ \alpha,\ 1)$	$\sqrt{\alpha^2 + 1}$	2	$A_1 + A_2 + B_1 + B_2$
	$Z^{(2)}$	$\pi/a(1,\ \alpha-1,\ 0)$	$\sqrt{\alpha^2 - 2\alpha + 2}$	2	$A_1 + A_2 + B_1 + B_2$
	$Z^{(3)}$	$\pi/a(0,\ \alpha-2,\ 1)$	$\sqrt{\alpha^2 - 4\alpha + 5}$	2	$A_1 + A_2 + B_1 + B_2$
	$S^{(1)}$	$\pi/a(\alpha,\ \alpha,\ 1)$	$\sqrt{2\alpha^2 + 1}$	2	$A_1 + A_2 + B_1 + B_2$
	$S^{(2)}$	$\pi/a(\alpha-1,\ \alpha-1,\ 0)$	$\sqrt{2}(1-\alpha)$	1	$B_1 + B_2$
	$U^{(1)}$	$\pi/a(1/4,\ 1/4,\ 1)$	$\sqrt{9/8}$	3	$A_1 + A_2 + 2B_1 + 2B_2$
	$U^{(2)}$	$\pi/a(5/4,\ -3/4,\ 1)$	$\sqrt{17/8}$	2	$A_1 + A_2 + B_1 + B_2$
	$K^{(1)}$	$\pi/a(3/4,\ 3/4,\ 0)$	$\sqrt{9/8}$	3	$A_1 + A_2 + 2B_1 + 2B_2$
	$K^{(2)}$	$\pi/a(3/4,\ -5/4,\ 1)$	$\sqrt{17/8}$	2	$A_1 + A_2 + B_1 + B_2$

4. Transmission Spectra

The transmission spectra of photonic crystals reflect their band structure directly. They are often used for the experimental characterization of real specimens. In this chapter, we will formulate the plane-wave expansion method for the numerical calculation of the transmittance and the Bragg reflectivity of 2D crystals. It will be demonstrated that the bandgaps and the uncoupled modes lead to opaque frequency ranges. It will also be shown that the small group velocity peculiar to the 2D and 3D crystals, that is, the group-velocity anomaly, is equivalent to a large effective refractive index. The refraction law on the surface of photonic crystals will be given.

4.1 Light Transmission and Bragg Reflection

Several methods are known for the numerical calculation of the transmission spectra of photonic crystals such as the spherical-wave expansion method [27, 30, 56, 57], the transfer matrix method [58], and the finite-difference time-domain (FDTD) method [59]. Although the spherical-wave expansion method is only applicable to those crystals composed of dielectric spheres and cylinders, its convergence property is quite good. The transfer matrix method can be applied to any periodic structures and many excellent calculations have been performed. The FDTD method may be the most general one, which is based on the numerical time-integration of Maxwell's equation. Although it is time-consuming (i.e., the computational time is long) and needs a large memory capacity, fairly accurate calculation is possible thanks to the tremendous development of computing facilities.

On the other hand, the plane-wave expansion method [47, 60] is based on the Fourier expansion of the internal field and the dielectric function, and is applicable to any 2D photonic crystal as long as the wave vector of the incident plane wave lies in the 2D x-y plane. Although its convergence becomes poor and it is not practical when the contrast of the dielectric constant is, say, larger than 4:1, it is very intuitive and has an advantage that the Bragg reflectivity can be obtained at the same time. Complex dielectric constants can be dealt with as well as real dielectric constants. The low-threshold lasing due to the group-velocity anomaly in 2D photonic crystals was demonstrated by this method [61]. An example of the configulation for the calculation is

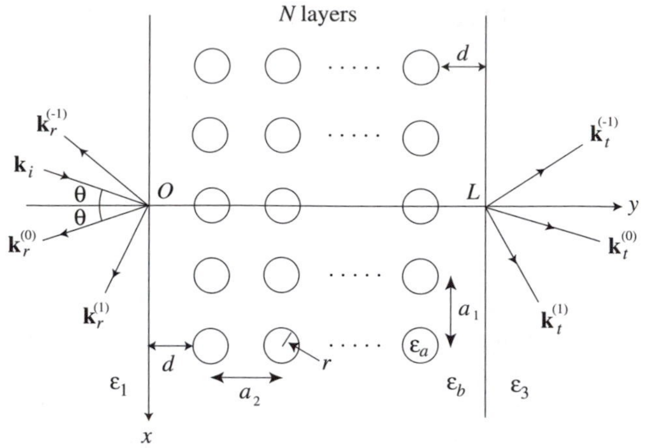

Fig. 4.1. Configuration for the calculation of the transmission and the Bragg reflection spectra (*top view*). See text for detail. (After [60])

shown in Fig. 4.1. The external plane wave in region 1 is incident on the left (front) surface of the specimen located in region 2, which consists of N layers of circular rods with a radius r_a and the background dielectric. The dielectric constant of the rods and the background are denoted by ε_a and ε_b, respectively. The lattice constant is denoted by a_1 in the x direction and a_2 in the y direction, and the distance between the surface and the first layer of the rods is denoted by d. When we deal with a square lattice, we assume the same value for a_1 and a_2. The dielectric constants of regions 1 and 3 are ε_1 and ε_3, respectively. The total thickness of the specimen is $L = (N-1)a_2 + 2(R+d)$.

Since photonic crystals may be regarded as periodic gratings, the electromagnetic field in region 1 is the superposition of the incident plane wave and the reflected Bragg waves, whereas that in region 3 is composed of the transmitted Bragg waves. Here, we regard the specular reflection and the straight transmission as the reflected and the transmitted Bragg waves of the zeroth order. The 2D wave vector of the incident plane wave is denoted by $\mathbf{k}_{i/\!/} = (k_1 \sin\theta,\ k_1 \cos\theta) = (k_x,\ k_{1y})$, where θ is the angle of incidence. $k_1 = \sqrt{\varepsilon_1}\omega/c$, where ω is the angular frequency of the incident field. The wave vectors of the reflected and the transmitted Bragg waves of order n, $\boldsymbol{k}_{r/\!/}^{(n)}$ and $\boldsymbol{k}_{t/\!/}^{(n)}$, are given by

$$k_{r,x}^{(n)} = k_{t,x}^{(n)} = k_x^{(n)} = k_x + G_n, \tag{4.1}$$

$$G_n = 2\pi n/a_1 \ (n = 0, \pm 1, \pm 2, \cdots), \tag{4.2}$$

$$k_{r,y}^{(n)} = \begin{cases} -\sqrt{k_1^2 - (k_x^{(n)})^2} & \text{if } k_1 \geq |k_x^{(n)}|, \\ -\mathrm{i}\sqrt{(k_x^{(n)})^2 - k_1^2} & \text{otherwise,} \end{cases} \tag{4.3}$$

$$k_{t,y}^{(n)} = \begin{cases} \sqrt{k_3^2 - (k_x^{(n)})^2} & \text{if } k_3 \geq |k_x^{(n)}|, \\ \mathrm{i}\sqrt{(k_x^{(n)})^2 - k_3^2} & \text{otherwise.} \end{cases} \tag{4.4}$$

Here, $k_3 = \sqrt{\varepsilon_3}\omega/c$. Because of the spatial periodicity along the x axis, the reflected and the transmitted Bragg waves receive a momentum along the x axis which is a multiple of $2\pi/a_1$ in (4.3) and (4.4).

As was explained in Sect. 2.2, the vector wave equations derived from Maxwell's equations can be reduced to two independent scalar equations when the wave vector lies in the x-y plane, or in other words, when the fields are independent of the z coordinate. These two modes are called the E polarization, for which the electric field is parallel to the z axis, and the H polarization, for which the magnetic field is parallel to the z axis, respectively. Because of the continuity condition for the tangential component of the electric or the magnetic field at each boundary, the fields in regions 1 and 3 are also characterized by these polarizations. In the next two sections, the details of the plane-wave expansion method will be explained. Those readers who are not interested in the details of the calculation may skip them and proceed to Sect. 4.4.

4.2 Field Equations

4.2.1 *E* Polarization

First, we consider the E polarization. The electric field in region 2 (E_{2z}) satisfies (2.57):

$$\mathcal{K}_E E_{2z}(\boldsymbol{r}_{/\!/}) \equiv \left\{ \frac{1}{\varepsilon(\boldsymbol{r}_{/\!/})} \left(\frac{\partial^2}{\partial x^2} + \frac{\partial^2}{\partial y^2} \right) + \frac{\omega^2}{c^2} \right\} E_{2z}(\boldsymbol{r}_{/\!/}) = 0. \tag{4.5}$$

The differencial operator $\mathcal{K}_E$ is defined by the first equality in the above equation. The electric fields in region 1 (E_{1z}) and region 3 (E_{3z}) are given by

$$E_{1z}(\boldsymbol{r}_{/\!/}) = E_0 \exp(\mathrm{i}\boldsymbol{k}_{i/\!/} \cdot \boldsymbol{r}_{/\!/}) + \sum_{n=-\infty}^{\infty} R_n^{(E)} \exp(\mathrm{i}\boldsymbol{k}_{r/\!/}^{(n)} \cdot \boldsymbol{r}_{/\!/}), \tag{4.6}$$

$$E_{3z}(\boldsymbol{r}_{/\!/}) = \sum_{n=-\infty}^{\infty} T_n^{(E)} \exp\left\{\mathrm{i}\boldsymbol{k}_{t/\!/}^{(n)} \cdot (\boldsymbol{r}_{/\!/} - \boldsymbol{L}_{/\!/})\right\}, \tag{4.7}$$

where E_0, $R_n^{(E)}$, and $T_n^{(E)}$ are the amplitudes of the electric field of the incident wave, the reflected Bragg waves, and the transmitted Bragg waves, respectively. $\boldsymbol{L}_{/\!/} = (0, L)$. Now, we introduce a boundary-value function $f_E(x, y)$:

$$f_E(x,y) \equiv \frac{1}{L} \sum_{n=-\infty}^{\infty} \left\{ y T_n^{(E)} + (L-y)\left(\delta_{n0} E_0 + R_n^{(E)}\right) \right\} \exp\left(\mathrm{i} k_x^{(n)} x\right), \tag{4.8}$$

where δ_{nm} is Kronecker's δ. Then, f_E is equal to E_{1z} or E_{3z} on each surface of the specimen:

$$f_E(x,0) = E_{1z}(x,0) \qquad \text{and} \qquad f_E(x,L) = E_{3z}(x,L). \tag{4.9}$$

Next, we define ψ_E by

$$\psi_E(x,y) \equiv E_{2z}(x,y) - f_E(x,y). \tag{4.10}$$

Then, from (4.5), (4.9), and (4.10), we obtain

$$\mathcal{K}_E \psi_E = -\mathcal{K}_E f_E, \tag{4.11}$$

$$\psi_E(x,0) = \psi_E(x,L) = 0. \tag{4.12}$$

By means of this mathematical trick, we have converted the boundary conditions for E_{2z} into the field equation of ψ_E [62]. Then, we can relate the internal field in region 2 to E_0 at the first stage of the calculation, thereby we can exclude the free oscillation and obtain the solution of the forced oscillation with the incident plane wave as the excitation source. The choice of f_E is arbitrary as long as (4.9) is satisfied.

Now, we expand $\psi_E(x,y)$ and $1/\varepsilon(x,y)$ in Fourier series:

$$\psi_E(x,y) = \sum_{n=-\infty}^{\infty} \sum_{m=1}^{\infty} A_{nm}^{(E)} \exp(\mathrm{i} k_x^{(n)} x) \sin \frac{m\pi}{L} y, \tag{4.13}$$

$$\frac{1}{\varepsilon(x,y)} = \sum_{n=-\infty}^{\infty} \sum_{m=-\infty}^{\infty} \kappa_{nm} \exp\left\{ \mathrm{i}\left(G_n x + \frac{m\pi}{L} y \right) \right\}. \tag{4.14}$$

Note that only sine functions appear in (4.13), since $\psi_E(x,0) = 0$. Substituting (4.8), (4.13), and (4.14) into (4.11) and comparing the coefficients of the independent Fourier components, we obtain the following equation after a simple but lengthy calculation.

$$\begin{aligned}
&\frac{\omega^2}{c^2} A_{nm}^{(E)} + \sum_{n'=-\infty}^{\infty} \sum_{m'=1}^{\infty} \left\{ (k_x^{(n')})^2 + \left(\frac{m'\pi}{L} \right)^2 \right\} \\
&\qquad\qquad \times \left(\kappa_{n-n',m+m'} - \kappa_{n-n',|m-m'|} \right) A_{n'm'}^{(E)} \\
&= -\frac{2\omega^2}{\pi c^2} \frac{(-1)^{m-1} T_n^{(E)} + R_n^{(E)} + \delta_{n0} E_0}{m} \\
&\quad + \frac{2}{\pi} \sum_{n'=-\infty}^{\infty} \left(k_x^{(n')} \right)^2 \sum_{m'=1}^{\infty} \left(\kappa_{n-n',|m-m'|} - \kappa_{n-n',m+m'} \right)
\end{aligned}$$

$$\times \frac{(-1)^{m'-1}T_{n'}^{(E)} + R_{n'}^{(E)} + \delta_{n'0}E_0}{m'}. \tag{4.15}$$

When we derived (4.15), the following two relations were used.

$$y = \frac{2L}{\pi}\sum_{m=1}^{\infty}\frac{(-1)^{m-1}}{m}\sin\frac{m\pi}{L}y, \tag{4.16}$$

$$1 = \frac{2}{\pi}\sum_{m=1}^{\infty}\frac{1-(-1)^{m}}{m}\sin\frac{m\pi}{L}y. \tag{4.17}$$

The remaining boundary conditions are the continuity of the x component of the magnetic field and the y component of the magnetic induction. The latter proves to be equivalent to the continuity of the z component of the electric field and does not yield an independent equation. Then, the former condition at both boundaries leads to

$$\pi\sum_{m=1}^{\infty} mA_{nm}^{(E)} = \left(\mathrm{i}Lk_{r,y}^{(n)} + 1\right)R_n^{(E)} - T_n^{(E)} + \delta_{n0}E_0\left(\mathrm{i}Lk_{1,y}+1\right), \tag{4.18}$$

$$\pi\sum_{m=1}^{\infty} m(-1)^m A_{nm}^{(E)} = R_n^{(E)} + \left(\mathrm{i}Lk_{t,y}^{(n)} - 1\right)T_n^{(E)} + \delta_{n0}E_0. \tag{4.19}$$

Equations (4.15), (4.18), and (4.19) determine the unknown coefficients, $A_{nm}^{(E)}$, $R_n^{(E)}$ and $T_n^{(E)}$. In the actual numerical calculation, we restrict the number of terms that appear in the Fourier expansions as we did for the photonic band calculation in Chap. 2, and solve the linear equations.

4.2.2 *H* Polarization

The H polarization can be treated in a similar manner. The magnetic fields in region 1 (H_{1z}) and in rigion 3 (H_{3z}) are given as follows.

$$H_{1z}(\boldsymbol{r}_{/\!/}) = H_0\exp\left(\mathrm{i}\boldsymbol{k}_{i/\!/}\cdot\boldsymbol{r}_{/\!/}\right) + \sum_n R_n^{(H)}\exp\left(\mathrm{i}\boldsymbol{k}_{r/\!/}^{(n)}\cdot\boldsymbol{r}_{/\!/}\right), \tag{4.20}$$

$$H_{3z}(\boldsymbol{r}_{/\!/}) = \sum_n T_n^{(H)}\exp\left\{\mathrm{i}\boldsymbol{k}_{t/\!/}^{(n)}\cdot\left(\boldsymbol{r}_{/\!/} - \boldsymbol{L}_{/\!/}\right)\right\}, \tag{4.21}$$

where H_0, $R_n^{(H)}$, and $T_n^{(H)}$ are the amplitudes of the magnetic field of the incident wave, the reflected Bragg waves, and the transmitted Bragg waves, respectively. The magnetic field in region 2 satisfies (2.58):

$$\begin{aligned}\mathcal{K}_H H_{2z}(\boldsymbol{r}_{/\!/}) &\equiv \left\{\frac{\partial}{\partial x}\frac{1}{\varepsilon(\boldsymbol{r}_{/\!/})}\frac{\partial}{\partial x} + \frac{\partial}{\partial y}\frac{1}{\varepsilon(\boldsymbol{r}_{/\!/})}\frac{\partial}{\partial y} + \frac{\omega^2}{c^2}\right\}H_{2z}(\boldsymbol{r}_{/\!/})\\ &= 0,\end{aligned} \tag{4.22}$$

where the differential operator $\mathcal{K}_H$ is defined by the first equality in the above equation. Now, we introduce another boundary-value function $f_H(x,y)$:

$$f_H(x,y) \equiv \frac{1}{L}\sum_{n=-\infty}^{\infty}\left\{yT_n^{(H)} + (L-y)(\delta_{n0}H_0 + R_n^{(H)})\right\}\exp\left(\mathrm{i}k_x^{(n)}x\right). \tag{4.23}$$

Then we have

$$f_H(x,0) = H_{1z}(x,0) \qquad \text{and} \qquad f_H(x,L) = H_{3z}(x,L). \tag{4.24}$$

Next, we define $\psi_H(x,y)$ by

$$\psi_H(x,y) \equiv H_{2z}(x,y) - f_H(x,y). \tag{4.25}$$

We thus obtain from (4.22), (4.24), and (4.25)

$$\mathcal{K}_H\psi_H = -\mathcal{K}_H f_H, \tag{4.26}$$

$$\psi_H(x,0) = \psi_H(x,L) = 0. \tag{4.27}$$

Now, we expand $\psi_H(x,y)$ in Fourier series as before:

$$\psi_H(x,y) = \sum_{n=-\infty}^{\infty}\sum_{m=1}^{\infty} A_{nm}^{(H)}\exp(\mathrm{i}k_x^{(n)}x)\sin\frac{m\pi}{L}y. \tag{4.28}$$

Substituting (4.23), (4.28), and (4.14) into (4.26), we obtain

$$\begin{aligned}
&\frac{\omega^2}{c^2}A_{nm}^{(H)} - \sum_{n'=-\infty}^{\infty}\sum_{m'=1}^{\infty}\left[\left(\frac{mm'\pi^2}{L^2} + k_x^{(n)}k_x^{(n')}\right)\kappa_{n-n',|m-m'|}\right.\\
&\qquad\qquad\left. + \left(\frac{mm'\pi^2}{L^2} - k_x^{(n)}k_x^{(n')}\right)\kappa_{n-n',m+m'}\right]A_{n'm'}^{(H)}\\
&= -\frac{2\omega^2}{\pi c^2}\frac{(-1)^{m-1}T_n^{(H)} + R_n^{(H)} + \delta_{n0}H_0}{m}\\
&\quad + \frac{2\pi m}{L^2}\sum_{n'=-\infty}^{\infty}(T_{n'}^{(H)} - R_{n'}^{(H)} - \delta_{n'0}H_0)\kappa_{n-n',m}\\
&\quad + \frac{2k_x^{(n)}}{\pi}\sum_{n'=-\infty}^{\infty}\sum_{m'=1}^{\infty}k_x^{(n')}(\kappa_{n-n',|m-m'|} - \kappa_{n-n',m+m'})\\
&\quad \times\frac{(-1)^{m'-1}T_{n'}^{(H)} + R_{n'}^{(H)} + \delta_{n'0}H_0}{m'}.
\end{aligned} \tag{4.29}$$

The remaining boundary condition is the continuity of the x component of the electric field, whereas the continuity of the y component of the electric displacement is equivalent to that of the z component of the magnetic field. The former boundary condition at $y = 0$ and $y = L$ leads to

$$\pi\varepsilon_1 \sum_{m=1}^{\infty} m A_{nm}^{(H)}$$
$$= \left(\mathrm{i}\varepsilon_2 L k_{r,y}^{(n)} + \varepsilon_1\right) R_n^{(H)} - \varepsilon_1 T_n^{(H)} + \delta_{n0} H_0 \left(\mathrm{i}\varepsilon_2 L k_{1,y} + \varepsilon_1\right), \tag{4.30}$$

$$\pi\varepsilon_3 \sum_{m=1}^{\infty} m(-1)^m A_{nm}^{(H)}$$
$$= \varepsilon_3 R_n^{(H)} + \left(\mathrm{i}\varepsilon_2 L k_{t,y}^{(n)} - \varepsilon_3\right) T_n^{(H)} + \delta_{n0}\varepsilon_3 H_0. \tag{4.31}$$

Equations (4.29), (4.30), and (4.31) determine the unknown coefficients, $A_{nm}^{(H)}$, $R_n^{(H)}$, and $T_n^{(H)}$.

4.3 Fourier Transform of the Dielectric Function

4.3.1 Square Lattice

The next task is to calculate the Fourier coefficients of the dielectric function, κ_{nm}. First, we calculate them for the configuration shown in Fig. 4.1. In order to do this, we extend $\varepsilon^{-1}(x, y)$ symmetrically to the region of negative y $(-L \le y \le 0)$. This extension, of course, does not make changes in $\varepsilon^{-1}(x, y)$ in the positive y region $(0 \le y \le L)$. Whether we extend it symmetrically or antisymmetrically to the negative y region is not critical. It is thus expressed as

$$\frac{1}{\varepsilon(x,y)} = \frac{1}{\varepsilon_\mathrm{b}} + \left(\frac{1}{\varepsilon_\mathrm{a}} - \frac{1}{\varepsilon_\mathrm{b}}\right) \sum_{j=1}^{2} \sum_{l=-\infty}^{\infty} \sum_{l'=0}^{N-1} S(\boldsymbol{r}_{/\!/} - \boldsymbol{u}_{j/\!/}(l, l')), \tag{4.32}$$

where $\boldsymbol{u}_{j/\!/}(l, l')$ is given by

$$\boldsymbol{u}_{1/\!/}(l, l') = (a_1 l,\ a_2 l' + r_\mathrm{a} + d), \tag{4.33}$$
$$\boldsymbol{u}_{2/\!/}(l, l') = (a_1 l,\ -a_2 l' - r_\mathrm{a} - d), \tag{4.34}$$

and

$$S(\boldsymbol{r}_{/\!/}) = \begin{cases} 1 & (|\boldsymbol{r}_{/\!/}| \le r_\mathrm{a}), \\ 0 & (|\boldsymbol{r}_{/\!/}| > r_\mathrm{a}). \end{cases} \tag{4.35}$$

Then,

$$\kappa_{nm} = \frac{1}{2a_1 L} \int_0^{a_1} \mathrm{d}x \int_{-L}^{L} \mathrm{d}y\, \frac{1}{\varepsilon(x,y)} \exp\left\{-\mathrm{i}\left(G_n x + \frac{m\pi}{L} y\right)\right\}$$
$$= \frac{1}{\varepsilon_\mathrm{b}} \delta_{n0}\delta_{m0} + \frac{1}{2a_1 L}\left(\frac{1}{\varepsilon_\mathrm{a}} - \frac{1}{\varepsilon_\mathrm{b}}\right) \sum_{j=1}^{2} \sum_{l=-\infty}^{\infty} \sum_{l'=0}^{N-1}$$

$$\times \int_0^{a_1} \mathrm{d}x \int_{-L}^{L} \mathrm{d}y \; S(\boldsymbol{r}_{/\!/} - \boldsymbol{u}_{j/\!/}(l,l')) \exp\left(-\mathrm{i}\boldsymbol{G}_{nm/\!/} \cdot \boldsymbol{r}_{/\!/}\right)$$

$$= \frac{1}{\varepsilon_\mathrm{b}} \delta_{n0}\delta_{m0} + \frac{1}{2a_1 L}\left(\frac{1}{\varepsilon_\mathrm{a}} - \frac{1}{\varepsilon_\mathrm{b}}\right) \sum_{j=1}^{2} \sum_{l'=0}^{N-1} \exp\left\{-\mathrm{i}\boldsymbol{G}_{nm/\!/} \cdot \boldsymbol{u}_{j/\!/}(0,l')\right\}$$

$$\times \int_{-\infty}^{\infty} \mathrm{d}x \int_{-\infty}^{\infty} \mathrm{d}y \; S(\boldsymbol{r}_{/\!/}) \exp\left(-\mathrm{i}\boldsymbol{G}_{nm/\!/} \cdot \boldsymbol{r}_{/\!/}\right), \tag{4.36}$$

where $\boldsymbol{G}_{nm/\!/} = (G_n, m\pi/L)$. The Fourier transform of $S(\boldsymbol{r}_{/\!/})$ is given by

$$\int_{-\infty}^{\infty} \mathrm{d}x \int_{-\infty}^{\infty} \mathrm{d}y \; S(\boldsymbol{r}_{/\!/}) \exp\left(-i\boldsymbol{G}_{nm/\!/} \cdot \boldsymbol{r}_{/\!/}\right) = \frac{2\pi r_\mathrm{a}}{G_{nm}} J_1(G_{nm} r_\mathrm{a}), \tag{4.37}$$

where J_1 is the Bessel function of the first order and

$$G_{nm} = |\boldsymbol{G}_{nm/\!/}| = \sqrt{G_n^2 + \left(\frac{m\pi}{L}\right)^2}. \tag{4.38}$$

We used (2.96) when we derived (4.37). Next, we calculate the summation over j and l' in (4.36). We define S_j $(j = 1, 2)$ by

$$S_j = \sum_{l'=0}^{N-1} \exp\left\{-\mathrm{i}\boldsymbol{G}_{nm/\!/} \cdot \boldsymbol{u}_{j/\!/}(0,l')\right\}. \tag{4.39}$$

We can show easily that

$$S_1 = \begin{cases} \exp\{-\mathrm{i}m\pi(r_\mathrm{a} + d)/L\} \dfrac{1 - \exp(-\mathrm{i}mNa_2\pi/L)}{1 - \exp(-\mathrm{i}ma_2\pi/L)} & (m \neq 0), \\ N & (m = 0), \end{cases} \tag{4.40}$$

$$S_2 = S_1^*. \tag{4.41}$$

Finally, we obtain

$$\kappa_{nm} = \begin{cases} \dfrac{f}{\varepsilon_\mathrm{a}} + \dfrac{1-f}{\varepsilon_\mathrm{b}} & (n = 0,\ m = 0) \\ 2f\left(\dfrac{1}{\varepsilon_\mathrm{a}} - \dfrac{1}{\varepsilon_\mathrm{b}}\right) \dfrac{J_1(G_{nm} r_\mathrm{a})}{G_{nm} r_\mathrm{a}} & (n \neq 0,\ m = 0) \\ \dfrac{2f}{N}\left(\dfrac{1}{\varepsilon_\mathrm{a}} - \dfrac{1}{\varepsilon_\mathrm{b}}\right) \cos\dfrac{m\pi}{2} & \\ \quad \times \dfrac{\sin[a_2 mN\pi/(2L)]}{\sin[a_2 m\pi/(2L)]} \dfrac{J_1(G_{nm} r_\mathrm{a})}{G_{nm} r_\mathrm{a}} & (m \neq 0) \end{cases} \tag{4.42}$$

where f is the filling factor given by

$$f = \frac{N\pi r_\mathrm{a}^2}{a_1 L}. \tag{4.43}$$

When the thickness L and the distance d are commensurate with the lattice constant a_2, i.e., if

$$r_\mathrm{a} + d = \frac{a_2}{2}, \qquad \text{and thus,} \qquad L = Na_2, \tag{4.44}$$

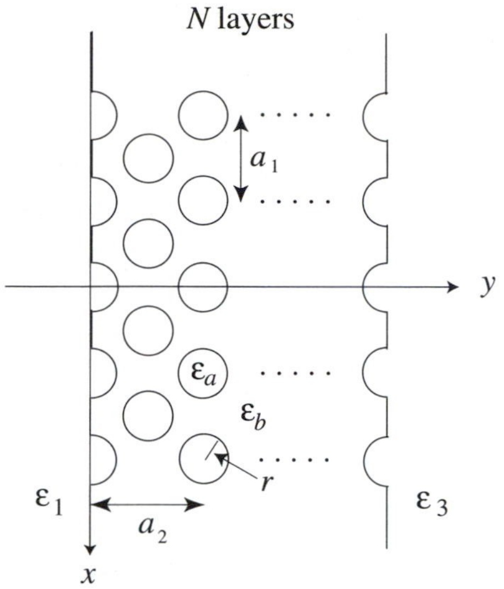

Fig. 4.2. Top view of the geometry of the 2D crystal for the numerical calculation of the transmission and the Bragg reflection spectra by means of the plane-wave expansion method

the Fourier coefficeints are simplified. For this case, κ_{nm} are given by

$$\kappa_{nm} = \begin{cases} \dfrac{f}{\varepsilon_{\mathrm{a}}} + \dfrac{1-f}{\varepsilon_{\mathrm{b}}} & (n=0,\ m=0) \\ 2f\left(\dfrac{1}{\varepsilon_{\mathrm{a}}} - \dfrac{1}{\varepsilon_{\mathrm{b}}}\right) \dfrac{J_1(G_{nm}r_{\mathrm{a}})}{G_{nm}r_{\mathrm{a}}} & (n \neq 0,\ m=0) \\ 2f\left(\dfrac{1}{\varepsilon_{\mathrm{a}}} - \dfrac{1}{\varepsilon_{\mathrm{b}}}\right) (-1)^j \dfrac{J_1(G_{nm}r_{\mathrm{a}})}{G_{nm}r_{\mathrm{a}}} & (m = 2jN) \\ 0 & (\text{otherwise}) \end{cases} \tag{4.45}$$

where j denotes a positive integer.

4.3.2 Hexagonal Lattice

When we calculate the transmittance of the hexagonal lattice or that of the square lattice in the Γ to M direction, we have to deal with a structure such as that shown in Fig. 4.2. For this structure, we assume that each surface crosses the center of the rods. Let us calculate the Fourier coefficients of the dielectric function of this structure. ε^{-1} is now given by

$$\frac{1}{\varepsilon(x,y)} = \frac{1}{\varepsilon_{\mathrm{b}}} + \left(\frac{1}{\varepsilon_{\mathrm{a}}} - \frac{1}{\varepsilon_{\mathrm{b}}}\right) \sum_{l=-\infty}^{\infty} \left[\sum_{l'=-N}^{N} S(\boldsymbol{r}_{/\!/} - \boldsymbol{v}_{1/\!/}(l,l')) + \sum_{l'=-N}^{N-1} S(\boldsymbol{r}_{/\!/} - \boldsymbol{v}_{2/\!/}(l,l')) \right]$$

$$= \sum_{n=-\infty}^{\infty} \sum_{m=-\infty}^{\infty} \kappa_{nm} \exp\left[\mathrm{i}\left(G_n x + \frac{m\pi}{L} y\right)\right], \tag{4.46}$$

where

$$\boldsymbol{v}_{1/\!/}(l, l') = (a_1 l,\ a_2 l'), \tag{4.47}$$

$$\boldsymbol{v}_{2/\!/}(l, l') = (a_1(l + 1/2),\ -a_2(l' + 1/2)). \tag{4.48}$$

Note that we extended ε^{-1} to the negative y region symmetrically as before. Then the Fourier coefficient κ_{nm} is given by

$$\begin{aligned}
&\kappa_{nm} \\
&= \frac{1}{2a_1 L} \int_0^{a_1} \mathrm{d}x \int_{-L}^{L} \mathrm{d}y \ \frac{1}{\varepsilon(x, y)} \exp\left\{-\mathrm{i}\left(G_n x + \frac{m\pi}{L} y\right)\right\} \\
&= \frac{1}{\varepsilon_\mathrm{b}} \delta_{n0}\delta_{m0} + \frac{1}{2a_1 L}\left(\frac{1}{\varepsilon_\mathrm{a}} - \frac{1}{\varepsilon_\mathrm{b}}\right) \sum_{l=-\infty}^{\infty} \\
&\quad \times \left[\sum_{l'=-N}^{N} \int_0^{a_1} \mathrm{d}x \int_{-L}^{L} \mathrm{d}y \ S(\boldsymbol{r}_{/\!/} - \boldsymbol{v}_{1/\!/}(l, l')) \exp\left(-\mathrm{i}\boldsymbol{G}_{nm/\!/} \cdot \boldsymbol{r}_{/\!/}\right)\right. \\
&\quad \left. + \sum_{l'=-N}^{N-1} \int_0^{a_1} \mathrm{d}x \int_{-L}^{L} \mathrm{d}y \ S(\boldsymbol{r}_{/\!/} - \boldsymbol{v}_{2/\!/}(l, l')) \exp\left(-\mathrm{i}\boldsymbol{G}_{nm/\!/} \cdot \boldsymbol{r}_{/\!/}\right)\right] \\
&= \frac{1}{\varepsilon_\mathrm{b}} \delta_{n0}\delta_{m0} + \frac{1}{2a_1 L}\left(\frac{1}{\varepsilon_\mathrm{a}} - \frac{1}{\varepsilon_\mathrm{b}}\right) \sum_{j=1}^{2} \sum_{l'=-N}^{N-1} \exp\left\{-\mathrm{i}\boldsymbol{G}_{nm/\!/} \cdot \boldsymbol{v}_{j/\!/}(0, l')\right\} \\
&\qquad \times \int_{-\infty}^{\infty} \mathrm{d}x \int_{-\infty}^{\infty} \mathrm{d}y \ S(\boldsymbol{r}_{/\!/}) \exp\left(-\mathrm{i}\boldsymbol{G}_{nm/\!/} \cdot \boldsymbol{r}_{/\!/}\right) \\
&= \begin{cases}
\dfrac{f}{\varepsilon_\mathrm{a}} + \dfrac{1-f}{\varepsilon_\mathrm{b}} & (n = 0,\ m = 0) \\
f\left(\dfrac{1}{\varepsilon_\mathrm{a}} - \dfrac{1}{\varepsilon_\mathrm{b}}\right)\left\{1 + (-1)^{n+j}\right\} \dfrac{J_1(G_{nm} r_\mathrm{a})}{G_{nm} r_\mathrm{a}} & \\
\qquad (m = jN, \ \text{and} \ \ n \neq 0 \text{ or } j \neq 0) & \\
0 & (\text{otherwise})
\end{cases}
\end{aligned} \tag{4.49}$$

where j is an integer. In (4.49),

$$f = \frac{2\pi r_\mathrm{a}^2}{a_1 a_2} \tag{4.50}$$

is the filling factor for the present structure, and G_{nm} is given by

$$G_{nm} = \sqrt{\left(\frac{2n\pi}{a}\right)^2 + \left(\frac{m\pi}{L}\right)^2}. \tag{4.51}$$

4.4 Some Examples

4.4.1 Square Lattice

An example of the transmission spectrum is presented in Fig. 4.3. This spectrum was calculated for the 2D square lattice whose photonic band structure was given in Fig. 3.4. We assumed that $a_1 = a_2 = a$ and the thickness L is commensurate with the lattice constant, i.e., (4.44) is satisfied. Hence, we used κ_{nm} in (4.45). The following parameters were used for the numerical calculation: $\varepsilon_{\mathrm{a}} = 1.0$ (air), $\varepsilon_{\mathrm{b}} = 2.1$ (methylpenthene polymer), $f = \pi r_{\mathrm{a}}^2/a^2 = 0.25$, and $N = 16$. The spectrum was calculated for the E polarization with 2700 Fourier components for the expansion in (4.13) $(-7 \leq n \leq 7,\ 1 \leq m \leq 180)$.

Since the wave vector of the incident plane wave is parallel to the Γ-X direction in our configuration (see Fig. 3.2), we concentrate on the band structure on the Δ point in Fig. 3.4(a). There are two bandgaps, one between the first and the second A modes, and the other between the second and the third A modes. The normalized frequency ranges of these bandgaps are listed in Table 4.1. On the other hand, there is a frequency range where only a B mode exists between the fourth and the fifth A modes. As was mentioned in Sect. 3.2, the B (uncoupled) mode is not excited by the incident plane wave because of the symmetry mismatch, and hence we expect that the transmittance at this frequency range should be low. This frequency range is also listed in Table 4.1. Corresponding to both bandgaps, two opaque frequency ranges are clearly observed in Fig. 4.3. The spectral ranges where the calculated transmittance is less than 0.1 are compared with those of the bandgaps

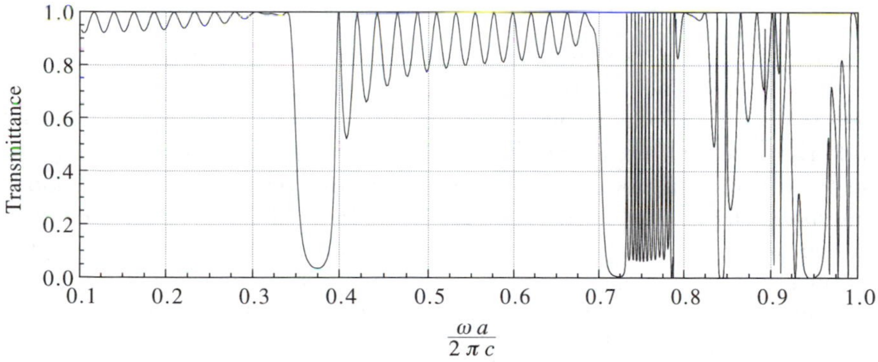

Fig. 4.3. Transmission spectrum for the E polarization in the Γ-X direction calculated for the square lattice composed of 16 layers of circular rods. The following parameters in Fig. 4.1 were used; $\varepsilon_{\mathrm{a}} = 1.0$ (air), $\varepsilon_{\mathrm{b}} = 2.1$ (methylpenthene polymer), $f = \pi r_{\mathrm{a}}^2/a^2 = 0.25$, $\varepsilon_1 = \varepsilon_3 = 1.0$ (air), and $\theta = 0$ (normal incidence). The abscissa is the normalized frequency. The corresponding band diagram is shown in Fig. 3.4. (After [46])

Table 4.1. Comparison of the bandgaps and the uncoupled (B) modes with the frequency ranges where the calculated transmittance is less than 0.1. The frequency is in units of $2\pi c/a$. (After [46])

Polarization	Bandgap	Uncoupled mode	Opaque range
E	0.353–0.387		0.360–0.388
	0.702–0.728		0.708–0.731
		0.927–0.962	0.927–0.961
H	0.354–0.401		0.356–0.404
	0.728–0.742		0.728–0.743
		0.923–0.943	0.923–0.948

in Table 4.1. The maximum deviation between them is 2%, and therefore we can conclude that their correspondence is good. As for the uncoupled mode around $\omega a/2\pi c = 0.94$, the calculated transmission spectrum is not very simple, and it has a plateau at $\omega a/2\pi c = 0.933$. However, this plateau disappears when the number of lattice layers is increased to 32. Therefore, this plateau originates from the finite nature of the sample dimension. Then the total width of the opaque range corresponds quite well to the region with one uncoupled mode.

Next, we examine the interference patterns. Their structures are simple for the lowest three A modes. We can estimate the effective refractive index n_{eff} for these modes from the interval $\Delta\omega$ of the local maxima. This is,

$$n_{\text{eff}} = \frac{a}{2L}\left(\frac{\Delta\omega a}{2\pi c}\right)^{-1} . \tag{4.52}$$

Similarly, it may be estimated from the dispersion curves using the relation $\mathrm{d}\omega/\mathrm{d}k = c/n_{\text{eff}}$. The results are listed in Table 4.2. The values obtained by these two methods coincide with each other quite well. Therefore we can conclude that the interference patterns observed in the calculated transmission spectrum originate from the band structure and the third lowest A mode that shows the group-velocity anomaly has a large effective refractive index. On the other hand, the interference pattern above $\omega a/2\pi c = 0.8$ is quite complicated. In particular, there are aperiodic and singular structures in the frequency ranges at $\omega a/2\pi c = 0.890 \sim 0.927$ and $0.962 \sim 1.0$, where two eigenmodes of type A with mutually different wave vectors are excited at the same time.

Now, we examine the transmission spectrum for the H polarization (Fig. 4.4). As in the case of the E polarization, the two bandgaps and the uncoupled (B) mode between the fourth and the fifth A modes correspond to the respective opaque frequency ranges quite well (see Table 4.1), although the bottom of the spectral dip around $\omega a/2\pi c = 0.93$ in Fig. 4.4 is not very

Table 4.2. Effective refractive index estimated from the interference pattern ($n^{\mathrm{i}}_{\mathrm{eff}}$) and the dispersion relation ($n^{\mathrm{d}}_{\mathrm{eff}}$). (After [46])

Polarization	Band	$n^{\mathrm{i}}_{\mathrm{eff}}$	$n^{\mathrm{d}}_{\mathrm{eff}}$
E	1st	1.34	1.36
	2nd	1.39	1.43
	3rd	6.81	7.13
H	1st	1.34	1.35
	2nd	1.38	1.38
	3rd	-	5.60

flat. However, it becomes wide and flat when the number of lattice layers is increased to 32. Hence, the origin of the incomplete dip is the finite thickness of the assumed specimen. On the other hand, the effective refractive index for the lowest three A modes was estimated as before, and listed in Table 4.2 except $n^{\mathrm{i}}_{\mathrm{eff}}$ for the third band, for which an aperiodic and singular interference pattern is observed. The correspondence between $n^{\mathrm{i}}_{\mathrm{eff}}$ and $n^{\mathrm{d}}_{\mathrm{eff}}$ for the first and the second bands is again quite good, and hence, we can conclude that the interference patterns observed for the H polarization also originate from the band structure. For the third A mode, the dispersion curve is convex downward near the Γ point, and most of its frequency range overlaps that of the fourth A mode, which means that there are two eigenmodes for a single ω for the most part of the third A branch. This feature is also found for the fourth and fifth A modes, and we find a complicated interference pattern at the corresponding frequency range, i.e., $\omega a/2\pi c = 0.742 \sim 1.0$.

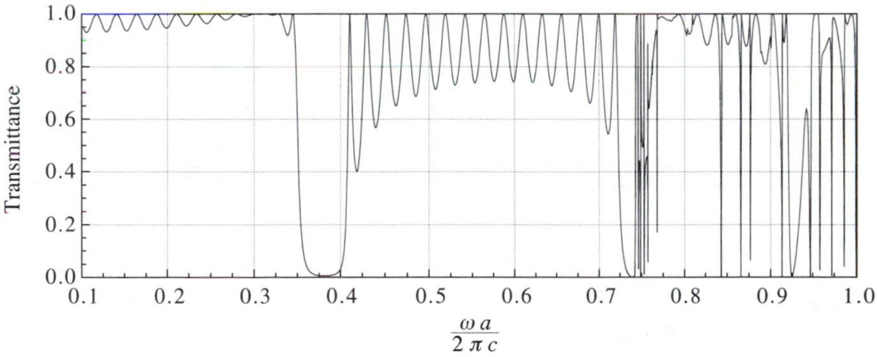

Fig. 4.4. Transmission spectrum for the H polarization in the Γ-X direction calculated for the 2D square lattice composed of 16 layers of circular air cylinders formed in a dielectric material. The same parameters as for Fig. 4.3 were used for numerical calculation. (After [46])

Then, our numerical calculation for both E and H polarizations has shown that the singular interference appears when the incident light excites two eigenmodes, and otherwise the interference pattern is quite normal and the effective refractive index estimated from it corresponds quite well to the photonic band structure.

4.4.2 Hexagonal Lattice

Figure 4.5 is another example of the transmission spectra that were calculated for the hexagonal lattice whose band diagrams and the symmetry of the eigenmodes are presented in Fig. 3.7. The spectra was calculated for the incident wave propagated in the Γ-X direction (see Fig. 3.6). The following parameters in Fig. 4.2 were used for the numerical calculation; $\varepsilon_{\mathrm{a}} = 1.0$ (air),

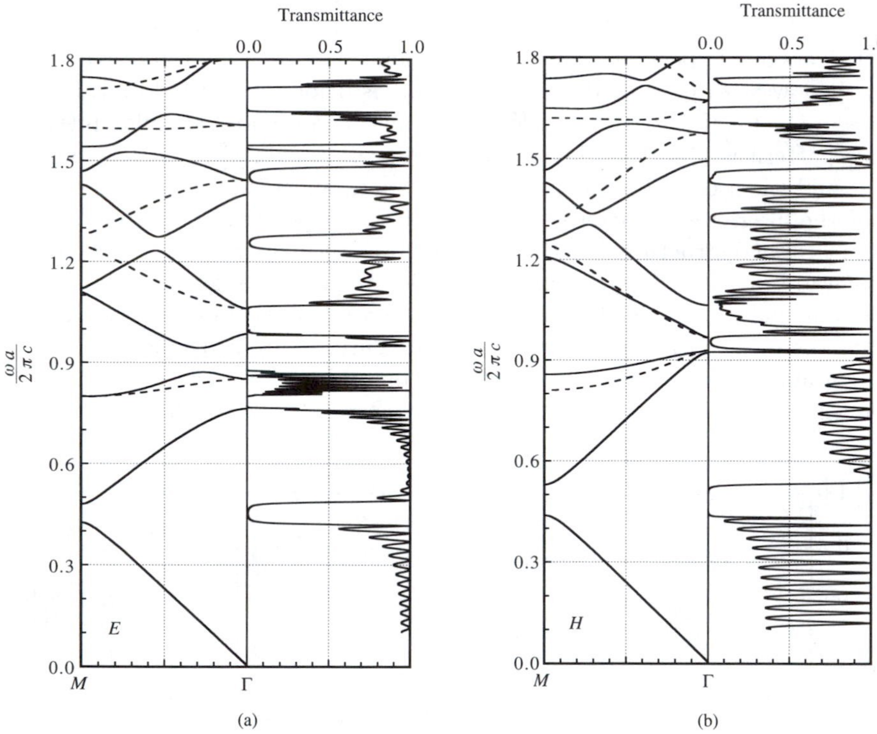

Fig. 4.5. Transmittance (*right-hand side*) and the dispersion relation (*left-hand side*) of a hexagonal lattice for (**a**) the E polarization and (**b**) the H polarization in the Γ-X direction. The ordinate is the normalized frequency. On the left-hand side of these figures, *solid lines* represent symmetric modes and *dashed lines* represent antisymmetric modes. The latter cannot be excited by the incident plane wave because of the mismatching of their spatial symmetry, and so they do not contribute to the light transmission. Opaque frequency regions are clearly observed in the transmission spectrum where no symmetric mode exists. (After [48])

$\varepsilon_{\mathrm{b}} = 2.72$ (PbO glass), $a_1 = a$, $a_2 = \sqrt{3}a$, $f = 2\pi r_{\mathrm{a}}^2/\sqrt{3}a^2 = 0.65$, $N = 16$, $\varepsilon_1 = \varepsilon_3 = 1.0$ (air), and $\theta = 0$ (normal incidence).

Now, we examine the calculated transmittance together with the dispersion relation in the Γ-X direction, which are shown in Fig. 4.5(a) (the E polarization) and Fig. 4.5(b) (the H polarization). The right-hand side of these figures shows the transmittance, while the left-hand side shows the dispersion relation, where solid lines and dashed lines denote the A and B modes, respectively. The antisymmetric B modes cannot be excited by the incident plane wave because of the mismatching of their spatial symmetry, and so they do not contribute to the light transmission as we mentioned previously. Opaque frequency regions are clearly observed in the transmission spectra for both polarizations where no symmetric A mode exists. This implies that 16 lattice layers are enough to reproduce the intrinsic optical properties of the specimen.

4.5 Refraction Law for Photonic Crystals

In this section, we examine how an incoming plane wave is refracted at the surface of the photonic crystal. Before we treat the case of the photonic crystal, we consider the refraction at the interface of two uniform materials with isotropic refractive indices, η_1 and η_2. We assume that a plane wave with an angular frequency ω is incident on the interface at an angle θ_1 (see Fig. 4.6(a)). We denote the wavelength of the incident wave in vacuum by λ_0. Then its wavelength in the materials λ_i is given by

$$\lambda_i = \frac{\lambda_0}{\eta_i}. \tag{4.53}$$

Snell's law tells us that the angles of reflection θ_1' and refraction θ_2 are given by,

$$\theta_1' = \theta_1, \tag{4.54}$$

$$\eta_1 \sin\theta_1 = \eta_2 \sin\theta_2. \tag{4.55}$$

These relations are derived from the fact that the equi-phase planes, which are drawn with fine lines in Fig. 4.6, of the relevant waves should have the same spacing along the interface.

Let us see this phenomenon from a slightly different point of view. Because the system that we are dealing with has translational symmetry in the direction parallel to the interface, the wave vector in that direction should be conserved when the refraction and the reflection take place. We denote the wave vectors of the incident, reflected, and refracted plane waves by $\boldsymbol{k}_1$, $\boldsymbol{k}_1'$, and $\boldsymbol{k}_2$, respectively. Their amplitudes are given by

$$|\boldsymbol{k}_1| = |\boldsymbol{k}_1'| = \frac{\eta_1\omega}{c}, \tag{4.56}$$

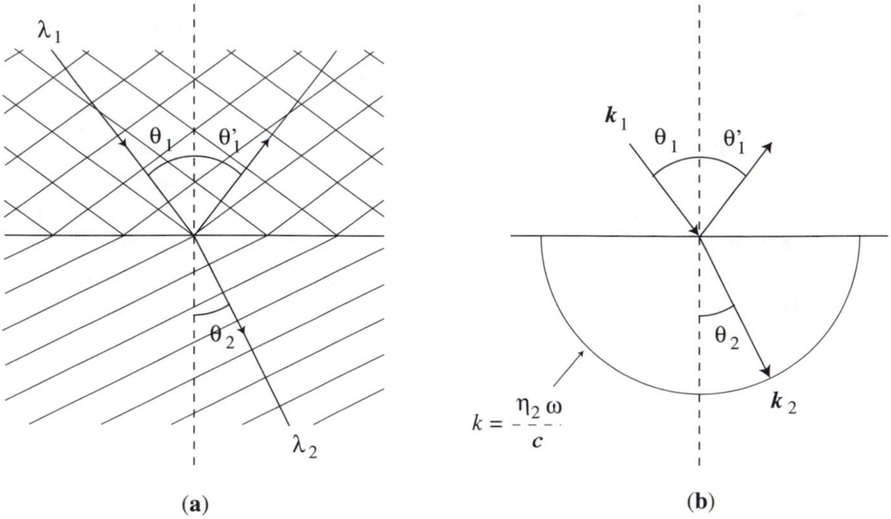

Fig. 4.6. Schematic illustration of the refraction at the surface of the uniform material.

$$|\boldsymbol{k}_2| = \frac{\eta_2 \omega}{c}. \tag{4.57}$$

The conservation of their parallel components leads to

$$\frac{\eta_1 \omega}{c} \sin\theta_1 = \frac{\eta_1 \omega}{c} \sin\theta_1' = \frac{\eta_2 \omega}{c} \sin\theta_2. \tag{4.58}$$

Thus, the same relations as (4.53) and (4.54) are obtained. The direction in which the radiational energy of the plane wave flows coincides with the direction of its wave vector. Let us examine this point somewhat rigorously. As we showed in Sect. 2.5, the energy velocity is equal to the group velocity in the photonic crystals. This is also true for the uniform materials, since we can regard the latter as a special case of the former without the spatial modification of the dielectric constant. For the present problem, the group velocity, and thus the energy velocity of the incident and refracted waves, $\boldsymbol{v}_1$ and $\boldsymbol{v}_2$ are given by

$$\boldsymbol{v}_i = \frac{\partial \omega}{\partial \boldsymbol{k}_i} = \frac{c}{\eta_i} \widehat{\boldsymbol{k}}_i, \tag{4.59}$$

where $\widehat{\boldsymbol{k}}_i$ denotes the unit vector parallel to $\boldsymbol{k}_i$. Hence, the direction of the energy flow coincides with the direction of the wave vector. Moreover, its direction is perpendicular to the equi-energy surface that is given by

$$\omega = \text{const.}, \tag{4.60}$$

which is denoted by a hemicircle in Fig. 4.6(b). With these facts in mind, let us proceed to the case of the photonic crystal.

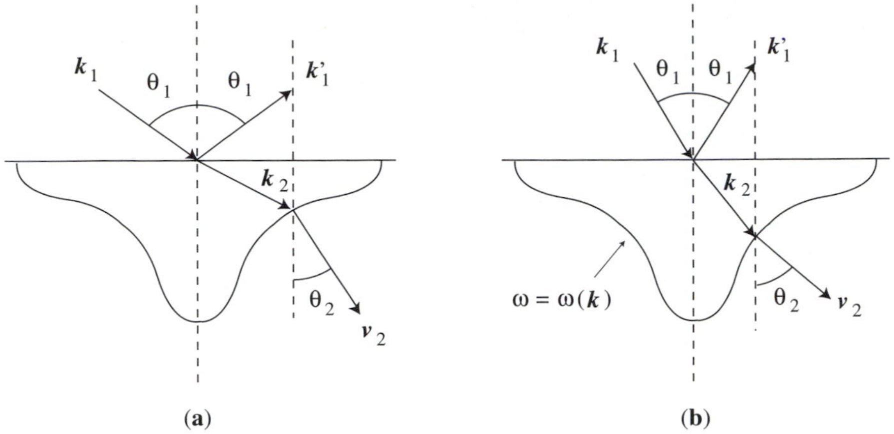

Fig. 4.7. Schematic illustration of refraction at the surface of the photonic crystal.

We assume that a plane wave with an angular frequency ω and a wave vector $\boldsymbol{k}$ is incident on the surface of the photonic crystal at an angle θ (Fig. 4.7(a)). The parallel component of the wave vector should be conserved, with allowance for the difference of a multiple of $2\pi/a$, as before. However, the equi-energy surface of the photonic crystal is not at all a spherical surface, and this leads to a marked difference. The wave vector of the refracted wave has its end point on the equi-energy surface given by the band structure, $\omega = \omega(\boldsymbol{k}_2)$, which is drawn with a solid curve in Fig. 4.7. This condition determines the wave vector of the refracted wave. Then, the propagation direction, that is the direction of the energy flow, which is denoted by an arrow, is perpendicular to the equi-energy surface and parallel to the group velocity of the refracted wave. Sometimes a very curious phenomenon occurs because of the complex dependence of ω on $\boldsymbol{k}_2$. For example, θ_2 in Fig. 4.7(b) is larger than that in Fig. 4.7(a), although θ_1 is smaller! Extraordinary angular dependence as well as frequency dependence in the refraction phenomena of the photonic crystals were observed and analyzed by Kosaka et al. [63, 64].

5. Optical Response of Photonic Crystals

In this chapter, we will derive a general formula for the optical response of photonic crystals by means of the Green's function given in Sect. 2.8. Then, it will be applied to four typical optical processes, i.e., dipole radiation, stimulated emission, sum-frequency generation, and free induction decay. This method can be readily extended to other kinds of nonlinear and coherent optical processes.

5.1 Solutions of Inhomogeneous Equations

We start from the following Maxwell equations:

$$\nabla \cdot \{\varepsilon_0 \varepsilon(\boldsymbol{r})\boldsymbol{E}(\boldsymbol{r},t) + \boldsymbol{P}_{\mathrm{ex}}(\boldsymbol{r},t)\} = 0, \tag{5.1}$$

$$\nabla \cdot \boldsymbol{H}(\boldsymbol{r},t) = 0, \tag{5.2}$$

$$\nabla \times \boldsymbol{E}(\boldsymbol{r},t) = -\mu_0 \frac{\partial}{\partial t}\boldsymbol{H}(\boldsymbol{r},t), \tag{5.3}$$

$$\nabla \times \boldsymbol{H}(\boldsymbol{r},t) = \frac{\partial}{\partial t}\{\varepsilon_0 \varepsilon(\boldsymbol{r})\boldsymbol{E}(\boldsymbol{r},t) + \boldsymbol{P}_{\mathrm{ex}}(\boldsymbol{r},t)\}. \tag{5.4}$$

The difference between this set of equations and (2.11)–(2.14) is the presence of $\boldsymbol{P}_{\mathrm{ex}}(\boldsymbol{r},t)$, which is an extrinsic polarization field that is not described by the dielectric function $\varepsilon(\boldsymbol{r})$. The nonlinear polarization of the photonic crystal and the dipole moments of impurity atoms induced by external fields are examples of such a polarization field. The magnetic permeability is assumed to be equal to that in free space as before. When we eliminate the magnetic field $\boldsymbol{H}(\boldsymbol{r},t)$ in (5.3) and (5.4), we obtain the inhomogeneous wave equation for $\boldsymbol{Q}(\boldsymbol{r},t)$ defined by (2.132):

$$-\left(\frac{1}{c^2}\frac{\partial^2}{\partial t^2} + \mathcal{H}\right)\boldsymbol{Q}(\boldsymbol{r},t) = \frac{1}{c^2\varepsilon_0\sqrt{\varepsilon(\boldsymbol{r})}}\frac{\partial^2}{\partial t^2}\boldsymbol{P}_{\mathrm{ex}}(\boldsymbol{r},t), \tag{5.5}$$

where the differential operator $\mathcal{H}$ is defined by (2.133).

We can obtain the solution of (5.5) as the convolution integral of Green's function given in (2.167) and the inhomogeneous term:

$$\boldsymbol{Q}(\boldsymbol{r},t) = \int_V \mathrm{d}\boldsymbol{r}' \int_{-\infty}^{\infty} \mathrm{d}t' \overleftrightarrow{G}(\boldsymbol{r},\boldsymbol{r}',t-t') \frac{1}{c^2\varepsilon_0\sqrt{\varepsilon(\boldsymbol{r}')}} \frac{\partial^2}{\partial t'^2} \boldsymbol{P}_{\mathrm{ex}}(\boldsymbol{r}',t'), \quad (5.6)$$

where V is the volume of the photonic crystal. It can be easily shown by using (2.162) that $\boldsymbol{Q}(\boldsymbol{r},t)$ given by this equation satisfies (5.5). Substituting (2.167) into (5.6), we obtain

$$\begin{aligned} &\boldsymbol{Q}(\boldsymbol{r},t) \\ &= -\frac{1}{\varepsilon_0 V} \sum_{\boldsymbol{k}n} \int_V \mathrm{d}\boldsymbol{r}' \int_{-\infty}^{t} \mathrm{d}t' \left\{ \frac{\sin \omega_{\boldsymbol{k}n}^{(\mathrm{T})}(t-t')}{\omega_{\boldsymbol{k}n}^{(\mathrm{T})}} \boldsymbol{Q}_{\boldsymbol{k}n}^{(\mathrm{T})}(\boldsymbol{r}) \otimes \boldsymbol{Q}_{\boldsymbol{k}n}^{(\mathrm{T})*}(\boldsymbol{r}') \right. \\ &\quad \left. +(t-t')\boldsymbol{Q}_{\boldsymbol{k}n}^{(\mathrm{L})}(\boldsymbol{r}) \otimes \boldsymbol{Q}_{\boldsymbol{k}n}^{(\mathrm{L})*}(\boldsymbol{r}') \right\} \frac{1}{\sqrt{\varepsilon(\boldsymbol{r}')}} \frac{\partial^2}{\partial t'^2} \boldsymbol{P}_{\mathrm{ex}}(\boldsymbol{r}',t'). \quad (5.7) \end{aligned}$$

In this equation, we assume that the extrinsic polarization $\boldsymbol{P}_{\mathrm{ex}}(\boldsymbol{r},t)$ was introduced adiabatically, i.e., $\boldsymbol{P}_{\mathrm{ex}}(\boldsymbol{r},-\infty) = 0$. When we carry out the t'-integral by parts, we obtain

$$\begin{aligned} &\boldsymbol{Q}(\boldsymbol{r},t) \\ &= -\frac{1}{\varepsilon_0 V} \sum_{\boldsymbol{k}n} \int_V \mathrm{d}\boldsymbol{r}' \int_{-\infty}^{t} \mathrm{d}t' \left\{ \cos \omega_{\boldsymbol{k}n}^{(\mathrm{T})}(t-t') \, \boldsymbol{Q}_{\boldsymbol{k}n}^{(\mathrm{T})}(\boldsymbol{r}) \otimes \boldsymbol{Q}_{\boldsymbol{k}n}^{(\mathrm{T})*}(\boldsymbol{r}') \right. \\ &\quad \left. +\boldsymbol{Q}_{\boldsymbol{k}n}^{(\mathrm{L})}(\boldsymbol{r}) \otimes \boldsymbol{Q}_{\boldsymbol{k}n}^{(\mathrm{L})*}(\boldsymbol{r}') \right\} \frac{1}{\sqrt{\varepsilon(\boldsymbol{r}')}} \frac{\partial}{\partial t'} \boldsymbol{P}_{\mathrm{ex}}(\boldsymbol{r}',t'). \quad (5.8) \end{aligned}$$

When we carry out the partial integration again, we have

$$\begin{aligned} \boldsymbol{Q}(\boldsymbol{r},t) &= -\frac{1}{\varepsilon_0 V} \sum_{\boldsymbol{k}n} \int_V \mathrm{d}\boldsymbol{r}' \left\{ \boldsymbol{Q}_{\boldsymbol{k}n}^{(\mathrm{T})}(\boldsymbol{r}) \otimes \boldsymbol{Q}_{\boldsymbol{k}n}^{(\mathrm{T})*}(\boldsymbol{r}') \right. \\ &\quad \left. +\boldsymbol{Q}_{\boldsymbol{k}n}^{(\mathrm{L})}(\boldsymbol{r}) \otimes \boldsymbol{Q}_{\boldsymbol{k}n}^{(\mathrm{L})*}(\boldsymbol{r}') \right\} \frac{1}{\sqrt{\varepsilon(\boldsymbol{r}')}} \boldsymbol{P}_{\mathrm{ex}}(\boldsymbol{r}',t) \\ &\quad +\frac{1}{\varepsilon_0 V} \sum_{\boldsymbol{k}n} \int_V \mathrm{d}\boldsymbol{r}' \int_{-\infty}^{t} \mathrm{d}t' \left\{ \omega_{\boldsymbol{k}n}^{(\mathrm{T})} \sin \omega_{\boldsymbol{k}n}^{(\mathrm{T})}(t-t') \right. \\ &\quad \left. \times \boldsymbol{Q}_{\boldsymbol{k}n}^{(\mathrm{T})}(\boldsymbol{r}) \otimes \boldsymbol{Q}_{\boldsymbol{k}n}^{(\mathrm{T})*}(\boldsymbol{r}') \right\} \frac{1}{\sqrt{\varepsilon(\boldsymbol{r}')}} \boldsymbol{P}_{\mathrm{ex}}(\boldsymbol{r}',t') \\ &= -\frac{\boldsymbol{P}_{\mathrm{ex}}(\boldsymbol{r},t)}{\varepsilon_0\sqrt{\varepsilon(\boldsymbol{r})}} + \frac{1}{\varepsilon_0 V} \sum_{\boldsymbol{k}n} \boldsymbol{Q}_{\boldsymbol{k}n}^{(\mathrm{T})}(\boldsymbol{r}) \int_V \mathrm{d}\boldsymbol{r}' \int_{-\infty}^{t} \mathrm{d}t' \\ &\quad \times \frac{\boldsymbol{Q}_{\boldsymbol{k}n}^{(\mathrm{T})*}(\boldsymbol{r}') \cdot \boldsymbol{P}_{\mathrm{ex}}(\boldsymbol{r}',t')}{\sqrt{\varepsilon(\boldsymbol{r}')}} \omega_{\boldsymbol{k}n}^{(\mathrm{T})} \sin \omega_{\boldsymbol{k}n}^{(\mathrm{T})}(t-t'), \quad (5.9) \end{aligned}$$

where we used (2.146). Finally, we obtain

$$
\begin{aligned}
&\boldsymbol{E}(\boldsymbol{r},t) + \frac{\boldsymbol{P}_{\mathrm{ex}}(\boldsymbol{r},t)}{\varepsilon_0\varepsilon(\boldsymbol{r})} \\
&= \frac{1}{\varepsilon_0 V}\sum_{\boldsymbol{k}n} \boldsymbol{E}_{\boldsymbol{k}n}^{(\mathrm{T})}(\boldsymbol{r}) \int d\boldsymbol{r}' \int_{-\infty}^{t} dt' \boldsymbol{E}_{\boldsymbol{k}n}^{(\mathrm{T})*}(\boldsymbol{r}') \cdot \boldsymbol{P}_{\mathrm{ex}}(\boldsymbol{r}',t') \\
&\qquad \times \omega_{\boldsymbol{k}n}^{(\mathrm{T})} \sin\omega_{\boldsymbol{k}n}^{(\mathrm{T})}(t-t'). \qquad (5.10)
\end{aligned}
$$

Here, we should note that (a) the role of the quasi-longitudinal modes that appear in the explicit expression of Green's function (2.167) is only to reconstruct the extrinsic polarization field and they do not contribute to the propagating radiation field, i.e., to the right-hand side of (5.10), and that (b) (5.1) holds, as it should, because of (2.144).

In the case of the E polarization in the 2D crystal, the wave equation with the extrinsic polarization $P_{z,\mathrm{ex}}$ is given by

$$
-\left(\frac{1}{c^2}\frac{\partial^2}{\partial t^2} + \mathcal{H}^{(2)}\right) Q_z(\boldsymbol{r}_{/\!/},t) = \frac{1}{c^2\varepsilon_0\sqrt{\varepsilon(\boldsymbol{r}_{/\!/})}}\frac{\partial^2}{\partial t^2}P_{z,\mathrm{ex}}(\boldsymbol{r}_{/\!/},t), \qquad (5.11)
$$

where Q_z and $\mathcal{H}^{(2)}$ are defined by (2.157) and (2.158). The solution of (5.11) is given by the convolution integral of Green's function $G^{(2)}$ in (2.172) and the inhomogeneous term:

$$
\begin{aligned}
&Q_z(\boldsymbol{r}_{/\!/},t) \\
&= \int_{V^{(2)}} \mathrm{d}\boldsymbol{r}'_{/\!/} \int_{-\infty}^{\infty} \mathrm{d}t' G^{(2)}(\boldsymbol{r}_{/\!/},\boldsymbol{r}'_{/\!/},t-t') \frac{1}{c^2\varepsilon_0\sqrt{\varepsilon(\boldsymbol{r}'_{/\!/})}}\frac{\partial^2}{\partial t^2}P_{z,\mathrm{ex}}(\boldsymbol{r}'_{/\!/},t') \\
&= -\frac{1}{\varepsilon_0 V^{(2)}}\sum_{\boldsymbol{k}_{/\!/}n} Q_{z,\boldsymbol{k}_{/\!/}n}(\boldsymbol{r}_{/\!/}) \int_{V^{(2)}} \mathrm{d}\boldsymbol{r}'_{/\!/} \int_{-\infty}^{t} \mathrm{d}t' \\
&\quad \times \frac{\sin\omega_{\boldsymbol{k}_{/\!/}n}^{(E)}(t-t')}{\omega_{\boldsymbol{k}_{/\!/}n}^{(E)}} \frac{1}{\sqrt{\varepsilon(\boldsymbol{r}')}}\frac{\partial^2}{\partial t^2}P_{z,\mathrm{ex}}(\boldsymbol{r}'_{/\!/},t'). \qquad (5.12)
\end{aligned}
$$

When we perform the partial integration twice, we have

$$
\begin{aligned}
&Q_z(\boldsymbol{r}_{/\!/},t) \\
&= -\frac{P_{z,\mathrm{ex}}(\boldsymbol{r}_{/\!/},t)}{\varepsilon_0\sqrt{\varepsilon(\boldsymbol{r}_{/\!/})}} + \frac{1}{\varepsilon_0 V^{(2)}}\sum_{\boldsymbol{k}_{/\!/}n} Q_{z,\boldsymbol{k}_{/\!/}n}(\boldsymbol{r}_{/\!/}) \int_{V^{(2)}} \mathrm{d}\boldsymbol{r}'_{/\!/} \int_{-\infty}^{t} \mathrm{d}t' \\
&\quad \times \frac{Q^*_{z,\boldsymbol{k}_{/\!/}n}(\boldsymbol{r}'_{/\!/})P_{z,\mathrm{ex}}(\boldsymbol{r}'_{/\!/},t')}{\sqrt{\varepsilon(\boldsymbol{r}'_{/\!/})}} \omega_{\boldsymbol{k}_{/\!/}n}^{(E)} \sin\omega_{\boldsymbol{k}_{/\!/}n}^{(E)}(t-t'), \qquad (5.13)
\end{aligned}
$$

where we have used the completeness of the eigenfunctions, (2.161). Finally, we obtain

$$E_z(\boldsymbol{r}_{/\!/}, t) + \frac{P_{z,\mathrm{ex}}(\boldsymbol{r}_{/\!/}, t)}{\varepsilon_0 \varepsilon(\boldsymbol{r}_{/\!/})}$$
$$= \frac{1}{\varepsilon_0 V^{(2)}} \sum_{\boldsymbol{k}_{/\!/} n} E_{z,\boldsymbol{k}_{/\!/} n}(\boldsymbol{r}_{/\!/}) \int_{V^{(2)}} \mathrm{d}\boldsymbol{r}'_{/\!/} \int_{-\infty}^{t} \mathrm{d}t' E^*_{z,\boldsymbol{k}_{/\!/} n}(\boldsymbol{r}'_{/\!/}) P_{z,\mathrm{ex}}(\boldsymbol{r}'_{/\!/}, t')$$
$$\times \omega^{(E)}_{\boldsymbol{k}_{/\!/} n} \sin \omega^{(E)}_{\boldsymbol{k}_{/\!/} n}(t - t'). \tag{5.14}$$

5.2 Dipole Radiation

In this section, we will deal with the radiation from an oscillating dipole moment. Suppose that there is a dipole moment oscillating at the frequency ω located at position $\boldsymbol{r}_0$ in the photonic crystal, and let $\boldsymbol{P}_{\mathrm{d}}(\boldsymbol{r}, t)$ be the external polarization of this problem:

$$\boldsymbol{P}_{\mathrm{d}}(\boldsymbol{r}, t) = \boldsymbol{d}\, \delta(\boldsymbol{r} - \boldsymbol{r}_0) \mathrm{e}^{(-\mathrm{i}\omega + \delta)t}. \tag{5.15}$$

Here, $\boldsymbol{d}$ is the magnitude of the dipole moment and δ is a positive infinitesimal. The function $\mathrm{e}^{\delta t}$ was introduced to assure the adiabatic switching of the extrinsic polarization. Equation (5.15) represents the positive frequency part of the dipole moment and the actual real field is given by the sum of $\boldsymbol{P}_{\mathrm{d}}(\boldsymbol{r}, t)$ and its complex conjugate. By substituting $\boldsymbol{P}_{\mathrm{d}}(\boldsymbol{r}, t)$ for $\boldsymbol{P}_{\mathrm{ex}}(\boldsymbol{r}, t)$ in (5.10), we obtain

$$\boldsymbol{E}_{\mathrm{d}}(\boldsymbol{r}, t) + \frac{\boldsymbol{P}_{\mathrm{d}}(\boldsymbol{r}, t)}{\varepsilon_0 \varepsilon(\boldsymbol{r})} = \frac{\mathrm{e}^{-\mathrm{i}\omega t}}{2\varepsilon_0 V} \sum_{\boldsymbol{k}n} \omega^{(\mathrm{T})}_{\boldsymbol{k}n} \boldsymbol{E}^{(\mathrm{T})}_{\boldsymbol{k}n}(\boldsymbol{r}) \left\{ \boldsymbol{E}^{(\mathrm{T})*}_{\boldsymbol{k}n}(\boldsymbol{r}_0) \cdot \boldsymbol{d} \right\}$$
$$\times \left(\frac{1}{\omega + \omega^{(\mathrm{T})}_{\boldsymbol{k}n} + \mathrm{i}\delta} - \frac{1}{\omega - \omega^{(\mathrm{T})}_{\boldsymbol{k}n} + \mathrm{i}\delta} \right). \tag{5.16}$$

The radiation of the electromagnetic energy is described by Poynting's vector $\boldsymbol{S}$ that is given by the vector product of the $\boldsymbol{E}$ and $\boldsymbol{H}$ fields. Here, we should note that the actual real fields are given by the sum of $\boldsymbol{E}$ and $\boldsymbol{E}^*$, and that of $\boldsymbol{H}$ and $\boldsymbol{H}^*$, respectively. Hence,

$$\boldsymbol{S}(\boldsymbol{r}, t) = \{\boldsymbol{E}(\boldsymbol{r}, t) + \boldsymbol{E}^*(\boldsymbol{r}, t)\} \times \{\boldsymbol{H}(\boldsymbol{r}, t) + \boldsymbol{H}^*(\boldsymbol{r}, t)\}. \tag{5.17}$$

The time-averaged Poynting vector $\overline{\boldsymbol{S}(\boldsymbol{r}, t)}$ is thus given by

$$\overline{\boldsymbol{S}(\boldsymbol{r}, t)} = \overline{\{\boldsymbol{E}_{\mathrm{d}}(\boldsymbol{r}, t) + \boldsymbol{E}^*_{\mathrm{d}}(\boldsymbol{r}, t)\} \times \{\boldsymbol{H}_{\mathrm{d}}(\boldsymbol{r}, t) + \boldsymbol{H}^*_{\mathrm{d}}(\boldsymbol{r}, t)\}}$$
$$= \left\{ \boldsymbol{E}_{\mathrm{d}}(\boldsymbol{r}, t) \times \boldsymbol{H}^*_{\mathrm{d}}(\boldsymbol{r}, t) + \boldsymbol{E}^*_{\mathrm{d}}(\boldsymbol{r}, t) \times \boldsymbol{H}_{\mathrm{d}}(\boldsymbol{r}, t) \right\}, \tag{5.18}$$

where the magnetic field $\boldsymbol{H}_{\mathrm{d}}(\boldsymbol{r}, t)$ is given by

$$\boldsymbol{H}_{\mathrm{d}}(\boldsymbol{r}, t) = \frac{1}{\mathrm{i}\omega\mu_0} \nabla \times \boldsymbol{E}_{\mathrm{d}}(\boldsymbol{r}, t). \tag{5.19}$$

Note that both $\boldsymbol{E}_{\mathrm{d}}$ and $\boldsymbol{H}_{\mathrm{d}}$ are proportional to $\mathrm{e}^{-\mathrm{i}\omega t}$, and thus, $\overline{\boldsymbol{S}(\boldsymbol{r},t)}$ in (5.18) does not depend on time. From (5.4) and (5.19), we have

$$\begin{aligned}\nabla\cdot\overline{\boldsymbol{S}(\boldsymbol{r},t)} &= \Big[\boldsymbol{H}_{\mathrm{d}}^*(\boldsymbol{r},t)\cdot\{\nabla\times\boldsymbol{E}_{\mathrm{d}}(\boldsymbol{r},t)\} - \boldsymbol{E}_{\mathrm{d}}(\boldsymbol{r},t)\cdot\{\nabla\times\boldsymbol{H}_{\mathrm{d}}^*(\boldsymbol{r},t)\} \\ &\qquad + \boldsymbol{H}_{\mathrm{d}}(\boldsymbol{r},t)\cdot\{\nabla\times\boldsymbol{E}_{\mathrm{d}}^*(\boldsymbol{r},t)\} - \boldsymbol{E}_{\mathrm{d}}^*(\boldsymbol{r},t)\cdot\{\nabla\times\boldsymbol{H}_{\mathrm{d}}(\boldsymbol{r},t)\}\Big] \\ &= \mathrm{i}\omega\{\boldsymbol{E}_{\mathrm{d}}^*(\boldsymbol{r},t)\cdot\boldsymbol{P}_{\mathrm{d}}(\boldsymbol{r},t) - \boldsymbol{E}_{\mathrm{d}}(\boldsymbol{r},t)\cdot\boldsymbol{P}_{\mathrm{d}}^*(\boldsymbol{r},t)\}. \qquad (5.20)\end{aligned}$$

When we substitute (5.16) into (5.20) and use the following relation,

$$\frac{1}{\omega-\omega_0\pm\mathrm{i}\delta} = \frac{\mathcal{P}}{\omega-\omega_0} \mp \pi\mathrm{i}\delta(\omega-\omega_0), \qquad (5.21)$$

where $\mathcal{P}$ denotes Cauchy's principal value, we obtain

$$\nabla\cdot\overline{\boldsymbol{S}(\boldsymbol{r},t)} = \frac{\pi\omega^2}{\varepsilon_0 V}\delta(\boldsymbol{r}-\boldsymbol{r}_0)\sum_{\boldsymbol{k}n}\left|\boldsymbol{d}\cdot\boldsymbol{E}_{\boldsymbol{k}n}^{(\mathrm{T})*}(\boldsymbol{r}_0)\right|^2\delta\left(\omega-\omega_{\boldsymbol{k}n}^{(\mathrm{T})}\right). \qquad (5.22)$$

When we denote a small volume which includes $\boldsymbol{r}_0$ by V_1 and its surface by S_1, the radiational energy U emitted per unit time by the oscillating dipole is given by the surface integral of the normal component of Poynting's vector, S_n, which can be transformed into a volume integral by means of Gauss's theorem:

$$\begin{aligned}U &= \int_{S_1}\mathrm{d}S\ \overline{S_n(\boldsymbol{r},t)} = \int_{V_1}\mathrm{d}\boldsymbol{r}\ \nabla\cdot\overline{\boldsymbol{S}(\boldsymbol{r},t)} \\ &= \frac{\pi\omega^2}{\varepsilon_0 V}\sum_{\boldsymbol{k}n}\left|\boldsymbol{d}\cdot\boldsymbol{E}_{\boldsymbol{k}n}^{(\mathrm{T})*}(\boldsymbol{r}_0)\right|^2\delta\left(\omega-\omega_{\boldsymbol{k}n}^{(\mathrm{T})}\right). \qquad (5.23)\end{aligned}$$

Because of the spatial variation of $\boldsymbol{E}_{\boldsymbol{k}n}^{(\mathrm{T})}(\boldsymbol{r})$, the radiated power generally dependes on the position of the oscillationg dipole moment.

We make an estimation of U by considering the density of states of the radiational eigenmodes of the photonic crystal, which we denote by $D(\omega)$. The number of eigenmodes whose eigenfrequency is in the interval $[\omega,\omega+\mathrm{d}\omega]$ is given by $D(\omega)\mathrm{d}\omega$. When we denote the averaged value of $\left|\boldsymbol{d}\cdot\boldsymbol{E}_{\boldsymbol{k}n}^{(\mathrm{T})*}(\boldsymbol{r}_0)\right|^2$ by an overline and convert the sum over $\boldsymbol{k}$ and n into an integral over the eigen-angular frequency, we obtain

$$\begin{aligned}U &\simeq \frac{\pi\omega^2}{\varepsilon_0 V}\overline{\left|\boldsymbol{d}\cdot\boldsymbol{E}^{(\mathrm{T})*}(\boldsymbol{r}_0)\right|^2}\int_0^\infty \mathrm{d}\omega' D(\omega')\delta(\omega-\omega') \\ &= \frac{\pi\omega^2}{\varepsilon_0 V}\overline{\left|\boldsymbol{d}\cdot\boldsymbol{E}^{(\mathrm{T})*}(\boldsymbol{r}_0)\right|^2}D(\omega). \qquad (5.24)\end{aligned}$$

The energy radiated by the oscillating dipole is thus approximately proportional to the density of states. If $D(\omega)$ is equal to zero in a certain frequency range, no dipole radiation takes place. The discussion given above is based on classical electromagnetic theory. We can also derive the same result by quantum mechanics, and we can conclude that no spontaneous emission of photons takes place when $D(\omega) = 0$.

In free space, the dispersion relation is

$$\omega = ck. \tag{5.25}$$

Hence, the number of eigenmodes in the volume V with eigenfrequencies less than ω_0, $N(\omega_0)$, is given by

$$\begin{aligned} N(\omega_0) &= 2 \times \sum_{ck<\omega_0} 1 = 2 \times \frac{V}{(2\pi)^3} \int_{k<\omega_0/c} \mathrm{d}\boldsymbol{k} \\ &= \frac{\omega_0^3 V}{3\pi^2 c^3}. \end{aligned} \tag{5.26}$$

We took into account that there are two polarizations for each $\boldsymbol{k}$ vector. $D(\omega)$ is thus given by

$$D(\omega) = \frac{\partial N(\omega)}{\partial \omega} = \frac{\omega^2 V}{\pi^2 c^3}. \tag{5.27}$$

In free space, the eigenmodes of the radiational field are plane waves:

$$\boldsymbol{E}_{\boldsymbol{k}}^{(\mathrm{T})}(\boldsymbol{r}) = \boldsymbol{E}_0 \mathrm{e}^{i\boldsymbol{k}\cdot\boldsymbol{r}}. \tag{5.28}$$

The magnitude of $\boldsymbol{E}_0$ is determined by the normalization condition, (2.145). For the present case, it is

$$|\boldsymbol{E}_0|^2 = 1. \tag{5.29}$$

Because the angle between $\boldsymbol{d}$ and $\boldsymbol{E}_0$ distributes with equal probability,

$$\overline{\left|\boldsymbol{d} \cdot \boldsymbol{E}^{(\mathrm{T})*}(\boldsymbol{r}_0)\right|^2} = \frac{1}{3}|\boldsymbol{d}|^2 \, |\boldsymbol{E}_0|^2 = \frac{|\boldsymbol{d}|^2}{3}. \tag{5.30}$$

Hence, we obtain

$$U = \frac{|\boldsymbol{d}|^2 \omega^4}{3\pi\varepsilon_0 c^3}. \tag{5.31}$$

The probability P to emit a photon with energy $\hbar\omega$ in unit time is thus obtained as

$$P = \frac{|\boldsymbol{d}|^2 \omega^3}{3\pi\varepsilon_0 \hbar c^3} \propto \omega D(\omega), \tag{5.32}$$

where $\boldsymbol{d}$ should be regarded as the transition dipole moment. This formula agrees with that mentioned in Chap. 1.

Now, we consider the case where there is a dissipation caused by a small non-zero imaginary part of the dielectric constant. In this case, all eigenmodes have a finite lifetime. This can be taken into account by replacing δ by a finite positive number $\gamma_{\boldsymbol{k}n}^{(\mathrm{T})}$ in the denominators in (2.166), and hence, in (5.16), since this replacement leads to

$$\overleftrightarrow{\mathrm{G}}(\boldsymbol{r}, \boldsymbol{r}', t) \to 0 \qquad (t \to \infty), \tag{5.33}$$

which is equivalent to saying that all eigenmodes vanish after a long period. In this case, we have

$$\boldsymbol{E}_{\mathrm{d}} \simeq -\frac{\boldsymbol{P}_{\mathrm{d}}(\boldsymbol{r},t)}{\varepsilon_0\varepsilon(\boldsymbol{r})} - \frac{\mathrm{e}^{-\mathrm{i}\omega t}}{2\varepsilon_0 V}\sum_{\boldsymbol{k}n}\frac{\omega_{\boldsymbol{k}n}^{(\mathrm{T})}\boldsymbol{E}_{\boldsymbol{k}n}^{(\mathrm{T})}(\boldsymbol{r})\left\{\boldsymbol{E}_{\boldsymbol{k}n}^{(\mathrm{T})*}(\boldsymbol{r}_0)\cdot\boldsymbol{d}\right\}}{\omega-\omega_{\boldsymbol{k}n}^{(\mathrm{T})}+\mathrm{i}\gamma_{\boldsymbol{k}n}^{(\mathrm{T})}}, \tag{5.34}$$

$$U \simeq \frac{\omega}{\varepsilon_0 V}\sum_{\boldsymbol{k}n}\frac{\omega_{\boldsymbol{k}n}^{(\mathrm{T})}\gamma_{\boldsymbol{k}n}^{(\mathrm{T})}\left|\boldsymbol{E}_{\boldsymbol{k}n}^{(\mathrm{T})*}(\boldsymbol{r}_0)\cdot\boldsymbol{d}\right|^2}{\left(\omega-\omega_{\boldsymbol{k}n}^{(\mathrm{T})}\right)^2+\gamma_{\boldsymbol{k}n}^{(\mathrm{T})2}}. \tag{5.35}$$

In (5.34), we neglected the contribution from non-resonant terms. These formulae will be used to investigate the localized defect modes in Chap. 6 and the photonic band structure with frequency-dependent dielectric constants in Chap. 7.

5.3 Stimulated Emission

Next, we consider stimulated emission.[1] Assume that impurity atoms with a population inversion are distributed in the photonic crystal. We denote their number density by $\varrho(\boldsymbol{r})$ and their polarizability by α. When an eigenmode $\boldsymbol{E}_{\boldsymbol{k}n}^{(\mathrm{T})}$ is propagated in the crystal, it induces an extrinsic polarization field given by

$$\boldsymbol{P}_{\mathrm{st}}^{(1)}(\boldsymbol{r},t) = \alpha\varrho(\boldsymbol{r})\boldsymbol{E}_{\boldsymbol{k}n}^{(\mathrm{T})}(\boldsymbol{r})\exp\{(-\mathrm{i}\omega+\delta)t\}. \tag{5.36}$$

ω will be set equal to $\omega_{\boldsymbol{k}n}^{(\mathrm{T})}$ at the end of the calculation in order to avoid an unphysical divergence. We assume that $\alpha\varrho(\boldsymbol{r})$ is sufficiently smaller than $\varepsilon_0\varepsilon(\boldsymbol{r})$ to justify the following perturbative calculation. The positive infinitesimal, δ, was introduced to assure adiabatic switching of the source term as before. The imaginary part of α is negative for atoms with population inversion, and it leads to light amplification, as will be shown below. If it is positive, it leads to light absorption. We assume the same spatial periodicity for $\varrho(\boldsymbol{r})$ as $\varepsilon(\boldsymbol{r})$. We also assume for simplicity that the lattice is simple cubic with a lattice constant a. Hence, $V = N_x N_y N_z a^3$, where N_x, N_y, N_z are positive integers that denote the number of unit cells in each direction included in the crystal volume. In order to calculate the light amplification factor in a unit length, we further assume that $\varrho(\boldsymbol{r})$ is non-zero only in the region $0 \le z \le an_z \equiv l$, where n_z is a positive integer (Fig. 5.1).

When we substitute $\boldsymbol{P}_{\mathrm{st}}^{(1)}$ for $\boldsymbol{P}_{\mathrm{ex}}$ in (5.10), the propagating part of the induced electric field, $\boldsymbol{E}^{(1)}$, is obtained:

---

[1] The same problem can be treated by photonic band calculation assuming an optical gain. See [65] and [66].

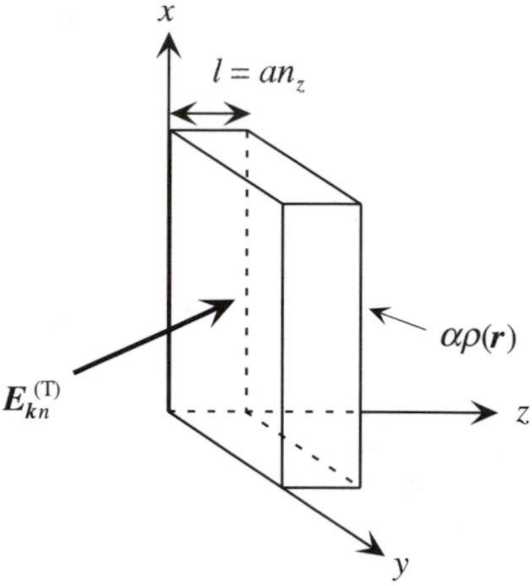

Fig. 5.1. Geometry for the analysis of stimulated emission. The lattice is assumed to be simple cubic with a lattice constant a. The crystal volume is $a^3N_xN_yN_z$. Impurity atoms with population inversion are distributed with the number density $\varrho(\boldsymbol{r})$ in the spatial region $0 \le z \le an_z \equiv l$

$$\begin{aligned}
&\boldsymbol{E}^{(1)}(\boldsymbol{r},t)\\
&= \frac{\alpha \mathrm{e}^{(-\mathrm{i}\omega+\delta)t}}{2\varepsilon_0 V} \sum_{\boldsymbol{k}'n'} \omega_{\boldsymbol{k}'n'}^{(\mathrm{T})} \boldsymbol{E}_{\boldsymbol{k}'n'}^{(\mathrm{T})}(\boldsymbol{r}) \int_V \mathrm{d}\boldsymbol{r}' \varrho(\boldsymbol{r}') \boldsymbol{E}_{\boldsymbol{k}'n'}^{(\mathrm{T})*}(\boldsymbol{r}') \cdot \boldsymbol{E}_{\boldsymbol{k}n}^{(\mathrm{T})}(\boldsymbol{r}')\\
&\times \left(\frac{1}{\omega + \omega_{\boldsymbol{k}'n'}^{(\mathrm{T})} + \mathrm{i}\delta} - \frac{1}{\omega - \omega_{\boldsymbol{k}'n'}^{(\mathrm{T})} + \mathrm{i}\delta} \right).
\end{aligned} \tag{5.37}$$

When we use (2.22) and (2.24), we have

$$\begin{aligned}
&\int_V \mathrm{d}\boldsymbol{r}' \varrho(\boldsymbol{r}') \boldsymbol{E}_{\boldsymbol{k}'n'}^{(\mathrm{T})*}(\boldsymbol{r}') \cdot \boldsymbol{E}_{\boldsymbol{k}n}^{(\mathrm{T})}(\boldsymbol{r}')\\
&= \sum_{j_1=0}^{N_x-1} \sum_{j_2=0}^{N_y-1} \sum_{j_3=0}^{n_z-1} \int_{V_0} \mathrm{d}\boldsymbol{r}' \varrho(\boldsymbol{r}') \boldsymbol{u}_{\boldsymbol{k}'n'}^{(\mathrm{T})*}(\boldsymbol{r}') \cdot \boldsymbol{u}_{\boldsymbol{k}n}^{(\mathrm{T})}(\boldsymbol{r}')\\
&\qquad \times \exp\{\mathrm{i}(\boldsymbol{k}-\boldsymbol{k}') \cdot (\boldsymbol{r}' + a\boldsymbol{j})\},
\end{aligned} \tag{5.38}$$

where $\boldsymbol{j} = (j_1, j_2, j_3)$ and V_0 is the volume of the unit cell. Because of the periodic boundary condition for $\boldsymbol{E}_{\boldsymbol{k}n}^{(\mathrm{T})}$ and $\boldsymbol{E}_{\boldsymbol{k}'n'}^{(\mathrm{T})}$, ak_xN_x, ak_yN_y, ak'_xN_x, and ak'_yN_y are all multiples of 2π. Hence, the summations over j_1 and j_2 in (5.38) are equal to zero unless $k_x = k'_x$ and $k_y = k'_y$. We thus have

$$\begin{aligned}
&\sum_{j_1=0}^{N_x-1} \sum_{j_2=0}^{N_y-1} \sum_{j_3=0}^{n_z-1} \exp\{\mathrm{i}a(\boldsymbol{k}-\boldsymbol{k}') \cdot \boldsymbol{j}\}\\
&= N_xN_y \delta_{k_xk'_x} \delta_{k_yk'_y} \frac{1-\exp(\mathrm{i}an_z\Delta k_z)}{1-\exp(\mathrm{i}a\Delta k_z)}
\end{aligned}$$

$$= N_x N_y \delta_{k_x k_x'} \delta_{k_y k_y'} \exp\{\mathrm{i}a(n_z - 1)\Delta k_z/2\} \frac{\sin(an_z \Delta k_z/2)}{\sin(a\Delta k_z/2)}, \tag{5.39}$$

where

$$\Delta k_z = k_z - k_z'. \tag{5.40}$$

If n_z is sufficiently large, the factor given by the ratio of the two sine functions in (5.39) is sharply peaked at $\Delta k_z = 0$ with the maximum value of n_z. If we take into consideration that the dominant contribution in (5.37) is given by those eigenmodes which have eigenfrequencies near to ω, we can conclude that the main contribution comes from modes with $k_z' \simeq k_z$ and $n' = n$ in the summation in (5.37) unless an accidental degeneracy of eigenfrequencies takes place.

We define the effective number density of the impurity atoms with respect to $\boldsymbol{E}_{\boldsymbol{k}n}^{(\mathrm{T})}$ by

$$F_1(\boldsymbol{k}n) \equiv \frac{1}{V_0} \int_{V_0} \mathrm{d}\boldsymbol{r}\, \varrho(\boldsymbol{r}) \left| \boldsymbol{E}_{\boldsymbol{k}n}^{(\mathrm{T})}(\boldsymbol{r}) \right|^2. \tag{5.41}$$

We thus obtain

$$\begin{aligned} \boldsymbol{E}^{(1)}(\boldsymbol{r}, t) \simeq{} & \frac{\alpha \omega_{\boldsymbol{k}n}^{(\mathrm{T})} N_x N_y V_0 F_1(\boldsymbol{k}n)}{2\varepsilon_0 V} \boldsymbol{E}_{\boldsymbol{k}n}^{(\mathrm{T})}(\boldsymbol{r}) \mathrm{e}^{(-\mathrm{i}\omega + \delta)t} \\ & \times \sum_{k_z'} \left(\frac{1}{\omega + \omega_{\boldsymbol{k}'n}^{(\mathrm{T})} + \mathrm{i}\delta} - \frac{1}{\omega - \omega_{\boldsymbol{k}'n}^{(\mathrm{T})} + \mathrm{i}\delta} \right) \\ & \times \exp\{\mathrm{i}a(n_z - 1)\Delta k_z/2\} \frac{\sin(an_z \Delta k_z/2)}{\sin(a\Delta k_z/2)}. \end{aligned} \tag{5.42}$$

In this equation, $k_x' = k_x$ and $k_y' = k_y$. Now, we denote the group velocity of $\boldsymbol{E}_{\boldsymbol{k}n}^{(\mathrm{T})}$ in the z direction by

$$v_{\mathrm{g}}(\boldsymbol{k}n) = \frac{\partial \omega_{\boldsymbol{k}n}^{(\mathrm{T})}}{\partial k_z}. \tag{5.43}$$

We convert the summation over k_z' in (5.42) into integration and change the variable from k_z' to $\omega_{\boldsymbol{k}'n}^{(\mathrm{T})}$:

$$\begin{aligned} \boldsymbol{E}^{(1)}(\boldsymbol{r}, t) \simeq{} & -\frac{\alpha \omega_{\boldsymbol{k}n}^{(\mathrm{T})} N_x N_y n_z V_0 F_1(\boldsymbol{k}n)}{2\varepsilon_0 V} \boldsymbol{E}_{\boldsymbol{k}n}^{(\mathrm{T})}(\boldsymbol{r}) \mathrm{e}^{(-\mathrm{i}\omega + \delta)t} \\ & \times \frac{aN_z}{2\pi} \int \frac{\mathrm{d}\omega_{\boldsymbol{k}'n}^{(\mathrm{T})}}{v_{\mathrm{g}}(\boldsymbol{k}'n)} \left\{ \frac{\mathcal{P}}{\omega - \omega_{\boldsymbol{k}'n}^{(\mathrm{T})}} - \pi \mathrm{i} \delta \left(\omega - \omega_{\boldsymbol{k}'n}^{(\mathrm{T})} \right) \right\}, \end{aligned} \tag{5.44}$$

where we neglected the non-resonant term, took $\Delta k_z = 0$ in the integrand, and used (5.21). Because the frequency range of the nth mode and the frequency dependence of its group velocity are unknown, we have to make a reasonable approximation for the integral in (5.44). We make an approximation that is exact for uniform materials, i.e., when the spatial modulation of the dielectric constant is absent. As will be shown later, this approximation is given by twice the imaginary part of the integral. We thus obtain

$$\boldsymbol{E}^{(1)}(\boldsymbol{r},t) \simeq \beta_{\boldsymbol{k}n} l \, \boldsymbol{E}_{\boldsymbol{k}n}^{(\mathrm{T})}(\boldsymbol{r}) \exp\{(-\mathrm{i}\omega + \delta)t\}, \tag{5.45}$$

where $\beta_{\boldsymbol{k}n}$ is defined as

$$\beta_{\boldsymbol{k}n} = \frac{\mathrm{i}\alpha\omega_{\boldsymbol{k}n}^{(\mathrm{T})} F_1(\boldsymbol{k}n)}{2\varepsilon_0 v_{\mathrm{g}}(\boldsymbol{k}n)}. \tag{5.46}$$

The induced electric field $\boldsymbol{E}^{(1)}(\boldsymbol{r},t)$ then makes another polarization field:

$$\boldsymbol{P}_{\mathrm{st}}^{(2)}(\boldsymbol{r},t) = \alpha\varrho(\boldsymbol{r})\boldsymbol{E}^{(1)}(\boldsymbol{r},t). \tag{5.47}$$

The induced field of the second order, $\boldsymbol{E}^{(2)}(\boldsymbol{r},t)$, can be calculated in the same way:

$$\boldsymbol{E}^{(2)}(\boldsymbol{r},t) \simeq \frac{1}{2}\beta_{\boldsymbol{k}n}^2 l^2 \boldsymbol{E}_{\boldsymbol{k}n}^{(\mathrm{T})}(\boldsymbol{r}) \exp\{(-\mathrm{i}\omega + \delta)t\}, \tag{5.48}$$

where a factor 1/2 was introduced by taking into account that $E^{(1)}$ increases linearly with the optical path length. In general, a factor $1/j!$ should be introduced for the jth order term. The total field that contains the induced fields to the infinite order and the original field is thus given by

$$\begin{aligned} \boldsymbol{E}(\boldsymbol{r},t) &= \sum_{j=0}^{\infty} \frac{1}{j!}\beta_{\boldsymbol{k}n}^j l^j \boldsymbol{E}_{\boldsymbol{k}n}^{(\mathrm{T})}(\boldsymbol{r}) \exp\left(-\mathrm{i}\omega_{\boldsymbol{k}n}^{(\mathrm{T})} t\right) \\ &= \boldsymbol{E}_{\boldsymbol{k}n}^{(\mathrm{T})}(\boldsymbol{r}) \exp\left(\beta_{\boldsymbol{k}n} l - \mathrm{i}\omega_{\boldsymbol{k}n}^{(\mathrm{T})} t\right). \end{aligned} \tag{5.49}$$

Hence, the real part of $\beta_{\boldsymbol{k}n}$,

$$\mathrm{Re}\,[\beta_{\boldsymbol{k}n}] = -\frac{\mathrm{Im}[\alpha]\omega_{\boldsymbol{k}n}^{(\mathrm{T})} F_1(\boldsymbol{k}n)}{2\varepsilon_0 v_{\mathrm{g}}(\boldsymbol{k}n)}, \tag{5.50}$$

is the amplitude amplification factor in the unit length. Hence, there is an enhancement effect when the group velocity of the eigenmode is small. In particular, the group-velocity anomaly that is peculiar to 2D and 3D crystals leads to a fairly large enhancement of stimulated emission. The enhancement originates from the long interaction time between the radiation field and the matter system due to the small group velocity [19]. The enhancement of stimulated emission sometimes leads to a large reduction of the lasing threshold. This point will be discussed in detail in Chap. 9.

β for Uniform Materials

For uniform materials, β can be evaluated rigorously. We will show here that the exact value coincides with the expression given by (5.46). We assume that the relative dielectric constant ε and the number density of the impurity atoms ϱ do not depend on $\boldsymbol{r}$. The polarization field due to the impurity atoms can be included in the new dielectric constant, $\varepsilon_{\mathrm{st}}$:

$$\varepsilon_0\varepsilon_{\rm st} = \varepsilon_0\varepsilon + \alpha\varrho. \tag{5.51}$$

We assume that the second term on the right-hand side is considerably smaller than the first term. The wave vector of a plane wave with angular frequency ω is thus given by

$$k = \frac{\sqrt{\varepsilon_{\rm st}}\omega}{c} \simeq \left(\sqrt{\varepsilon} + \frac{\alpha\varrho}{2\varepsilon_0\sqrt{\varepsilon}}\right)\frac{\omega}{c}. \tag{5.52}$$

The group velocity is

$$v_{\rm g} = \frac{\partial\omega}{\partial k} \simeq \frac{c}{\sqrt{\varepsilon}}. \tag{5.53}$$

β is thus given by

$$\beta = \frac{{\rm i}\alpha\varrho\omega}{2\varepsilon_0\sqrt{\varepsilon}c} = \frac{{\rm i}\alpha\varrho\omega}{2\varepsilon_0\varepsilon v_{\rm g}}. \tag{5.54}$$

On the other hand, the eigenmode in the uniform material is a plane wave:

$$\boldsymbol{E}_{\boldsymbol{k}}^{(\rm T)}(\boldsymbol{r}) = \boldsymbol{E}_0{\rm e}^{{\rm i}\boldsymbol{k}\cdot\boldsymbol{r}}. \tag{5.55}$$

Since the normalization condition is

$$\varepsilon|\boldsymbol{E}_0|^2 = 1, \tag{5.56}$$

F_1 is given by

$$F_1(\boldsymbol{k}) = \varrho|\boldsymbol{E}_0|^2 = \frac{\varrho}{\varepsilon}. \tag{5.57}$$

From (5.46), we have

$$\beta = \frac{{\rm i}\alpha\omega}{2\varepsilon_0 v_{\rm g}} \times \frac{\varrho}{\varepsilon}, \tag{5.58}$$

which agrees with (5.54).

5.4 Sum-Frequency Generation

5.4.1 Three-Dimensional Case

In this section, we will deal with sum-frequency generation as a typical example of nonlinear optical phenomena. We assume that the photonic crystal is composed of materials with second-order nonlinearity. Then, we introduce a position-dependent second-order susceptibility tensor $\overleftrightarrow{\chi}^{(2)}(\boldsymbol{r})$ that has the same spatial periodicity as $\varepsilon(\boldsymbol{r})$. We assume for simplicity, as in the previous section, that the lattice is simple cubic. In order to clarify the phase-matching condition, we further assume that $\overleftrightarrow{\chi}^{(2)}(\boldsymbol{r})$ is non-zero only in the region $0 \leq z \leq an_z$, where n_z is a positive integer. We will examine the generation of the sum-frequency component when two eigenmodes, $\boldsymbol{E}^{(\mathrm{T})}_{\boldsymbol{k}_1 n_1}(\boldsymbol{r})$ and $\boldsymbol{E}^{(\mathrm{T})}_{\boldsymbol{k}_2 n_2}(\boldsymbol{r})$, are propagated in the crystal.

The nonlinear polarization $\boldsymbol{P}_{\mathrm{NL}}(\boldsymbol{r},t)$ induced by the two electric fields acts as the external polarization $\boldsymbol{P}_{\mathrm{ex}}$ in (5.10), which is given by the next equation for the present problem:

$$\begin{aligned}\boldsymbol{P}_{\mathrm{NL}}(\boldsymbol{r},t) = \frac{1}{2}A^2 \overleftrightarrow{\chi}^{(2)}(\boldsymbol{r}) : \boldsymbol{E}^{(\mathrm{T})}_{\boldsymbol{k}_1 n_1}(\boldsymbol{r}) \boldsymbol{E}^{(\mathrm{T})}_{\boldsymbol{k}_2 n_2}(\boldsymbol{r}) \\ \times \exp\left\{\left(-\mathrm{i}\omega^{(\mathrm{T})}_{\boldsymbol{k}_1 n_1} - \mathrm{i}\omega^{(\mathrm{T})}_{\boldsymbol{k}_2 n_2} + \delta\right) t\right\}. \end{aligned} \tag{5.59}$$

In this equation A denotes the amplitude of the two eigenmodes, where we assume that the two modes have the same amplitude for simplicity, and the symbol ":" denotes the product of $\overleftrightarrow{\chi}^{(2)}(\boldsymbol{r})$ with $\boldsymbol{E}^{(\mathrm{T})}_{\boldsymbol{k}_1 n_1}(\boldsymbol{r})$ and $\boldsymbol{E}^{(\mathrm{T})}_{\boldsymbol{k}_2 n_2}(\boldsymbol{r})$ as a tensor. Note that we wrote the positive frequency part of the polarization field. The actual real field is given by the sum of the above field and its complex conjugate. The same holds for the electric field derived below.

Substituting $\boldsymbol{P}_{\mathrm{NL}}(\boldsymbol{r},t)$ for $\boldsymbol{P}_{\mathrm{ex}}(\boldsymbol{r},t)$ in (5.10), we obtain the sum-frequency component of the electric field. The propagating part of the induced field, $\boldsymbol{E}_{\mathrm{NL}}(\boldsymbol{r},t)$, can be calculated like (5.37). When we define the effective nonlinear susceptibility with respect to the relevant fields, F_2, by

$$\begin{aligned} &F_2(\boldsymbol{k}n, \boldsymbol{k}_1 n_1, \boldsymbol{k}_2 n_2) \\ &= \frac{1}{V_0}\int_{V_0} \mathrm{d}\boldsymbol{r}\, \boldsymbol{E}^{(\mathrm{T})*}_{\boldsymbol{k}n}(\boldsymbol{r}) \cdot \overleftrightarrow{\chi}^{(2)}(\boldsymbol{r}) : \boldsymbol{E}^{(\mathrm{T})}_{\boldsymbol{k}_1 n_1}(\boldsymbol{r}) \boldsymbol{E}^{(\mathrm{T})}_{\boldsymbol{k}_2 n_2}(\boldsymbol{r}), \end{aligned} \tag{5.60}$$

$\boldsymbol{E}_{\mathrm{NL}}(\boldsymbol{r},t)$ is given by

$$\begin{aligned} &\boldsymbol{E}_{\mathrm{NL}}(\boldsymbol{r},t) \\ &= \frac{A^2 V_0 \exp\left\{\left(-\mathrm{i}\omega^{(\mathrm{T})}_{\boldsymbol{k}_1 n_1} - \mathrm{i}\omega^{(\mathrm{T})}_{\boldsymbol{k}_2 n_2} + \delta\right) t\right\}}{4\varepsilon_0 V} \sum_{\boldsymbol{k}n} \omega^{(\mathrm{T})}_{\boldsymbol{k}n} \boldsymbol{E}^{(\mathrm{T})}_{\boldsymbol{k}n}(\boldsymbol{r}) \\ &\times \sum_{j_1=0}^{N_x-1}\sum_{j_2=0}^{N_y-1}\sum_{j_3=0}^{n_z-1} \exp\{\mathrm{i}a(\boldsymbol{k}_1+\boldsymbol{k}_2-\boldsymbol{k})\cdot\boldsymbol{j}\} F_2(\boldsymbol{k}n, \boldsymbol{k}_1 n_1, \boldsymbol{k}_2 n_2) \\ &\times \left(\frac{1}{\omega^{(\mathrm{T})}_{\boldsymbol{k}_1 n_1} + \omega^{(\mathrm{T})}_{\boldsymbol{k}_2 n_2} + \omega^{(\mathrm{T})}_{\boldsymbol{k}n} + \mathrm{i}\delta} - \frac{1}{\omega^{(\mathrm{T})}_{\boldsymbol{k}_1 n_1} + \omega^{(\mathrm{T})}_{\boldsymbol{k}_2 n_2} - \omega^{(\mathrm{T})}_{\boldsymbol{k}n} + \mathrm{i}\delta}\right). \end{aligned} \tag{5.61}$$

The summation over $\boldsymbol{j}$ can be calculated like (5.39):

$$\begin{aligned} &\sum_{j_1=0}^{N_x-1}\sum_{j_2=0}^{N_y-1}\sum_{j_3=0}^{n_z-1} \exp\left\{\mathrm{i}a\big(\boldsymbol{k}_1+\boldsymbol{k}_2-\boldsymbol{k}\big)\cdot\boldsymbol{j}\right\} \\ &= N_x N_y \overline{\delta}_{k_x, k_{1x}+k_{2x}} \overline{\delta}_{k_y, k_{1y}+k_{2y}} \\ &\qquad \times \exp\{\mathrm{i}a(n_z-1)\Delta k_z/2\} \frac{\sin(a n_z \Delta k_z/2)}{\sin(a\Delta k_z/2)}, \end{aligned} \tag{5.62}$$

where

$$\Delta k_z = k_{1z} + k_{2z} - k_z. \tag{5.63}$$

$\overline{\delta}$ is Kronecker's delta with allowance for the Umklapp process, that is,

$$\overline{\delta}_{kk'} = \begin{cases} 1 & (k' = k + 2\pi j/a, \text{ where } j \text{ is an integer}), \\ 0 & (\text{otherwise}). \end{cases} \tag{5.64}$$

The function

$$\frac{\sin(an_z\Delta k_z/2)}{\sin(a\Delta k_z/2)} \tag{5.65}$$

has a sharp peak at

$$\Delta k_z = \frac{2\pi j}{a}, \tag{5.66}$$

if n_z is sufficiently large. As a result, the summation in (5.62) has a large value only when

$$\boldsymbol{k} = \boldsymbol{k}_1 + \boldsymbol{k}_2 + \boldsymbol{G}, \tag{5.67}$$

where $\boldsymbol{G}$ is a reciprocal lattice vector. We say that the phase-matching condition is satisfied when crystalline momentum is conserved. In a uniform material, the phase-matching condition is equivalent to the conservation of momentum, i.e., it is given by

$$\boldsymbol{k} = \boldsymbol{k}_1 + \boldsymbol{k}_2. \tag{5.68}$$

This is a special case of (5.67) for which $\boldsymbol{G} = 0$. In photonic crystals, $\boldsymbol{G}$ is arbitrary, and therefore, the phase-matching condition is relaxed considerably compared with uniform materials. Generally speaking, the phase-matching is much easier for the former than the latter.

For the summation over $\boldsymbol{k}$ and n in (5.61), we make the same approximation as we did for (5.42). We thus set

$$\left(\frac{1}{\omega^{(\mathrm{T})}_{\boldsymbol{k}_1 n_1} + \omega^{(\mathrm{T})}_{\boldsymbol{k}_2 n_2} + \omega^{(\mathrm{T})}_{\boldsymbol{k}n} + \mathrm{i}\delta} - \frac{1}{\omega^{(\mathrm{T})}_{\boldsymbol{k}_1 n_1} + \omega^{(\mathrm{T})}_{\boldsymbol{k}_2 n_2} - \omega^{(\mathrm{T})}_{\boldsymbol{k}n} + \mathrm{i}\delta} \right)$$
$$\to 2\pi\mathrm{i}\delta\left(\omega^{(\mathrm{T})}_{\boldsymbol{k}_1 n_1} + \omega^{(\mathrm{T})}_{\boldsymbol{k}_2 n_2} - \omega^{(\mathrm{T})}_{\boldsymbol{k}n} \right). \tag{5.69}$$

We assume for simplicity that there is only one eigenmode, $\boldsymbol{E}^{(\mathrm{T})}_{\overline{\boldsymbol{k}}\overline{n}}$, for which

$$\overline{k}_x = k_{1x} + k_{2x} + \frac{2\pi p}{a} \qquad (p : \text{integer}), \tag{5.70}$$

$$\overline{k}_y = k_{1y} + k_{2y} + \frac{2\pi q}{a} \qquad (q : \text{integer}), \tag{5.71}$$

$$\overline{k}_z \approx k_{1z} + k_{2z} + \frac{2\pi j}{a}, \tag{5.72}$$

$$\omega_{\overline{kn}}^{(\mathrm{T})} = \omega_{\boldsymbol{k}_1 n_1} + \omega_{\boldsymbol{k}_2 n_2}. \tag{5.73}$$

When we convert the summation over k_z in (5.61) to the integration and change the variable from k_z to $\omega_{\boldsymbol{k}n}$, we obtain

$$\begin{aligned} \boldsymbol{E}_{\mathrm{NL}}(\boldsymbol{r},t) \simeq\ & \frac{\mathrm{i}aA^2\omega_{\overline{kn}}^{(\mathrm{T})} F_2(\overline{\boldsymbol{k}n}, \boldsymbol{k}_1 n_1, \boldsymbol{k}_2 n_2)}{4\varepsilon_0 v_{\mathrm{g}}(\overline{\boldsymbol{k}n})} \\ & \times \boldsymbol{E}_{\overline{kn}}^{(\mathrm{T})} \exp\left(-\mathrm{i}\omega_{\overline{kn}}^{(\mathrm{T})} t\right) \\ & \times \exp\left\{\mathrm{i}a(n_z - 1)\Delta\bar{k}_z/2\right\} \frac{\sin a n_z \Delta\bar{k}_z/2}{\sin a\Delta\bar{k}_z/2}, \end{aligned} \tag{5.74}$$

where

$$\Delta\bar{k}_z = k_{1z} + k_{2z} - \bar{k}_z, \tag{5.75}$$

and $v_{\mathrm{g}}(\overline{\boldsymbol{k}n})$ is the group velocity in the z direction. If there is more than one mode, $\boldsymbol{E}_{\mathrm{NL}}$ is given by the sum of the contribution from them. The intensity of the electric field of the sum frequency is given by

$$\begin{aligned} |\boldsymbol{E}_{\mathrm{NL}}(\boldsymbol{r},t)|^2 \simeq\ & \frac{a^2 A^4 \omega_{\overline{kn}}^{(\mathrm{T})2} \left|F_2(\overline{\boldsymbol{k}n}, \boldsymbol{k}_1 n_1, \boldsymbol{k}_2 n_2)\right|^2}{16\varepsilon_0^2 v_{\mathrm{g}}^2(\overline{\boldsymbol{k}n})} \\ & \times \left|\boldsymbol{E}_{\overline{kn}}^{(\mathrm{T})}\right|^2 \frac{\sin^2 a n_z \Delta\bar{k}_z/2}{\sin^2 a\Delta\bar{k}_z/2}. \end{aligned} \tag{5.76}$$

Here, we should note that (1) the field intensity (the Poynting vector) is proportional to v_{g}^{-2} (v_{g}^{-1}), and thus we can expect its enhancement at the photonic band edges where v_{g} tends to zero and for the group-velocity anomaly, (2) the phase-matching condition is, as we mentioned before, satisfied when $\Delta\bar{k}_z = 0$, i.e., when the crystalline momentum is conserved, and (3) the effective nonlinear susceptibility $F_2(\overline{\boldsymbol{k}n}, \boldsymbol{k}_1 n_1, \boldsymbol{k}_2 n_2)$ may vanish for particular combinations of $\boldsymbol{k}_1 n_1$ and $\boldsymbol{k}_2 n_2$ in highly symmetric photonic crystals because of the symmetry of the relevant wave functions, and this leads to an additional selection rule besides that due to the crystallographic symmetry of the host crystal. We should also note that these three features are common to other nonlinear optical processes in the photonic crystals for which the phase-matching condition is imposed by the momentum conservation law.

5.4.2 Two-Dimensional Case

In this section, we derive the expression for sum-frequency generation in the 2D square lattice. Some numerical results will be presented in the next section. We assume that the crystal is composed of materials with second-order nonlinearity as before. We introduce the position-dependent second-order susceptibility $\chi^{(2)}(\boldsymbol{r}_{/\!/})$ that has the same spatial periodicity as $\varepsilon(\boldsymbol{r}_{/\!/})$. We also assume that the wave vectors of two incident waves, which are eigenmodes of the 2D photonic crystal, lie in the x-y plane. Among the two polarizations, we will deal with the E polarization in this section. As for the H polarization, the full vector treatment given in the previous section is appropriate, since it has two components of the electric field.

The z component of the nonlinear polarization $P_{z,NL}$ is given by

$$P_{z,NL}(\boldsymbol{r}_{/\!/},t) = \frac{1}{2}A^2\chi^{(2)}(\boldsymbol{r}_{/\!/})E_{z,\boldsymbol{k}_{1/\!/}n_1}(\boldsymbol{r}_{/\!/})E_{z,\boldsymbol{k}_{2/\!/}n_2}(\boldsymbol{r}_{/\!/}) \times \exp\left\{\left(-\mathrm{i}\omega^{(E)}_{\boldsymbol{k}_{1/\!/}n_1} - \mathrm{i}\omega^{(E)}_{\boldsymbol{k}_{2/\!/}n_2} + \delta\right)t\right\}. \quad (5.77)$$

Generally speaking, $\chi^{(2)}$ is a tensor of rank 3, and the x and y components of the polarization field, $P_{x,NL}$ and $P_{y,NL}$, may be non-zero. However, we assume that they are equal to zero for simplicity. This situation can be realized if we use, for example, a nonlinear crystal of $3m$-C_{3v} symmetry such as $LiNbO_3$ with its crystalline c axis parallel to the z axis. Then, the E polarization corresponds to the extraordinary wave in this uniaxial crystal. We will omit the suffix z hereafter. In order to clarify the phase-matching condition, we further assume that $\chi^{(2)}(\boldsymbol{r}_{/\!/})$ is non-zero only in the region $0 \le y \le a n_y$, where n_y is a positive integer.

The effective nonlinear susceptibility for the 2D crystal, $F_2^{(2)}$ is defined by

$$F_2^{(2)}(\boldsymbol{k}_{/\!/}n, \boldsymbol{k}_{1/\!/}n_1, \boldsymbol{k}_{2/\!/}n_2) = \frac{1}{V_0^{(2)}}\int_{V_0^{(2)}} \mathrm{d}\boldsymbol{r}_{/\!/}\chi^{(2)}(\boldsymbol{r}_{/\!/})E^*_{\boldsymbol{k}_{/\!/}n}(\boldsymbol{r}_{/\!/})E_{\boldsymbol{k}_{1/\!/}n_1}(\boldsymbol{r}_{/\!/})E_{\boldsymbol{k}_{2/\!/}n_2}(\boldsymbol{r}_{/\!/}), \quad (5.78)$$

where $V_0^{(2)}$ denotes the volume of the 2D unit cell. Following the same procedure as we did for the 3D case, we obtain the propagating part of the induced electric field, $E_{\mathrm{NL}}(\boldsymbol{r}_{/\!/},t)$:

$$E_{\mathrm{NL}}(\boldsymbol{r}_{/\!/},t) \simeq \frac{\mathrm{i}aA^2\omega^{(E)}_{\overline{\boldsymbol{k}}_{/\!/}\overline{n}}F_2^{(2)}(\overline{\boldsymbol{k}}_{/\!/}\overline{n}, \boldsymbol{k}_{1/\!/}n_1\boldsymbol{k}_{2/\!/}n_2)}{4\varepsilon_0 v_{\mathrm{g}}(\overline{\boldsymbol{k}}_{/\!/}\overline{n})} \times E_{\overline{\boldsymbol{k}}_{/\!/}\overline{n}}(\boldsymbol{r})\exp\left(-\mathrm{i}\omega^{(E)}_{\overline{\boldsymbol{k}}_{/\!/}\overline{n}}t\right) \times \exp\left\{\mathrm{i}a(n_y-1)\Delta\overline{k}_y/2\right\}\frac{\sin\left(an_y\Delta\overline{k}_y/2\right)}{\sin\left(a\Delta\overline{k}_y/2\right)}, \quad (5.79)$$

where

$$\overline{k}_x = k_{1x} + k_{2x} + \frac{2\pi j}{a} \qquad (j : \text{integer}), \quad (5.80)$$

$$\omega^{(E)}_{\overline{\boldsymbol{k}}_{/\!/}\overline{n}} = \omega^{(E)}_{\boldsymbol{k}_{1/\!/}n_1} + \omega^{(E)}_{\boldsymbol{k}_{2/\!/}n_2}, \quad (5.81)$$

$$\Delta\overline{k}_y = k_{1y} + k_{2y} - \overline{k}_y, \quad (5.82)$$

and $v_g(\overline{\boldsymbol{k}}_{/\!/}\overline{n})$ is the group velocity in the y direction. We assumed for simplicity that there is only one eigenmode that satisfied (5.80) and (5.81). The intensity of the induced nonlinear electric field is thus given by

$$|E_{\rm NL}(\boldsymbol{r}_{/\!/},t)|^2 \simeq \frac{a^2 A^4 \omega^{(E)2}_{\overline{\boldsymbol{k}}_{/\!/}\overline{n}} \left| E_{\overline{\boldsymbol{k}}_{/\!/}\overline{n}}(\boldsymbol{r}_{/\!/}) \right|^2}{16\varepsilon_0^2 v_g^2(\overline{\boldsymbol{k}}_{/\!/}\overline{n})} \times \frac{\sin^2(an_y \Delta\overline{k}_y/2)}{\sin^2(a\Delta\overline{k}_y/2)} |F_2^{(2)}(\overline{\boldsymbol{k}}_{/\!/}\overline{n}, \boldsymbol{k}_{1/\!/}n_1, \boldsymbol{k}_{2/\!/}n_2)|^2 . \quad (5.83)$$

$E_{\rm NL}$ for Uniform Materials

We will examine the result of the approximation given in (5.79) for the case of a uniform material. We take $\varepsilon(\boldsymbol{r}_{/\!/})$ and $\chi^{(2)}(\boldsymbol{r}_{/\!/})$ to be independent of $\boldsymbol{r}_{/\!/}$. We further assume that $\boldsymbol{k}_{1/\!/} = \boldsymbol{k}_{2/\!/}$ for simplicity. We take the y axis to be parallel to $\boldsymbol{k}_{1/\!/}$, i.e., $\boldsymbol{k}_{1/\!/} = (0, k_1)$ $(k_1 > 0)$. This plane wave will be referred to as the fundamental wave. On the other hand, the sum-frequency component will be referred to as the second harmonic, since it has twice as large a frequency as the fundamental wave. We denote the angular frequency of the fundamental wave by ω. Note that the wave vector of the second harmonics, $\boldsymbol{k}_{/\!/}$ is also parallel to the y axis because of the conservation of the x component of the crystalline momentum.

Since the eigenfunctions are plane wave in the uniform material, we have

$$E_{z,\boldsymbol{k}_{1/\!/}}(\boldsymbol{r}_{/\!/}) = \frac{1}{\sqrt{\varepsilon_\omega}} e^{ik_1 y} , \quad (5.84)$$

$$E_{z,\boldsymbol{k}_{/\!/}}(\boldsymbol{r}_{/\!/}) = \frac{1}{\sqrt{\varepsilon_{2\omega}}} e^{iky} , \quad (5.85)$$

$$v_g(2\omega) = \frac{c}{\sqrt{\varepsilon_{2\omega}}} , \quad (5.86)$$

where ε_ω and $\varepsilon_{2\omega}$ are the dielectric constants at ω and 2ω, respectively, and $v_g(2\omega)$ is the group velocity of the second harmonics. In (5.84) and (5.85), we took the extended zone scheme, and the suffix to distinguish the band was omitted. Because the Umklapp process is absent in the uniform material, (5.78) leads to

$$F_2^{(2)}(\boldsymbol{k}_{/\!/}, \boldsymbol{k}_{1/\!/}, \boldsymbol{k}_{1/\!/}) = \frac{\chi^{(2)}}{V_0^{(2)} \varepsilon_\omega \sqrt{\varepsilon_{2\omega}}} \int_{V_0^{(2)}} d\boldsymbol{r}_{/\!/} \exp\{i(2k_1 - k)y\}$$
$$= \frac{\chi^{(2)}[\exp\{i(2k_1-k)a\} - 1]}{ia\varepsilon_\omega \sqrt{\varepsilon_{2\omega}}(2k_1 - k)} . \quad (5.87)$$

Substituting these equations into (5.79), we obtain

$$E_{2\omega}(\boldsymbol{r}_{/\!/},t) \simeq \frac{i\omega\chi^{(2)}A^2 l}{2c\varepsilon_0\varepsilon_\omega\sqrt{\varepsilon_{2\omega}}} e^{i(ky-2\omega t)} e^{il\Delta k/2} \frac{\sin(l\Delta k/2)}{l\Delta k/2} , \quad (5.88)$$

where

$$\Delta k = 2k_1 - k \qquad \text{and} \qquad l = an_y. \tag{5.89}$$

On the other hand, from the first equality in (5.12), we have

$$\begin{aligned} &E_{2\omega}(\boldsymbol{r}_{/\!/}, t) \\ &= -\frac{4\omega^2}{c^2\varepsilon_0\varepsilon_{2\omega}}\int_{V^{(2)}} \mathrm{d}\boldsymbol{r}' \int_{-\infty}^{\infty} \mathrm{d}t' G^{(2)}(\boldsymbol{r}, \boldsymbol{r}', t-t') P_{z,\mathrm{ex}}(\boldsymbol{r}', t') \\ &= -\frac{2\omega^2\chi^{(2)}A^2}{c^2\varepsilon_0\varepsilon_\omega\varepsilon_{2\omega}}\int_{0\le y'\le l} \mathrm{d}\boldsymbol{r}' \int_{-\infty}^{\infty} \mathrm{d}t' G^{(2)}(\boldsymbol{r}, \boldsymbol{r}', t-t') \mathrm{e}^{2\mathrm{i}(k_1 y' - \omega t')}. \end{aligned} \tag{5.90}$$

When we use (2.170), we obtain

$$\begin{aligned} &\int_{-\infty}^{\infty} \mathrm{d}t' G^{(2)}(\boldsymbol{r}, \boldsymbol{r}', t-t') \mathrm{e}^{-2\mathrm{i}\omega t'} \\ &= \mathrm{e}^{-2\mathrm{i}\omega t}\int_{-\infty}^{\infty} \mathrm{d}t' G^{(2)}(\boldsymbol{r}, \boldsymbol{r}', t') \mathrm{e}^{2\mathrm{i}\omega t'} \\ &= \frac{c^2}{V^{(2)}}\mathrm{e}^{-2\mathrm{i}\omega t}\sum_{\boldsymbol{k}_{/\!/}} \frac{\exp(\mathrm{i}\boldsymbol{k}_{/\!/}\cdot\boldsymbol{r}_{/\!/})\exp(-\mathrm{i}\boldsymbol{k}_{/\!/}\cdot\boldsymbol{r}'_{/\!/})}{(2\omega - \omega_{\boldsymbol{k}_{/\!/}} + \mathrm{i}\delta)(2\omega + \omega_{\boldsymbol{k}_{/\!/}} + \mathrm{i}\delta)}. \end{aligned} \tag{5.91}$$

For the summation over $\boldsymbol{k}_{/\!/}$ in this equation, k_x must be equal to zero because of the conservation of the x component of the momentum, which originates from the spatial integral in (5.90). Since $\omega_{\boldsymbol{k}_{/\!/}} \approx 2\omega$ contribute dominantly to (5.91), we set,

$$\omega_{\boldsymbol{k}_{/\!/}} = \omega_k = \begin{cases} ck/\sqrt{\varepsilon_{2\omega}} & (k > 0), \\ -ck/\sqrt{\varepsilon_{2\omega}} & (k < 0). \end{cases} \tag{5.92}$$

Since

$$\int_{0\le y'\le l} \mathrm{d}\boldsymbol{r}' \mathrm{e}^{\mathrm{i}(2k_1 - k)y'} = aN_x\frac{\exp\{\mathrm{i}(2k_1 - k)l\} - 1}{\mathrm{i}(2k_1 - k)}, \tag{5.93}$$

we have

$$\begin{aligned} E_{2\omega}(\boldsymbol{r}_{/\!/}, t) = -\frac{2\omega^2\chi^{(2)}aN_xA^2\mathrm{e}^{-2\mathrm{i}\omega t}}{\varepsilon_0\varepsilon_\omega\varepsilon_{2\omega}V^{(2)}}\sum_k \frac{\mathrm{e}^{\mathrm{i}ky}}{(2\omega - \omega_k + \mathrm{i}\delta)(2\omega + \omega_k + \mathrm{i}\delta)} \\ \times\frac{\exp\{\mathrm{i}(2k_1 - k)l\} - 1}{\mathrm{i}(2k_1 - k)}. \end{aligned} \tag{5.94}$$

When we transform the summation into the integration over k and change the variable from k to $\Omega = \omega_k$, we have

$$\begin{aligned} &E_\omega(\boldsymbol{r}_{/\!/}, t) \\ &= -\frac{\omega^2\chi^{(2)}a^2A^2N_xN_y\mathrm{e}^{-2\mathrm{i}\omega t}}{\mathrm{i}c\varepsilon_0\varepsilon_\omega\sqrt{\varepsilon_{2\omega}}\pi V^{(2)}} \end{aligned}$$

$$\times\left\{\int_0^{\infty} \mathrm{d}\Omega \frac{\mathrm{e}^{\mathrm{i}k^{(+)}y}}{(2\omega-\Omega+\mathrm{i}\delta)(2\omega+\Omega+\mathrm{i}\delta)} \frac{\mathrm{e}^{\mathrm{i}(2k_1-k^{(+)})l}-1}{2k_1-k^{(+)}}\right.$$
$$\left.-\int_{\infty}^{0} \mathrm{d}\Omega \frac{\mathrm{e}^{\mathrm{i}k^{(-)}y}}{(2\omega-\Omega+\mathrm{i}\delta)(2\omega+\Omega+\mathrm{i}\delta)} \frac{\mathrm{e}^{\mathrm{i}(2k_1-k^{(-)})l}-1}{2k_1-k^{(-)}}\right\}, \quad (5.95)$$

where

$$k^{(\pm)} = \pm\frac{\sqrt{\varepsilon_{2\omega}}\Omega}{c}, \quad (5.96)$$

and we have taken into consideration that there are two branches with positive and negative k as shown in (5.92). We change the sign of the variable Ω in the second integral and modify the path of integration to enclose the upper half of the complex Ω plane for $y > l$. Using the residue theorem, we thus obtain

$$E_{2\omega}(\boldsymbol{r}_{/\!/}, t) = \frac{\omega^2 \chi^{(2)} A^2 \mathrm{e}^{-2\mathrm{i}\omega t}}{\mathrm{i}c\varepsilon_0\varepsilon_\omega\sqrt{\varepsilon_{2\omega}}\pi} \times 2\pi\mathrm{i}\frac{\mathrm{e}^{\mathrm{i}ky}}{4\omega}\frac{\mathrm{e}^{\mathrm{i}l\Delta k}-1}{\Delta k}, \quad (5.97)$$

which agrees with (5.88). For the case of general photonic crystals, the rigorous evaluation is not possible since the path of the integration is not the whole real axis and the analyticity of the eigenfunctions is unknown.

5.5 SHG in the Square Lattice

In this section, we will apply our method to second harmonic generation (SHG) in a square lattice of circular air-rods formed in a $LiNbO_3$ crystal. The geometry is shown in Fig. 5.2, which is the top view of the square lattice composed of identical cylinders with radius r_a. The dielectric constants of the cylinder and the background are denoted by ε_1 and ε_2, respectively. We assume that the crystalline c axis of this uniaxial crystal of the $3m$-C_{3v} symmetry is aligned along the z axis. Then, the incident fundamental waves of the E polarization yield only the z component of the nonlinear polarization field, and therefore, the method given in the previous section is applicable to the present problem. Similar situations can be realized for crystals of the 4-C_4, $4mm$-C_{4v}, 3-C_3, 6-C_6, and $6mm$-C_{6v} symmetries as well.

Figure 5.3 shows the photonic band structure of the square lattice, where we assume that $r_\mathrm{a}/a = 0.3$, $\varepsilon_1 = 1.0$, and $\varepsilon_2 = 4.9912$, which is the dielectric constant of $LiNbO_3$ at 532 nm. The calculation was performed by the plane-wave expansion method. The number of the basis plane waves was 289 and the accuracy was estimated as better than 1%. The symmetries of the eigenfunctions are also shown in Fig. 5.3.

Now, we assume for simplicity that $\boldsymbol{k}_{1/\!/} = (k_x, k_y)$ and $\boldsymbol{k}_{2/\!/} = (-k_x, k_y)$ $(k_y > 0)$. Then, the phase-matching condition is satisfied when

$$\overline{k}_x = 0 \quad \text{and} \quad \overline{k}_y = 2k_y - \frac{2j\pi}{a} \quad (j = 0,\ 1). \quad (5.98)$$

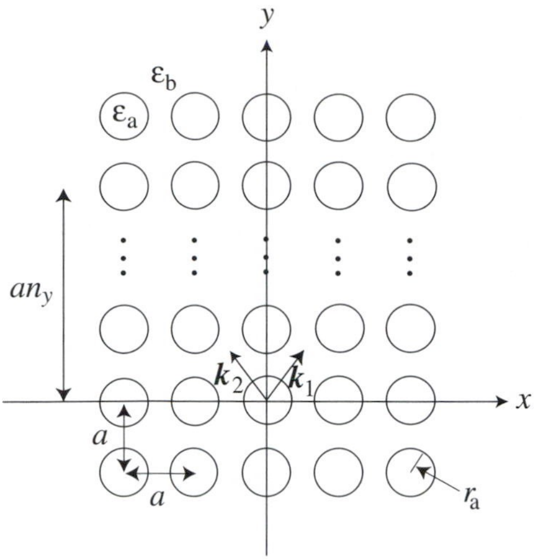

Fig. 5.2. Geometry for sum-frequency generation in a two-dimensional crystal. This figure is the top view of a square lattice with a lattice constant a composed of identical cylinders with a radius R. The dielectric constants of the cylinder and the background are denoted by ε_1 and ε_2, respectively. The position-dependent second-order susceptibility $\chi^{(2)}(\boldsymbol{r}_{/\!/})$ is assumed to be non-zero at $0 \leq y \leq an_y$ in this example. Two incident eigenmodes of E polarization with wave vectors $\boldsymbol{k}_{1/\!/}$ and $\boldsymbol{k}_{2/\!/}$, and angular frequencies $\omega^{(E)}_{\boldsymbol{k}_{1/\!/}n_1}$ and $\omega^{(E)}_{\boldsymbol{k}_{2/\!/}n_2}$ generate the sum-frequency component. (After [67])

Here, we consider two examples. In case 1, two incident waves in the lowest branch induce the second harmonic in the second lowest branch on the Δ point; $\omega a/2\pi c \simeq 0.14$ for the fundamental wave and $j = 0$. In case 2, the second harmonic in the fourth branch on the Δ point with negative $\overline{k}_y$ is induced by similar incident waves; $\omega a/2\pi c \simeq 0.25$ and $j = 1$. For both cases, the effective nonlinear susceptibility $F^{(2)}_2(\overline{\boldsymbol{k}}_{/\!/}\overline{n}, \boldsymbol{k}_{1/\!/}1, \boldsymbol{k}_{2/\!/}1)$ ($\overline{n} = 2$ or 4) is non-zero. On the other hand, the third branch does not contribute to SHG since $F^{(2)}_2(\overline{\boldsymbol{k}}_{/\!/}3, \boldsymbol{k}_{1/\!/}1, \boldsymbol{k}_{2/\!/}1) = 0$. This is because the wave function of the third branch on the Δ point is antisymmetric under the mirror reflection on the y-z plane, whereas the induced nonlinear polarization for the second harmonic is symmetric. As is seen in this example, $F^{(2)}_2(\overline{\boldsymbol{k}}_{/\!/}\overline{n}, \boldsymbol{k}_{1/\!/}n_1, \boldsymbol{k}_{2/\!/}n_2)$ vanishes for particular combinations of the initial and the final states in highly symmetric crystals, and this gives a new type of selection rules that are not relevant to the crystallographic symmetry of the host crystal.

Figure 5.4 shows the group velocity of the second and the fourth branches on the Δ point together with their dispersion curves. The former was calculated by means of the Hellmann–Feynman theorem. (The detail was described in Sect. 2.5.) As is clear from this figure, the group velocity of both bands tends to zero at the Γ and X points. In addition, the group velocity of the fourth band is small over the whole branch due to the group-velocity anomaly.

Now, we take the angular frequency of the fundamental wave ω such that $\omega a/2\pi c = 0.138$ for case 1 and $\omega a/2\pi c = 0.253$ for case 2. Then $|v_{\mathrm{g}}|/c$ at 2ω is 0.114 and 0.0740, respectively. The corresponding data points are denoted by solid circles in Fig. 5.4, where we have to note that $\overline{k}_y < 0$ for case 2. Figure 5.5 shows two contours of the lowest branch for both cases where the dielectric constant of the host crystal ε_2 is assumed to be 4.6483, which is that of $LiNbO_3$ at 1064 nm. In order to use this value consistently with the assumed angular frequency of the fundamental waves, the lattice constant should be about 0.15 mm for case 1 and 0.27 mm for case 2. From the dispersion relations of the second and the fourth bands, the phase-matching condition is fulfilled when $ak_y/2\pi = 0.245$ for case 1 and $ak_y/2\pi = 0.413$ for case 2, respectively. Note that the phase-matching is impossible for the E polarization in a uniform $LiNbO_3$ crystal because the dielectric constant at 2ω is larger than that at ω.

The average intensity of the electric field of the second harmonic, $\overline{|E_{\mathrm{NL}}(\boldsymbol{r}_{/\!/}, t)|^2}$, is given by

$$\overline{|E_{\mathrm{NL}}(\boldsymbol{r}_{/\!/}, t)|^2} = \frac{\pi^2 A^4}{\varepsilon_0^2} \left(\frac{\omega a}{2\pi c}\right)^2 \left(\frac{c}{v_{\mathrm{g}}}\right)^2 \overline{|E_{\overline{\boldsymbol{k}}_{/\!/}\overline{n}}(\boldsymbol{r}_{/\!/})|^2}$$
$$\times |F_2^{(2)}(\overline{\boldsymbol{k}}_{/\!/}\overline{n}, \boldsymbol{k}_{1/\!/}1, \boldsymbol{k}_{2/\!/}1)|^2 \frac{\sin^2(a n_y \Delta \overline{k}_y/2)}{\sin^2(a\Delta \overline{k}_y/2)}, \qquad (5.99)$$

where

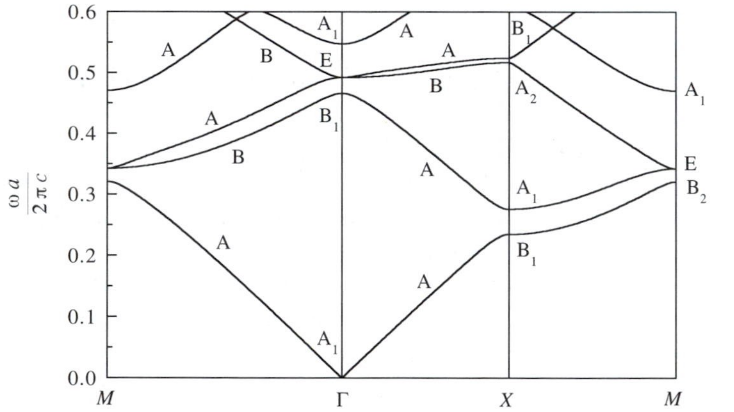

Fig. 5.3. Dispersion relation of the square lattice of circular air-rods formed in a $LiNbO_3$ crystal. The ordinate is the normalized angular frequency. The following values were assumed: $r_{\mathrm{a}}/a = 0.3$, $\varepsilon_1 = 1.0$, and $\varepsilon_2 = 4.9912$, which is the dielectric constant of $LiNbO_3$ at 532 nm. The symmetries of the eigenfunctions are also shown. (After [67])

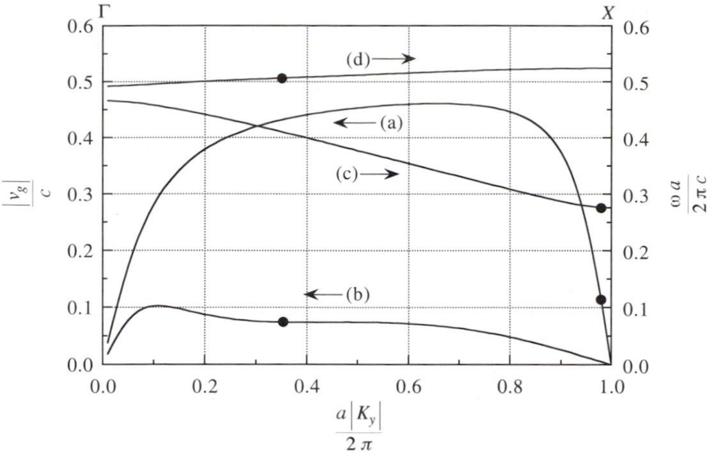

Fig. 5.4. Group velocity of (*a*) second branch and (*b*) fourth branch on the Δ point. Curves (*c*) and (*d*) are their dispersion relations. The same parameters as for Fig. 5.3 were assumed. The data points corresponding to case 1 and case 2 are denoted by *solid circles* (see text). (After [67])

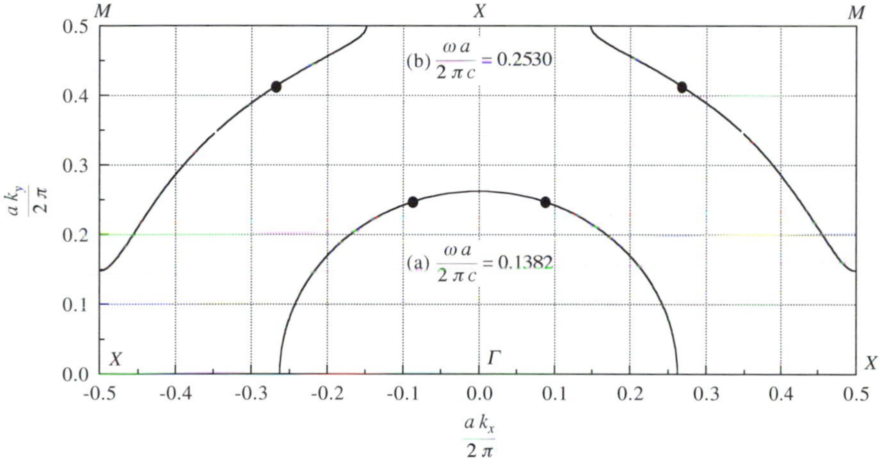

Fig. 5.5. Contours of the lowest branch for (*a*) case 1: $\omega a/2\pi c = 0.138$ and (*b*) case 2: $\omega a/2\pi c = 0.253$. The same parameters as for Fig. 5.3 were assumed except that $\varepsilon_2 = 4.6483$, which is the dielectric constant of $LiNbO_3$ at 1064 nm. The data points for which the phase-matching condition for SHG is fulfilled are denoted by *solid circles*. (After [67])

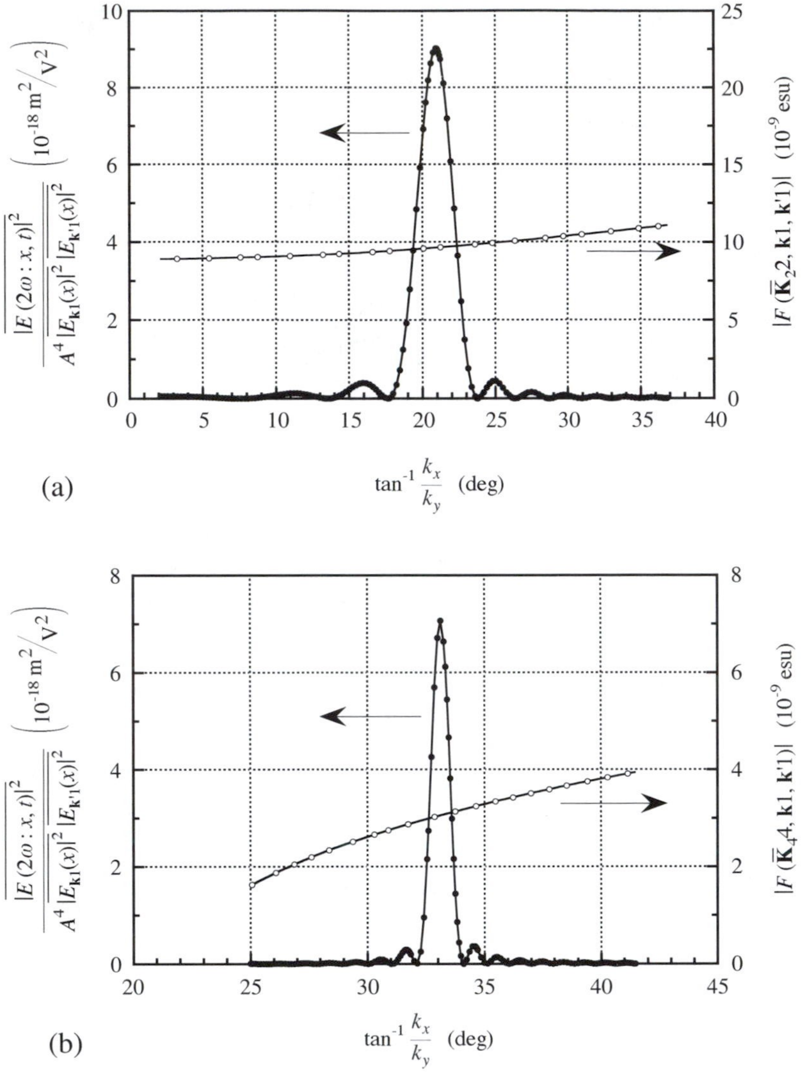

Fig. 5.6. Angular dependence of the average intensity of the electric field of the second harmonic (*solid circles*) and the effective nonlinear susceptibility (*open circles*) for (**a**) case 1 and (**b**) case 2. n_y is assumed to be 100, and therefore, the sample thickness an_y is 15 mm or 27 mm, respectively. The abscissa is the angle of incidence of the fundamental waves. (After [67])

$$\Delta \overline{k}_y = 2k_y - \overline{k}_y. \tag{5.100}$$

The results of the numerical calculation are shown in Fig. 5.6, where n_y was assumed to be 100 and $\chi^{(2)}$ of $LiNbO_3$ was taken as 1.94×10^{-7} (esu) [68]. The abscissa is the angle of incidence of the fundamental waves, and the angular dependence of $|F_2^{(2)}(\overline{\boldsymbol{k}}_{/\!/}\overline{n}, \boldsymbol{k}_{1/\!/}1, \boldsymbol{k}_{2/\!/}1)|$ is also shown. In this figure, we find the sharp peaks of $\overline{|E_{NL}(\boldsymbol{r}_{/\!/}, t)|^2}$ at the phase-matched angles, which are 20.9° for case 1 (Fig. 5.6(a)) and 33.1° for case 2 (Fig. 5.6(b)). We also observe the fairly weak angular variation of the effective nonlinear susceptibility around the peak. The sharpness of the peak depends on n_y. For a very small n_y, the peak becomes broad and the angular variation of $|F_2^{(2)}|$ may somewhat affect the peak position. On the other hand, the height of the peak is proportional to n_y^2. Therefore, the peak height increases quadratically with the sample thickness an_y. It was assumed to be about 15 mm (case 1) and 27 mm (case 2).

5.6 Free Induction Decay

In this section, we will treat free induction decay (FID) as an example of the coherent optical processes in the photonic crystal. We assume that impurity atoms with two electronic levels, which will be denoted by e (excited state) and g (ground state), are distributed in the crystal with the number density $n(\mathbf{r})$. We assume that both ground and excited states are non-degenerate, and hence their wave functions are real. Their transition dipole moment denoted by $\boldsymbol{\mu}$ is also real. We introduce their position- and time-dependent density matrix $\widehat{\varrho}(\boldsymbol{r}, t)$ and examine its temporal evolution when an external electric field, $\boldsymbol{E}_{\mathrm{ex}}(\boldsymbol{r}, t)$, is applied.

The Hamiltonian for one impurity atom, $\widehat{\mathcal{H}}$, is given by

$$\widehat{\mathcal{H}} = \widehat{\mathcal{H}}_0 - \widehat{\boldsymbol{\mu}} \cdot \boldsymbol{E}_{\mathrm{ex}}(\boldsymbol{r}, t), \tag{5.101}$$

where $\widehat{\mathcal{H}}_0$ is the unperturbed Hamiltonian and $\widehat{\boldsymbol{\mu}}$ is the dipole operator of the impurity atom. The equation of motion of the density matrix is

$$\mathrm{i}\hbar \frac{\partial \widehat{\varrho}}{\partial t} = [\widehat{\mathcal{H}}_0 - \widehat{\boldsymbol{\mu}} \cdot \boldsymbol{E}_{\mathrm{ex}}, \widehat{\varrho}]. \tag{5.102}$$

Hence,

$$\frac{\partial \varrho_{\mathrm{ee}}}{\partial t} = \frac{1}{\mathrm{i}\hbar} \left(\varrho_{\mathrm{eg}} - \varrho_{\mathrm{eg}}^*\right) \boldsymbol{\mu} \cdot \boldsymbol{E}_{\mathrm{ex}} - \frac{\varrho_{\mathrm{ee}}}{T_1}, \tag{5.103}$$

$$\frac{\partial \varrho_{\mathrm{eg}}}{\partial t} = -\mathrm{i}\Omega \varrho_{\mathrm{eg}} + \frac{1}{\mathrm{i}\hbar} \left(2\varrho_{\mathrm{ee}} - 1\right) \boldsymbol{\mu} \cdot \boldsymbol{E}_{\mathrm{ex}} - \frac{\varrho_{\mathrm{eg}}}{T_2}, \tag{5.104}$$

where we introduced the longitudinal and transverse relaxation time, T_1 and T_2, to take into consideration the population and phase relaxation processes. In these equations, Ω is the transition frequency,

$$\Omega = \frac{E_{\mathrm{e}} - E_{\mathrm{g}}}{\hbar}, \tag{5.105}$$

where E_{e} (E_{g}) is the energy of the excited (ground) state, and we used the following relations.

$$\varrho_{\mathrm{ge}} = \varrho_{\mathrm{eg}}^{*}, \tag{5.106}$$

$$\boldsymbol{\mu}_{\mathrm{eg}} = \boldsymbol{\mu}_{\mathrm{ge}} = \boldsymbol{\mu}. \tag{5.107}$$

We assumed $\boldsymbol{\mu}_{\mathrm{ee}} = \boldsymbol{\mu}_{\mathrm{gg}} = 0$ for simplicity.

The polarization due to the impurity atoms, which is denoted by $\boldsymbol{P}_{\mathrm{FID}}$, is proportional to the trace of the dipole operator times the density matrix:

$$\begin{aligned}\boldsymbol{P}_{\mathrm{FID}}(\boldsymbol{r},t) &= n(\boldsymbol{r})\mathrm{Tr}[\widehat{\boldsymbol{\mu}}\widehat{\varrho}(\boldsymbol{r},t)] \\ &= n(\boldsymbol{r})\boldsymbol{\mu}\left\{\varrho_{\mathrm{eg}}(\boldsymbol{r},t) + \varrho_{\mathrm{eg}}^{*}(\boldsymbol{r},t)\right\}.\end{aligned} \tag{5.108}$$

We assume that the system was driven by an eigenmode of the photonic crystal for $t \leq 0$:

$$\boldsymbol{E}_{\mathrm{ex}}(\boldsymbol{r},t) = A\boldsymbol{E}_{\boldsymbol{k}n}^{(\mathrm{T})}(\boldsymbol{r}) \exp\left(-\mathrm{i}\omega_{\boldsymbol{k}n}^{(\mathrm{T})} t\right). \tag{5.109}$$

A is the amplitude of the excitation field.

Under the assumption of weak excitation and rotating wave approximation, we obtain

$$\varrho_{\mathrm{eg}}(\boldsymbol{r},0) \simeq \frac{A}{2\left(T_2^{-1} - \mathrm{i}\Delta\omega\right)} \boldsymbol{\mu} \cdot \boldsymbol{E}_{\boldsymbol{k}n}^{(\mathrm{T})}(\boldsymbol{r}). \tag{5.110}$$

For $t > 0$, the external field is switched off, and hence,

$$\frac{\partial}{\partial t}\varrho_{\mathrm{eg}}(\boldsymbol{r},t) = -\mathrm{i}\Omega\varrho_{\mathrm{eg}}(\boldsymbol{r},t) - \frac{\varrho_{\mathrm{eg}}(\boldsymbol{r},t)}{T_2}. \tag{5.111}$$

Then, we obtain

$$\begin{aligned}&\varrho_{\mathrm{eg}}(\boldsymbol{r},t) \\ &= \exp\left\{-\left(\mathrm{i}\Omega + \frac{1}{T_2}\right)t\right\}\varrho_{\mathrm{eg}}(\boldsymbol{r},0) \\ &\simeq \frac{A\boldsymbol{\mu} \cdot \boldsymbol{E}_{\boldsymbol{k}n}^{(\mathrm{T})}(\boldsymbol{r})}{2\left(T_2^{-1} - \mathrm{i}\Delta\omega\right)} \exp\left\{-\left(\mathrm{i}\Omega + \frac{1}{T_2}\right)t\right\},\end{aligned} \tag{5.112}$$

where $\Delta\omega = \omega_{\boldsymbol{k}n}^{(\mathrm{T})} - \Omega$. Therefore, from (5.108)

$$\begin{aligned}\boldsymbol{P}_{\mathrm{FID}}(\boldsymbol{r},t) = &\frac{An\boldsymbol{\mu}\left\{\boldsymbol{\mu} \cdot \boldsymbol{E}_{\boldsymbol{k}n}^{(\mathrm{T})}(\boldsymbol{r})\right\}}{2\left(T_2^{-1} - \mathrm{i}\Delta\omega\right)} \exp\left\{-\left(\mathrm{i}\Omega + \frac{1}{T_2}\right)t\right\} \\ &+\mathrm{c.c.}\end{aligned} \tag{5.113}$$

Substituting (5.113) for $\boldsymbol{P}_{\mathrm{ex}}(\boldsymbol{r},t)$ in (5.10), we obtain the electric field for FID, $\boldsymbol{E}_{\mathrm{FID}}(\boldsymbol{r},t)$, which oscillates with the atomic angular frequency Ω :

$$\begin{aligned}
&\boldsymbol{E}_{\mathrm{FID}}(\boldsymbol{r},t)+\frac{4\pi\boldsymbol{P}_{\mathrm{FID}}(\boldsymbol{r},t)}{\varepsilon(\boldsymbol{r})}\\
&\simeq\frac{An\pi\omega_{\boldsymbol{k}n}^{(\mathrm{T})}\boldsymbol{E}_{\boldsymbol{k}n}^{(\mathrm{T})}(\boldsymbol{r})}{\mathrm{i}V\left(T_2^{-1}-\mathrm{i}\Delta\omega\right)^2}\left\langle\left|\boldsymbol{\mu}\cdot\boldsymbol{E}_{\boldsymbol{k}n}^{(\mathrm{T})}(\boldsymbol{r}')\right|^2\right\rangle\\
&\qquad\times\exp\left\{-\left(\mathrm{i}\Omega+\frac{1}{T_2}\right)t\right\},
\end{aligned}\tag{5.114}$$

where

$$\left\langle\left|\boldsymbol{\mu}\cdot\boldsymbol{E}_{\boldsymbol{k}n}^{(\mathrm{T})}(\boldsymbol{r}')\right|^2\right\rangle=\frac{1}{V}\int_V\mathrm{d}\boldsymbol{r}'\left|\boldsymbol{\mu}\cdot\boldsymbol{E}_{\boldsymbol{k}n}^{(\mathrm{T})}(\boldsymbol{r}')\right|^2.\tag{5.115}$$

Other kinds of coherent optical processes such as four-wave mixing can be treated in a similar manner.

6. Defect Modes in Photonic Crystals

As mentioned in Chap. 1, one of the most important properties of photonic crystals is the emergence of localized defect modes in the gap frequency region when a disorder is introduced to their periodic dielectric structure. In this chapter, we will present a numerical method to calculate the eigenfrequency and the eigenfunction of the defect mode that is based on the numerical simulation of the excitation process of the defect mode by a virtual oscillating dipole moment located near the dielectric defect. After we derive a general expression for the frequency dependence of the excited electromagnetic field and describe how to calculate it by the FDTD (finite-difference time-domain) method, we will apply our method to three examples, and show excellent agreement with experiments.

6.1 General Properties

First, the presence of a photonic bandgap is essential for the localized modes to appear. In order to see this point, let us assume that the defect mode (denoted by d) and a Bloch wave (denoted by $\boldsymbol{k}n$) have the same eigenfrequency:

$$\omega_{\mathrm{d}} = \omega_{\boldsymbol{k}n}. \tag{6.1}$$

We denote the eigenfunction of the Bloch wave by $\boldsymbol{E}_{\boldsymbol{k}n}(\boldsymbol{r})$ as before and that of the defect mode by $\boldsymbol{E}_{\mathrm{d}}(\boldsymbol{r})$. Because of the presence of the structural disorder, the Bloch wave is no longer an eigenmode of the photonic crystal. It is mixed with the defect mode to form two genuine eigenstates:

$$\boldsymbol{E}_1(\boldsymbol{r}) = p_1 \boldsymbol{E}_{\mathrm{d}}(\boldsymbol{r}) + q_1 \boldsymbol{E}_{\boldsymbol{k}n}(\boldsymbol{r}), \tag{6.2}$$

$$\boldsymbol{E}_2(\boldsymbol{r}) = p_2 \boldsymbol{E}_{\mathrm{d}}(\boldsymbol{r}) + q_2 \boldsymbol{E}_{\boldsymbol{k}n}(\boldsymbol{r}). \tag{6.3}$$

Since $\boldsymbol{E}_{\boldsymbol{k}n}(\boldsymbol{r})$ is an extended state, neither $\boldsymbol{E}_1(\boldsymbol{r})$ nor $\boldsymbol{E}_2(\boldsymbol{r})$ is a localized mode. This fact contradicts the initial assumption. Thus, the eigenfrequency of the defect mode should be in the photonic bandgap.

The experimental evidence that the defect mode is spatially localized was given by McCall et al. [35]. They investigated a two-dimensional (2D) square array of circular dielectric cylinders with a disorder in the microwave region. As Fig. 6.1 shows, one cylinder located at the origin was removed

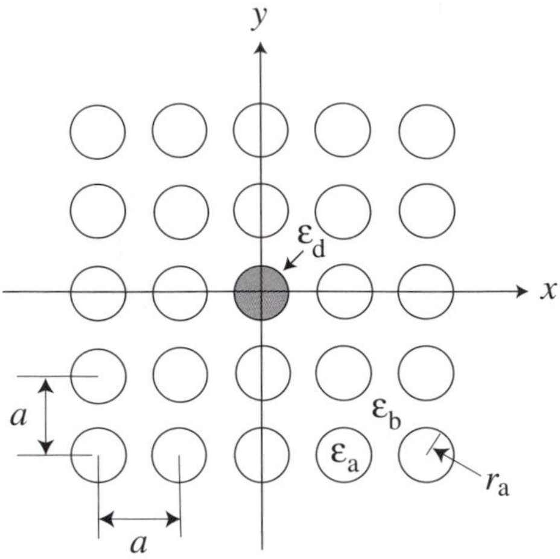

Fig. 6.1. Top view of the 2D square array of circular dielectric cylinders that was investigated by McCall et al. [35]. ε_a, ε_b, and ε_d denote the dielectric constants of the circular cylinders, the background, and the defect cylinder, respectively. a and r_a stand for the lattice constant and the radius of the cylinders. Their experimental values were as follows: $a = 1.27$ cm, $r_a = 0.48$ cm, $\varepsilon_a = 9.0$, and $\varepsilon_b = 1.0$. (After [69])

and this part acted as the defect. They observed quite a sharp peak in the transmission spectrum in the gap frequency region (Fig. 6.2), whose frequency was identified as that of a localized defect mode. The spatial distribution of the electric field at the peak frequency showed a strong localization (Fig. 6.3).

The eigenfrequencies and the eigenfunctions of the defect modes can be calculated by several methods. In early studies, the plane-wave expansion with the supercell method was frequently used. In this method, the structure shown in Fig. 6.1 is regarded as a unit cell, which is referred to as the supercell, and the band calculation by the plane-wave expansion method is performed for this unit cell structure. Then, a band with a very small width appears in the gap frequency region if a localized defect mode is present. Exactly speaking, this is an impurity-band calculation, since we assume a periodic array of the defect structure. Because the wave functions localized at adjacent defects hardly overlap each other when the supercell is sufficiently large, the bandwidth of the impurity band is usually quite small. Hence, we can obtain the eigenfrequency of the defect mode almost without ambiguity. However, we need a very large number of plane waves to perform an accurate calculation when the supercell is large. Generally speaking, the CPU (central processing unit) time necessary for a computer to diagonalize the matrix like the one that appears on the left-hand side of (2.32) is proportional to the cube of its dimension, and hence, to the cube of the number of the plane waves used for the expansion. This fact sometimes imposes a serious constraint on

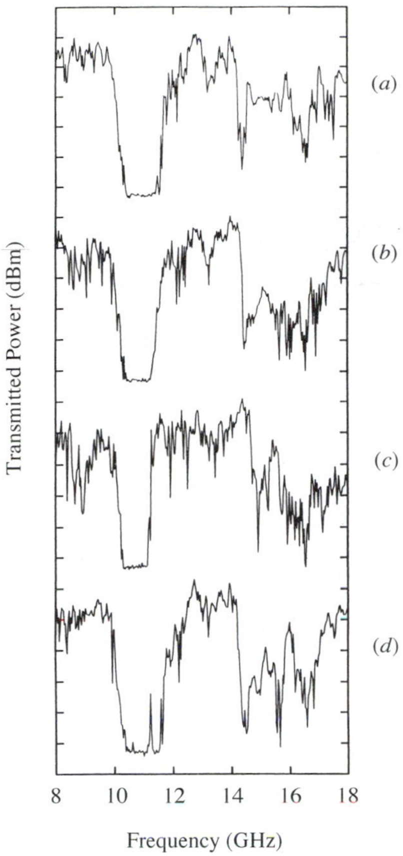

Fig. 6.2. Transmission spectra of the 9×18 array of cylinders. The incident wave comes along (*a*) the [1,0], (*b*) the [2,1], and (*c*) the [1,1] direction of the regular crystal without the defect. In (*d*), the orientation of the incident wave is the same as in (*a*), but a single cylinder located at the center of the array is removed. Note the sharp extra peak in the gap located at 11.2 GHz signifying a localized defect mode. Each vertical division is 5 dB. (After [35])

the accuracy of the numerical results especially for three-dimensional (3D) crystals.

Later, it was found that the real-space methods are generally more efficient than the plane-wave expansion. In the following, one such mehod will be described in detail. This is based on the numerical simulation of the radiation process of a virtual oscillating dipole moment embedded in the photonic crystal [69–72]. The numerical simulation is performed by the FDTD (finite-difference time-domain) method. Advantages of this method are that (1) the CPU time is comparatively small, (2) the algorithm is suitable for vector and parallel computing, and hence, the CPU time can be reduced by a consider-

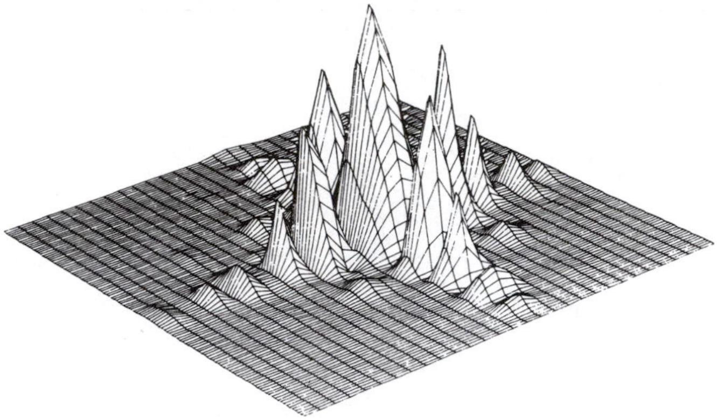

Fig. 6.3. Energy distribution of a localized defect mode observed for the square array of circular dielectric rods with one rod removed. (After McCall et al. [35])

able amount by using a vector processor or a parallel machine, (3) the spatial symmetry of the eigenmodes is rigorously taken into account, and (4) the frequency dependence and the imaginary part of the dielectric constant can naturally be dealt with if necessary. We consider these points in detail with some examples.

6.2 Principle of Calculation

In this section, we will first derive a general formula describing the radiation from an oscillating dipole moment located near a disorder introduced to the regular structure of the photonic crystal. We will then proceed to the description of the FDTD calcualtion.

Maxwell's equations for the present problem are given by

$$\nabla \cdot \{\varepsilon_0 \varepsilon(\boldsymbol{r}) \boldsymbol{E}(\boldsymbol{r}, t) + \boldsymbol{P}_{\mathrm{d}}(\boldsymbol{r}, t)\} = 0, \tag{6.4}$$

$$\nabla \cdot \boldsymbol{H}(\boldsymbol{r}, t) = 0, \tag{6.5}$$

$$\nabla \times \boldsymbol{E}(\boldsymbol{r}, t) = -\mu_0 \frac{\partial}{\partial t} \boldsymbol{H}(\boldsymbol{r}, t), \tag{6.6}$$

$$\nabla \times \boldsymbol{H}(\boldsymbol{r}, t) = \frac{\partial}{\partial t} \{\varepsilon_0 \varepsilon(\boldsymbol{r}) \boldsymbol{E}(\boldsymbol{r}, t) + \boldsymbol{P}_{\mathrm{d}}(\boldsymbol{r}, t)\}, \tag{6.7}$$

where $\boldsymbol{P}_{\mathrm{d}}(\boldsymbol{r}, t)$ is the polarization field of the oscillating dipole and its explicit form is given by (5.15). The difference between this set of equations and (5.1)–(5.4) is that $\varepsilon(\boldsymbol{r})$ now has the disorder introduced by the dielectric defect.

In Sect. 5.2, we derived the expression for the dipole radiation in the regular photonic crystal using Green's function. The same method can be applied to the present problem, if we take into account the presence of the defect mode in addition to the extended Bloch states. We assume that the angular frequency of the oscillating dipole ω is close to ω_d and that the latter is isolated in a photonic bandgap. Then we can neglect the contribution from all other extended eigenmodes, and (5.34) gives the electric field of the present problem:

$$\boldsymbol{E}(\boldsymbol{r},t) \simeq -\frac{\omega_\mathrm{d}\left\{\boldsymbol{E}_\mathrm{d}^*(\boldsymbol{r}_0)\cdot\boldsymbol{d}\right\}\boldsymbol{E}_\mathrm{d}(\boldsymbol{r})\exp(-\mathrm{i}\omega t)}{2\varepsilon_0 V(\omega-\omega_\mathrm{d}+\mathrm{i}\gamma)}, \tag{6.8}$$

where we have introduced the decay rate of the defect mode γ. $\boldsymbol{E}_\mathrm{d}(\boldsymbol{r})$ is normalized such that

$$\int_V \varepsilon(\boldsymbol{r})|\boldsymbol{E}_\mathrm{d}(\boldsymbol{r})|^2\mathrm{d}\boldsymbol{r} = V. \tag{6.9}$$

By calculating Poynting's vector, we obtain the expression for the electromagnetic energy emitted by the dipole in a unit time, U:

$$U \simeq \frac{\omega\omega_\mathrm{d}\gamma|\boldsymbol{d}\cdot\boldsymbol{E}_\mathrm{d}^*(\boldsymbol{r}_0)|^2}{\varepsilon_0 V\left\{(\omega-\omega_\mathrm{d})^2+\gamma^2\right\}}. \tag{6.10}$$

The derivation of (6.8) and (6.10) is similar to that of (5.34) and (5.35). Equation (6.10) implies that we can obtain ω_d as the resonance frequency if we can evaluate the frequency dependence of U. Moreover, (6.8) shows that the excited field is proportional to the eigenfunction. We will perform this task by solving the wave equation derived from the above Maxwell equations numerically. The numerical method is called the finite-difference time-domain (FDTD) method.

We will deal with the E polarization, since McCall et al. investigated this case experimentally. (The H polarization can be dealt with in a similar manner; see Sect. 7.3) Then the wave equation for the z component (parallel to the rod axis) of the electric field, $E_z(\boldsymbol{r}_{/\!/},t)$, is given by

$$\frac{\varepsilon(\boldsymbol{r}_{/\!/})}{c^2}\frac{\partial^2 E_z}{\partial t^2} - \frac{\partial^2 E_z}{\partial x^2} - \frac{\partial^2 E_z}{\partial y^2} = \frac{d\omega^2}{c^2}\delta(x-x_0)\delta(y-y_0)\exp(-\mathrm{i}\omega t), \tag{6.11}$$

where (x_0, y_0) denotes the 2D position of the dipole moment, and d is the amplitude of the oscillating dipole. We discretize this differential equation to obtain a difference equation, and solve the latter numerically with the initial condition $E_z = \partial E_z/\partial t = 0$ at $t = 0$. The derivatives of the second order in (6.11) are approximated as follows:

$$\frac{\partial^2 E_z(\boldsymbol{r}_{/\!/},t)}{\partial x^2} \simeq \frac{E_z(x+\Delta x,y,t)-2E_z(x,y,t)+E_z(x-\Delta x,y,t)}{(\Delta x)^2}, \tag{6.12}$$

$$\frac{\partial^2 E_z(\boldsymbol{r}_{/\!/},t)}{\partial y^2} \simeq \frac{E_z(x,y+\Delta y,t)-2E_z(x,y,t)+E_z(x,y-\Delta y,t)}{(\Delta y)^2}, \tag{6.13}$$

$$\frac{\partial^2 E_z(\boldsymbol{r}_{/\!/},t)}{\partial t^2} \simeq \frac{E_z(x,y,t+\Delta t)-2E_z(x,y,t)+E_z(x,y,t-\Delta t)}{(\Delta t)^2}, \tag{6.14}$$

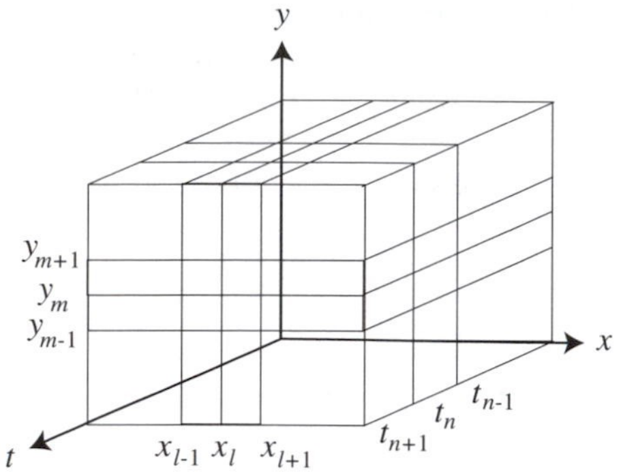

Fig. 6.4. Spatio-temporal mesh for the FDTD calculation

where Δx, Δy, and Δt are the differences between adjacent mesh points (see Fig. 6.4). Substituting (6.12)–(6.14) into (6.11), we obtain

$$\begin{aligned} & E_z(x_l, y_m, t_{n+1}) \\ \approx\ & 2E_z(x_l, y_m, t_n) - E_z(x_l, y_m, t_{n-1}) \\ & + \frac{c^2(\Delta t)^2\{E_z(x_{l+1}, y_m, t_n) - 2E_z(x_l, y_m, t_n) + E_z(x_{l-1}, y_m, t_n)\}}{\varepsilon(x_l, y_m)(\Delta x)^2} \\ & + \frac{c^2(\Delta t)^2\{E_z(x_l, y_{m+1}, t_n) - 2E_z(x_l, y_m, t_n) + E_z(x_l, y_{m-1}, t_n)\}}{\varepsilon(x_l, y_m)(\Delta y)^2} \\ & + \frac{d\omega^2(\Delta t)^2\delta_{ll_0}\delta_{mm_0}\exp(-\mathrm{i}\omega t_n)}{\varepsilon(x_l, y_m)\Delta x\Delta y} . \end{aligned} \tag{6.15}$$

Here, x_l denotes the lth mesh point in the x direction, and so on. In the last term on the right-hand side of (6.15), two delta functions were replaced by Kronecker deltas, where (l_0, m_0) denotes the position of the dipole. Note that

$$\int \mathrm{d}x \int \mathrm{d}y \longrightarrow \Delta x \Delta y \sum_{lm} . \tag{6.16}$$

Therefore, the replacement of $\delta(x - x_0)\delta(y - y_0)$ by $\delta_{ll_0}\delta_{mm_0}/\Delta x\Delta y$ gives the correct value. Since the right-hand side of (6.15) is all written with values at time $t = t_n$ and t_{n-1}, the temporal evolution of the electric field can be calculated iteratively. This allows the computer to do the vector processing and the parallel computing, which results in a tremendous acceleration of the calculation.

In the following three sections, we will treat lossless crystals with real $\varepsilon(\boldsymbol{r})$. Then, the shape of the resonance curve in the steady state, (6.10), should be modified from a Lorentzian to a delta function. However, our numerical method assumes an abrupt excitation of the defect modes and deals with transient but nearly steady states. The shape of the resonance curve for this case is also well represented by (6.10).

6.3 Point Defects in a Square Lattice

In this section, we will treat the experiments reported by McCall et al. [35]. The geometry for the numerical calculation is shown in Fig. 6.1. This is the top view of a 2D square lattice composed of circular dielectric rods. One of them is replaced by one with a different dielectric constant, which we call the defect rod hereafter. In Fig. 6.1, ε_a, ε_b, and ε_d, denote the dielectric constants of the circular rods, the background, and the defect rod, respectively. a and r_a represent the lattice constant and the radius of the rods. Note that we assumed the same radius for the defect rod as for the rest of the lattice. For the numerical calculation, the following values were assumed according to the experimental condition of McCall et al.; $\varepsilon_a = 9.0$, $\varepsilon_b = 1.0$, $a = 1.27$ cm, and $r_a = 0.48$ cm. ε_d was varied from 1.0 to 20.0.

12×12 unit cells, which we call the supercell hereafter, were taken into account in our numerical calculation, and the periodic boundary condition was imposed on its boundary. Because the numerically obtained wave functions of the defect modes, which will be shown below, are well localized and have a very small amplitude on the boundary, the supercell of this size was sufficient for our purpose. In the actual calculation, we can assume a certain symmetry for each localized mode since it is an irreducible representation of the C_{4v} point group to which the assumed square lattice of the circular rods belongs. Therefore, we can impose a respective boundary condition on one eighth of the supercell for the one-dimensional representation, i.e., the A_1, A_2, B_1, or B_2 mode, and on one-quarter of the supercell for the two-dimensional representation, i.e., the E mode. Then, it was sufficient to treat the above small area for the numerical calculation and the computational task was reduced. In our calculation, every unit cell was divided into 40×40 parts, and one period of the oscillation was divided into 640 steps in order to discretize the wave equation. The further decrease of the size of the spatial and temporal meshes did not give an apparent change in the eigenfrequencies of the defect modes.

The photonic band structure of the regular lattice was previously presented in Fig. 2.4, which showed the presence of three bandgaps. Since the experiment by McCall et al. was concerned with a totally symmetric A_1 mode in the second gap for the case of $\varepsilon_d = 1.0$, we examine this mode first. Figure 6.5 shows the calculated electromagnetic energy radiated by the oscillating dipole moment located at the center of the defect rod. $\square$, $\circ$, $\diamond$, and $\bullet$ denote the accumulated electromagnetic energy after 10, 20, 35, and 50 cycles of the oscillation, respectively. In this figure, we observe a clear resonance at $\omega a/2\pi c = 0.466$, or $\omega/2\pi = 11.0$ GHz, which is quite close to the experimental observation, i.e., $\omega/2\pi = 11.2$ GHz.

Figure 6.6 shows the distribution of the electric field at the resonance, i.e., at $\omega a/2\pi c = 0.466$ after 100 cycles of the oscillation. The abscissa is the distance from the center of the defect rod along the nearest-neighbor direction, i.e., the (1, 0) direction, normalized by the lattice constant. As

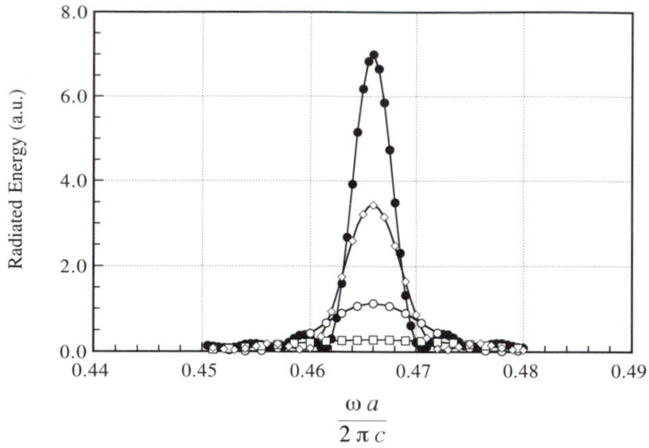

Fig. 6.5. Electromagnetic energy radiated by an oscillating dipole moment located at the center of the defect rod as a function of the oscillation frequency. □, ∘, ⋄, and • denote the accumulated electromagnetic energy after 10, 20, 35, and 50 cycles of the oscillation, respectively. The abscissa represents the normalized frequency. The same parameters as for Fig. 6.1 were used. ε_{d}, was assumed here to be 1.0 according to the experimental condition of McCall et al. [35]. A resonance at $\omega a/2\pi c = 0.466$ (or $\omega/2\pi = 11.2$ GHz) is clearly observed. (After [69])

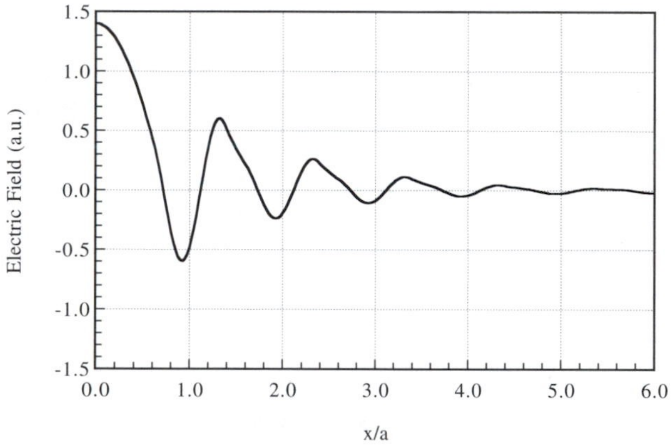

Fig. 6.6. Electric field excited by the oscillating dipole moment after 100 cycles of the oscillation at $\omega a/2\pi c = 0.466$. The abscissa is the distance from the center of the defect rod along the nearest-neighbor direction normalized by the lattice constant. This electric field can be regarded as that of the localized eigenmode created by the defect rod. (After [69])

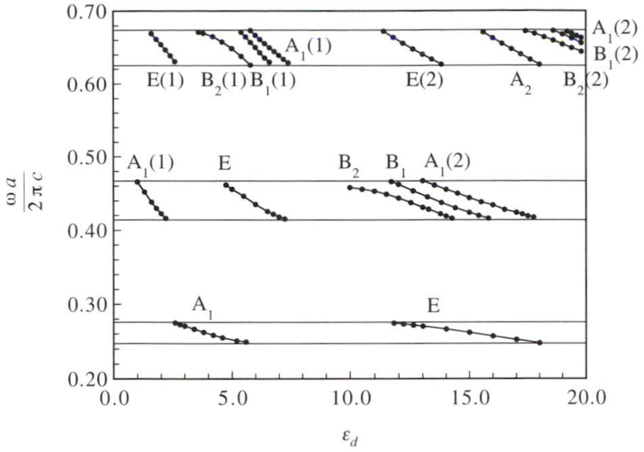

Fig. 6.7. Eigenfrequencies of the localized defect modes as a function of the dielectric constant of the defect rod ε_d. The ordinate is the normalized frequency and *horizontal lines* represent the boundaries of the photonic bandgaps. The symmetry of the localized modes is also shown, where the numbers in the parentheses are given in order of ε_d when there are more than one mode of the same symmetry in the same gap. (After [69])

was shown in (6.8), we can regard this distribution as the eigenfunction of the defect mode. We find in this figure that the defect mode is localized at the defect rod as we expected and the amplitude of the electric field on the boundary of the supercell, i.e., at $x = 6a$ is negligible. Hence, we may consider that the interference between the neighboring supercells or, in other words, the impurity-band effect, is small for our calculation. In addition, we confirmed that we obtain the same distribution of the electric field apart from its magnitude even when the oscillating dipole is located at $(a, 0)$ instead of $(0, 0)$. This implies that the induced electromagnetic field belongs to a single eigenmode. The CPU time that was necessary to obtain this figure with a super computer was about 12 s. The CPU time can be further reduced by the optimization of the parameters used when we discretize the wave equation. We can also apply this method to the same kind of problems for 3D lattices, as was shown by Painter et al. [73] and Hwang et al. [74].

Next, Fig. 6.7 shows the ε_d dependence of the eigenfrequencies of the localized modes for all symmetries. In this figure, horizontal lines represent the boundaries of the photonic bandgaps and the numbers in parentheses are given in order of ε_d when there are more than one modes of the same symmetry in the same gap. In addition to the A_1 mode [the $A_1(1)$ mode in the second gap] that was investigated by McCall et al., we observe that localized eigenstates of all possible symmetries appear when we change ε_d from 1.0 to 20.0. The eigenfrequency of each mode generally decreases with increasing ε_d, because the defect mode "feels" a large dielectric constant on average when ε_d is large, and this leads to a simple scaling of the eigenfrequency as $\omega_\mathrm{d} \propto 1/\sqrt{\varepsilon}$ that can be found from the left-hand side of (6.11).

Figure 6.8 shows the 2D distribution of the electric fields of the defect modes in the second gap. The maximum of each electric field is normalized to unity in this figure. For all defect modes, the eigenfunctions show their peculiar symmetry and are localized around the defect rod located at the origin. For the doubly degenerate E mode, one eigenfunction is a replica of the other given by a 90-degree rotation. Similarly, Fig. 6.9 shows the distribution of the electric fields of the defect modes in the first gap. We can also calculate those in the third gap quite easily, though we do not show them here. Note that the spatial variation of the electric fields is slower for modes in the first gap than for those in the second gap, which is a natural consequence of the relative magnitudes of their eigenfrequencies. Also, note that because of the mismatching of the spatial symmetry, we can excite neither A_2 nor B_2 mode by an incident plane wave pointed in the (1, 0) direction, and neither A_2 nor B_1 mode by that pointed in the (1, 1) direction.

We would like to conclude this section by making two remarks. First, we have treated a point defect, i.e., one defect rod in an otherwise regular lattice, for numerical calculation. We can treat localized modes on line defects, and calculate their dispersion relations as well in the framework of the present method by simply changing the dielectric structure and the boundary condition of the supercell, as will be shown in Sect. 6.5. The band structures of regular crystals can also be calculated. Secondly, we can also treat frequency-dependent dielectric constants, which is usually difficult for those methods which solve eigenvalue problems to obtain the eigenstates. This point will be described in detail in Chap. 7.

6.4 Point Defects in a Hexagonal Lattice

Smith et al. [75] investigated the defect modes in a 2D hexagonal lattice, and we examine their case here [71]. The geometry for the present numerical calculation is shown in Fig. 6.10. This is the top view of a supercell in a 2D hexagonal lattice composed of circular dielectric rods. One of the rods in the supercell is replaced by one with a different radius, which we call the defect rod. In Fig. 6.10, ε_a and ε_b denote the dielectric constants of the circular rods and the background, respectively, a is the lattice constant, r_a and r_d represent the radii of the regular and defect rods, and L is the distance between the centers of the defect rods in adjacent supercells. Note that we assumed the same dielectric constant for the defect rod as for the regular rods this time.

In the actual calculation, we could utilize the spatial symmetry of the eigenmodes to reduce the amount of the computational task as we did in the last section. Because the assumed crystal has C_{6v} spatial symmetry, each localized eigenmode is attributed to one of its irreducible representations,

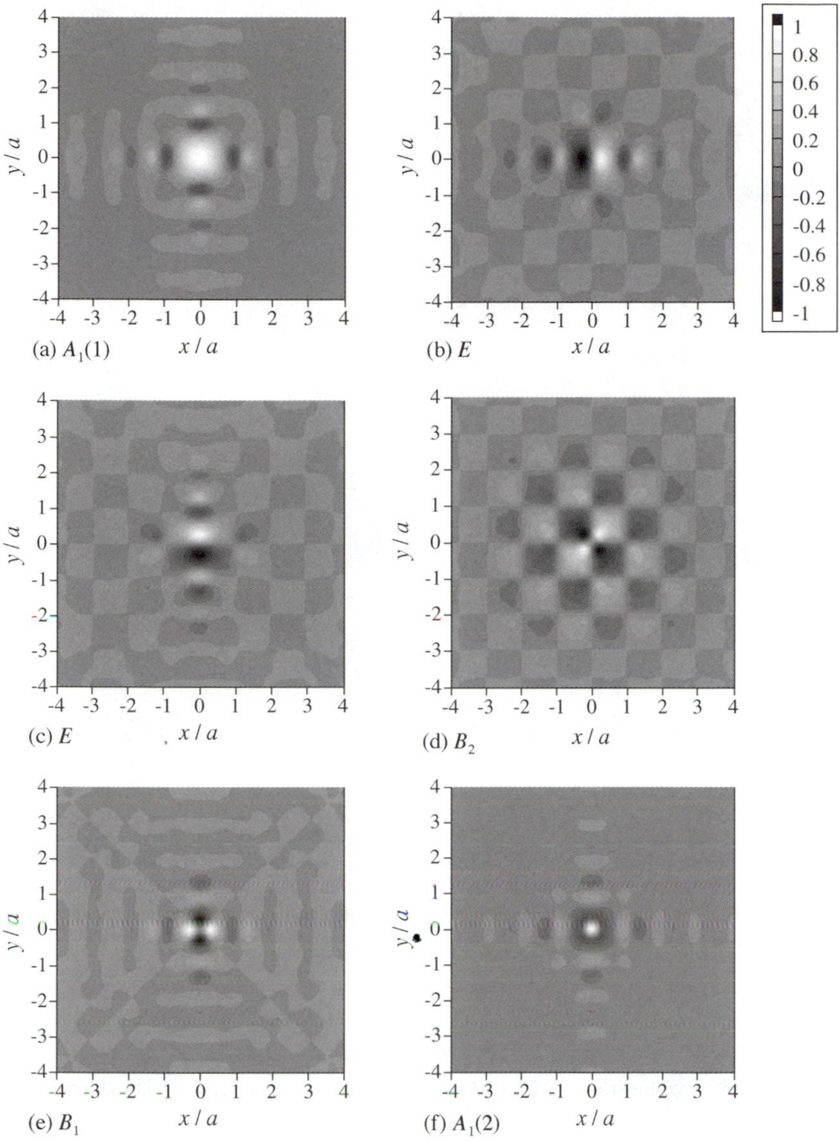

Fig. 6.8. Distribution of the electric fields of the (**a**) $A_1(1)$, (**b**) (**c**) E, (**d**) B_2, (**e**) B_1, and (**f**) $A_1(2)$ modes in the second gap. The maximum of each electric field is normalized to unity. For all defect modes, the eigenfunctions show their peculiar symmetry and are localized around the defect rod located at the origin. For the doubly degenerate E mode, one eigenfunction is a replica of the other given by a 90-degree rotation. (After [69])

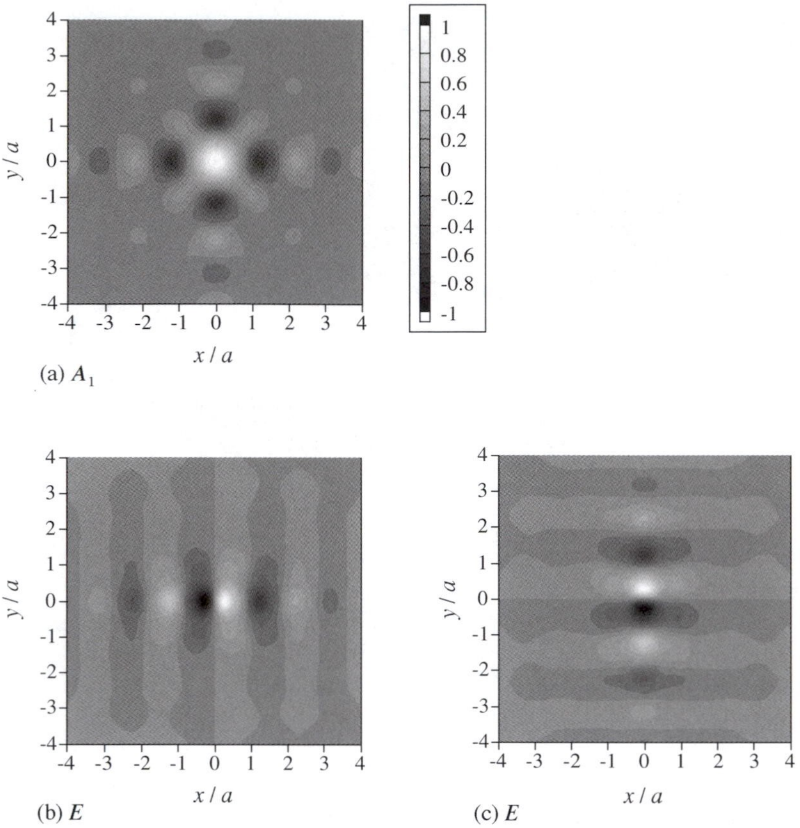

Fig. 6.9. Distribution of the electric fields of the (**a**) A_1 and (**b**) (**c**) E modes in the first gap. The maximum of each electric field is normalized to unity again. Note that the spatial variation of the electric fields is slower for modes in the first gap than for those in the second gap. (After [69])

i.e., four one-dimensional representations (A_1, A_2, B_1, and B_2) and two two-dimensional representations (E_1 and E_2). Then, for the one-dimensional representations, it was sufficient to deal with only one twelfth of the supercell. As for the two-dimensional representations, one quarter of the supercell was sufficient.

Since the experiment by Smith et al. was concerned with a totally symmetric A_1 mode for the case of $r_{\mathrm{d}} = 0$, we examine this mode first. Figure 6.11 shows the electromagnetic energy radiated by an oscillating dipole moment located at the origin as a function of the oscillation frequency. □, ○, ◇, and • denote the accumulated electromagnetic energy after 10, 20, 35, and 50 cycles of the oscillation, respectively. The following parameters were used according to the experimental condition of Smith et al.: $a = 1.27$ cm, $r_{\mathrm{a}} = 0.48$ cm, $\varepsilon_{\mathrm{a}} = 9.0$, $\varepsilon_{\mathrm{b}} = 1.0$, and $r_{\mathrm{d}} = 0$ cm. The supercell with $L = 16a$ was consid-

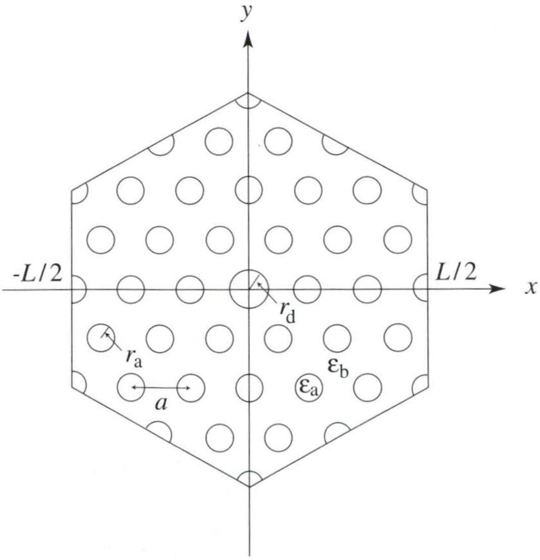

Fig. 6.10. Top view of a supercell of the 2D hexagonal array of circular rods that was assumed for the numerical calculation. ε_a, and ε_b denote the dielectric constants of the circular rods and the background, respectively. a is the lattice constant. r_a and r_d represent the radii of the regular and defect rods, respectively. L denotes the distance between the defect rods in two adjacent supercells. (After [71])

ered here. A resonance at $\omega a/2\pi c = 0.468$ (or $\omega/2\pi = 11.1$ GHz) is clearly observed in Fig. 6.11, which is quite close to the experimental observation, i.e., $\omega/2\pi = 11.23$ GHz. The deviation is only 1.2%. We should also note that a resonance curve after a larger number of the oscillation cycles is sharper. This is because the system approaches a pure state that is described by $\boldsymbol{E}_d(\boldsymbol{r})$ after an initial transient response caused by the abrupt introduction of the oscillating dipole moment at $t = 0$. Figure 6.12 shows the distribution of the electric field radiated by the oscillating dipole moment after 100 cycles of the oscillation at $\omega a/2\pi c = 0.468$. The same parameters as for Fig. 6.11 were used. The maximum of the electric field is normalized to unity in this figure.

Next, we examine the problem discussed by Feng and Arakawa [76]. They assumed that $r_a/a = 0.2$, $\varepsilon_a = 13.0$, and $\varepsilon_b = 1.0$ in Fig. 6.10. They also assumed a supercell with $L = 3a$. Figure 6.13 shows the photonic band structure and the state density of the regular hexagonal lattice of the circular rods for the E polarization. This figure shows the presence of a large bandgap from $\omega a/2\pi c = 0.26$ to 0.45.

Feng and Arakawa [76] showed by the plane-wave expansion method with 2269 basis plane waves that several defect modes appear when the radius of the defect rod, r_d, is varied from $r_d = 0$ to 0.5; and we consider this point here. Figure 6.14 shows the eigenfrequency of the defect modes as a

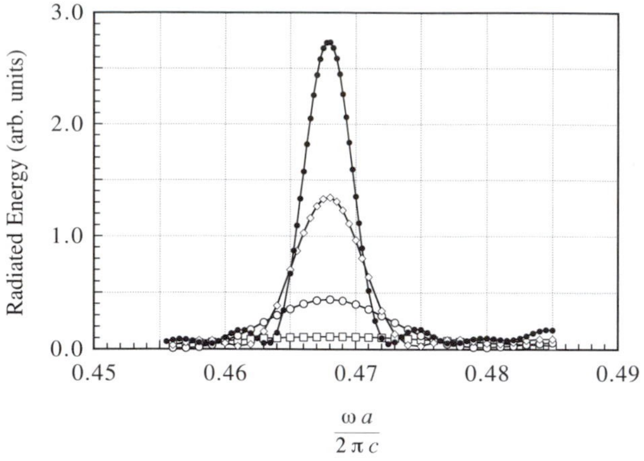

Fig. 6.11. Electromagnetic energy radiated by an oscillating dipole moment located at the center of the supercell as a function of the oscillation frequency. □, ○, ◇, and • denote the accumulated electromagnetic energy after 10, 20, 35, and 50 cycles of the oscillation, respectively. The abscissa represents the normalized frequency, where ω, a, and c stand for the angular frequency, the lattice constant, and the light velocity in vacuum, respectively. The following parameters were used according to the experimental condition of Smith et al. [75]: $a = 1.27$ cm, $r_a = 0.48$ cm, $\varepsilon_a = 9.0$, $\varepsilon_b = 1.0$, and $r_d = 0$ cm (see Fig. 6.10). A supercell with $L = 16a$ was assumed. A resonance at $\omega a/2\pi c = 0.468$ (or $\omega/2\pi = 11.1$ GHz) is clearly observed. (After [71])

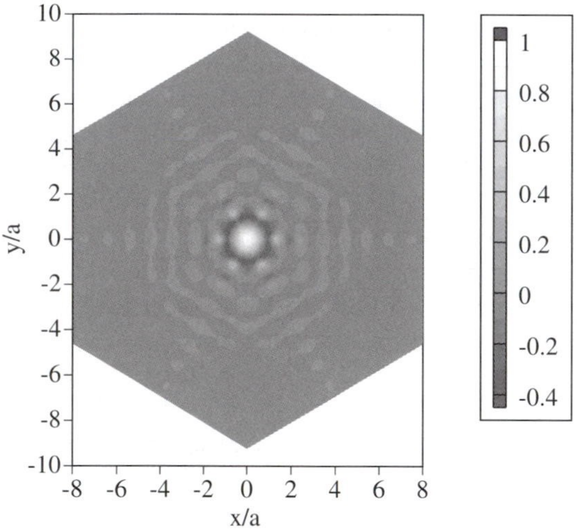

Fig. 6.12. Distribution of the electric field radiated by the oscillating dipole moment after 100 cycles of the oscillation at $\omega a/2\pi c = 0.468$. The same parameters as for Fig. 6.11 were used. The maximum of the electric field is normalized to unity. This electric field can be regarded as that of the localized eigenmode created by the defect. It is really localized at the origin and shows the A_1 symmetry of the C_{6v} point group. (After [71])

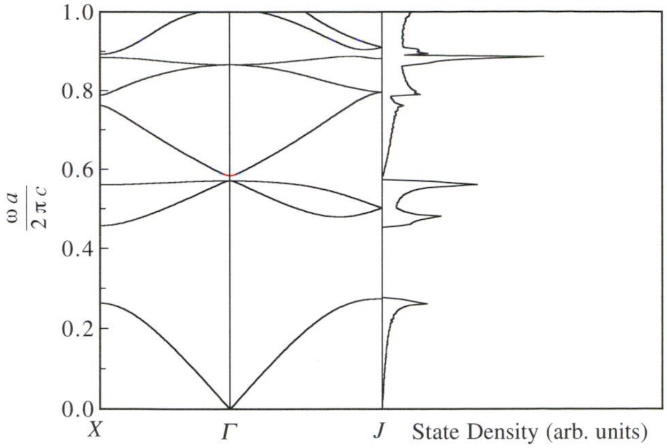

Fig. 6.13. Photonic band structure and the state density of the regular hexagonal lattice of the circular rods for the E polarization. The ordinate is the normalized eigenfrequency. According to the previous calculation by Feng and Arakawa [76], the following values were assumed: $r_a/a = 0.2$ cm, $\varepsilon_a = 13.0$, and $\varepsilon_b = 1.0$. (After [71])

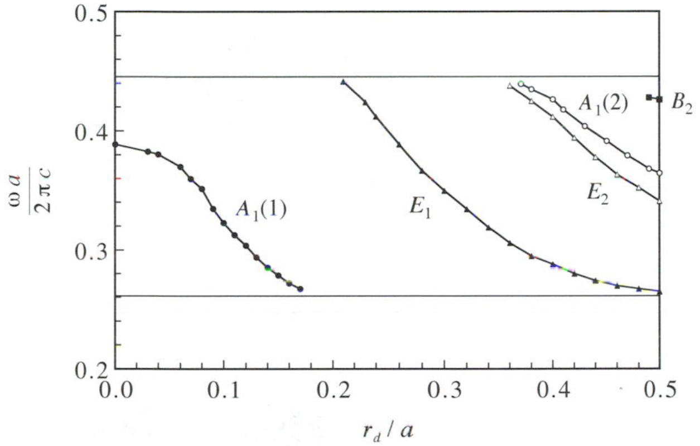

Fig. 6.14. Eigenfrequency of the localized defect modes as a function of the radius of the defect rod, r_d. The ordinate is the normalized frequency and horizontal lines represent the boundaries of the photonic bandgap. The symmetries of the localized modes are also shown, where the number in parentheses is given in order of r_d when there are more than one mode of the same symmetry. (After [71])

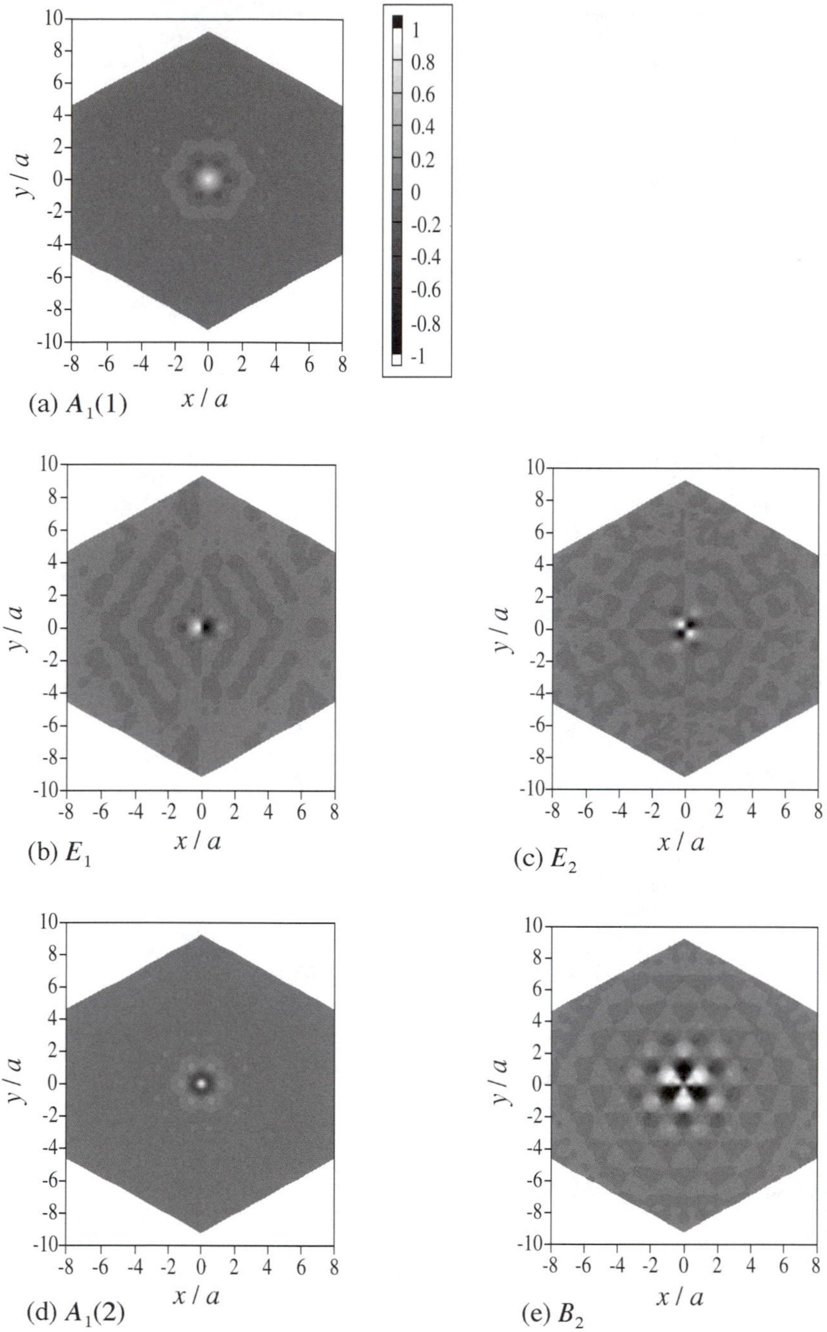

Fig. 6.15. Distribution of the electric fields of the (**a**) $A_1(1)$, (**b**) E_1, (**c**) E_2, (**d**) $A_1(2)$, and (**e**) B_2 modes. The maximum of each electric field is normalized to unity. For all defect modes, the eigenfunctions show their peculiar symmetries and are localized around the defect rod located at the origin. (After [71])

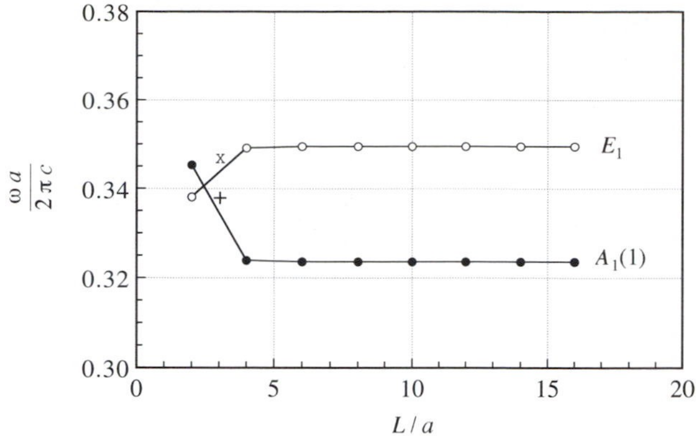

Fig. 6.16. L dependence of the eigenfrequencies, which were calculated with the same parameters as for Fig. 6.14 and with $r_\mathrm{d}/a = 0.1$ for the $A_1(1)$ mode and $r_\mathrm{d}/a = 0.3$ for the E_1 mode. The results of Feng and Arakawa [76] are denoted by $+$ for the $A_1(1)$ mode and by $\times$ for the E_1 mode. (After [71])

function of r_d, where a supercell with $L = 16a$ was used for our numerical calculation. The ordinate is the normalized frequency and the horizontal lines represent the boundaries of the photonic bandgap. The symmetries of the defect modes are also shown, where the number in parentheses is given in order of r_d when there are more than one mode of the same symmetry. The eigenfunctions are presented in Fig. 6.15. The maximum of the electric field is normalized to unity as before. For all defect modes, the eigenfunctions show their peculiar symmetries and are localized around the defect rod located at the origin. For the doubly degenerate E_1 and E_2 modes, only one eigenfunction is shown. When we compared our results with those by the plane-wave expansion method, it was found that the overall feature is common to both calculations except that they differ quantitatively from each other by 2–4%.

Now, we examine the reason for the quantitative disagreement between the two calculations. The L dependence of the eigenfrequencies is presented in Fig. 6.16, where $\bullet$ and $\circ$ represent the results of our calculation for the $A_1(1)$ and E_1 modes, and $+$ and $\times$ represent those by the plane-wave expansion method. Strictly speaking, the latter depend on the wave vector in the first Brillouin zone because it was an impurity-band calculation, but the widths of the obtained impurity bands were not very wide and we plot the centers of the impurity bands as representative points in Fig. 6.16. This figure shows that (1) the eigenfrequencies converge fairly rapidly and they scarcely change when L/a is more than 6, (2) but for L/a less than 6, there is a considerable change, and (3) the results by the plane-wave expansion method are consistent with our results for small L. Thus, we should note that the small supercell brings about not only the width but also the shift of the impurity bands. Therefore, if we are interested in isolated and well localized defect modes, we should assume a somewhat large supercell.

6.5 Line Defects in a Square Lattice

In this section, we will deal with the structure shown in Fig. 6.17, where the spacing between the adjacent layers around the x axis is different from the rest. This part, which we refer to as the line defect, brings about the eigenmodes localized in the y direction. Since the structure has translational symmetry in the x direction, those localized eigenmodes satisfy the following Bloch condition:

$$E_z(x + a, y, t) = \mathrm{e}^{\mathrm{i}ka} E_z(x, y, t). \tag{6.17}$$

That is, each of them is characterized by a wave number k in the x direction. They thus form a 1D impurity band.

It is essential for our method described in the previous sections that the eigenfrequency of the relevant localized mode is isolated since it is obtained from the resonance structure in the spectrum of the dipole radiation. If it is close to other eigenmodes, we cannot distinguish it from them. Therefore, our method does not seem, at first glance, to be applicable to the problem of the impurity band, for which the spectrum of the localized eigenmodes forms a continuum. However, this is *not* true. The area surrounded by a

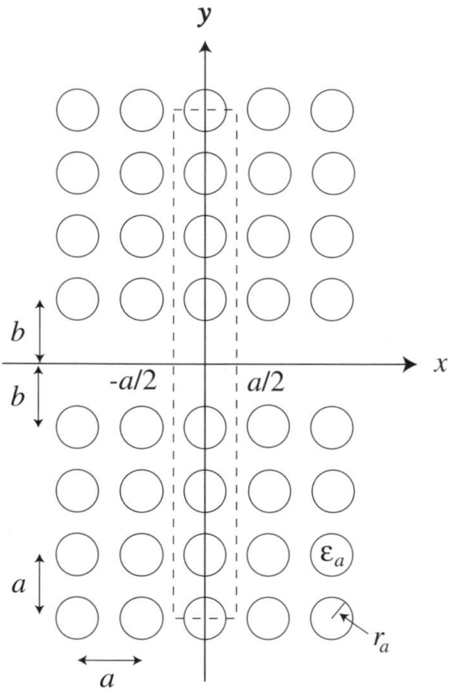

Fig. 6.17. Top view of the 2D square array of circular rods that was assumed for the numerical calculation (see text for details.) (After [70])

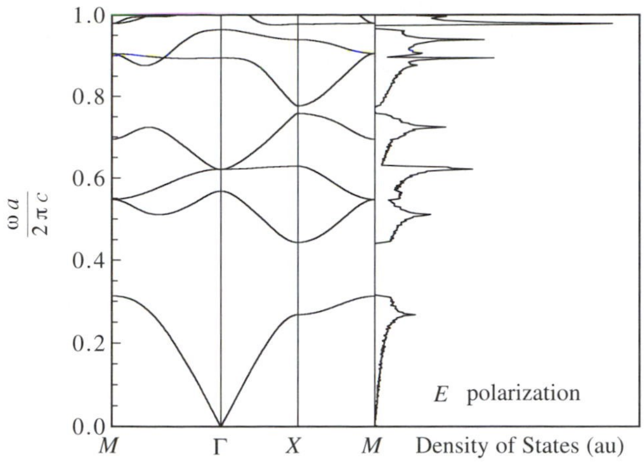

Fig. 6.18. Photonic band structure of the regular square lattice of the circular rods for E polarization. The ordinate is the normalized frequency. According to the experimental condition of Lin et al. [77], the following values were assumed: $a = 1.27$ mm, $r_a = 0.255$ mm, $\varepsilon_a = 10.0$. (After [70])

broken line in Fig. 6.17 is the supercell on which the numerical calculation was carried out. The number of rods included in the supercell was more than 6. We imposed a boundary condition on $E_z(x, y, t)$ given by (6.17). Then, we could extract the contribution to the radiated electromagnetic field from the particular eigenmodes with the designated wave vector. In what follows, we will show that this simple procedure works quite well and that excellent agreement with the experimental result is obtained.

Now, Fig. 6.18 shows the photonic band structure of the regular (i.e., $2b = a$) square lattice. According to the experimental condition of Lin et al. [77] the following values were assumed: $a = 1.27$ mm, $r_a = 0.255$ mm, $\varepsilon_a = 10.0$. Figure 6.18 shows the presence of three bandgaps. We will deal with the lowest one below.

As will be shown below, the localized mode at the Γ point ($k = 0$) of the 1D impurity band found by Lin et al. was a B_2 mode of the C_{2v} point group. First we examine this case, for which $b = 1.5a$. Figure 6.19 shows the frequency dependence of the electromagnetic energy radiated by the oscillating dipole moment located 0.2 mm from the origin in the y direction. $\square$, $\circ$, $\diamond$, and $\bullet$ denote the accumulated electromagnetic energy after 10, 20, 35, and 50 cycles of the oscillation, respectively. In this figure, we observe a clear resonance at $\omega a/2\pi c = 0.3602$, or $\omega/2\pi = 85.03$ GHz, which is quite close to the experimental observation, that is, $\omega/2\pi = 85.5$ GHz. The discrepancy between them is less than 0.6%. On the other hand, the inset of Fig. 6.19 shows the calculated dispersion relation of the 1D impurity band where the abscissa represents the normalized wave vector in the x direction. In the experiments carried out by Lin et al., transmission spectra of the specimen

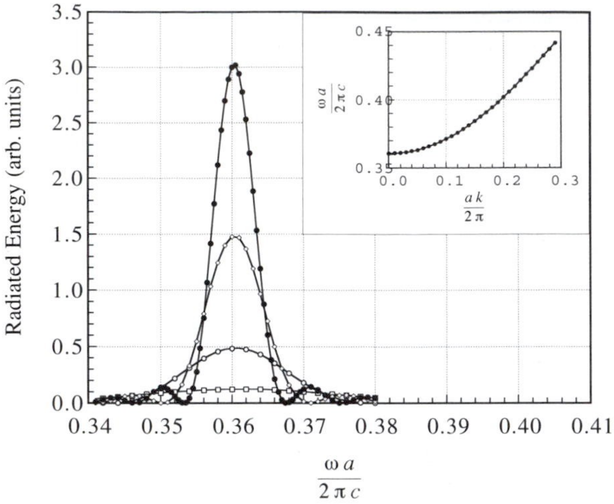

Fig. 6.19. Electromagnetic energy radiated by the dipole moment as a function of the oscillation frequency. □, ○, ◇, and • denote the accumulated electromagnetic energy after 10, 20, 35, and 50 cycles of the oscillation, respectively. The same parameters as for Fig. 6.18 were used. b and k were assumed to be $1.5a$ and 0. The inset shows the dispersion relation of the 1D impurity band. Its abscissa is the normalized wave vector in the x direction. (After [70])

were measured with microwaves at normal incidence, i.e., in the y direction. Therefore, they observed the localized mode at the Γ point. If the angle of incidence is tilted, then this dispersion curve can be observed.

Next, Fig. 6.20 shows the b dependence of the eigenfrequencies of the localized modes at the Γ point where open and solid circles represent the A_1 and B_2 modes, respectively. Modes of other symmetries, i.e., A_2 and B_1 symmetries, were not found in this parameter range. In this figure, horizontal lines represent the edges of the first and second bands at the X point in the 2D Brillouin zone (see Fig. 6.18), between which the localized modes with $k = 0$ can exist. The number in parentheses in Fig. 6.20 is a band index in order to distinguish branches with the same symmetry. Figures 6.21 and 6.22 show the 2D distribution of the electric fields of the localized modes of the A_1 and B_2 symmetries, respectively. The maximum of each electric field is normalized to unity in these figures. Note that the number of nodes of the eigenfunctions along the y axis increases with increasing band index. Also note that the A_1 (B_2) modes are symmetric (antisymmetric) about the x axis.

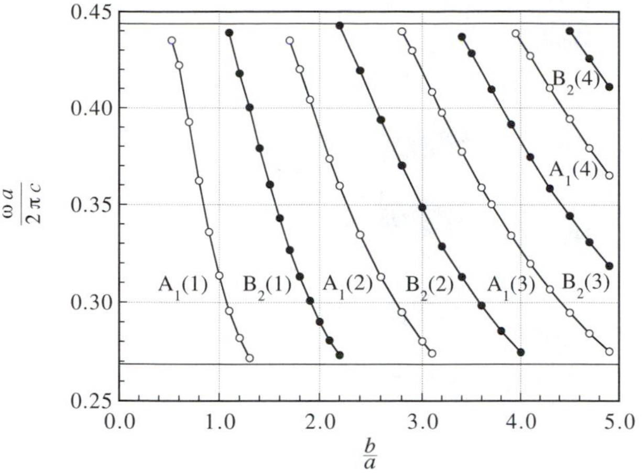

Fig. 6.20. Eigenfrequencies at the Γ point of the 1D impurity bands as a function of the width b of the defect structure. The ordinate is the normalized frequency, and horizontal lines represent the edges of the first and the second bands of the x point in the 2D Brillouin zone. The symmetry of the localized modes is also shown, where the number in parentheses is the band index that distinguishes branches with the same symmetry. (After [70])

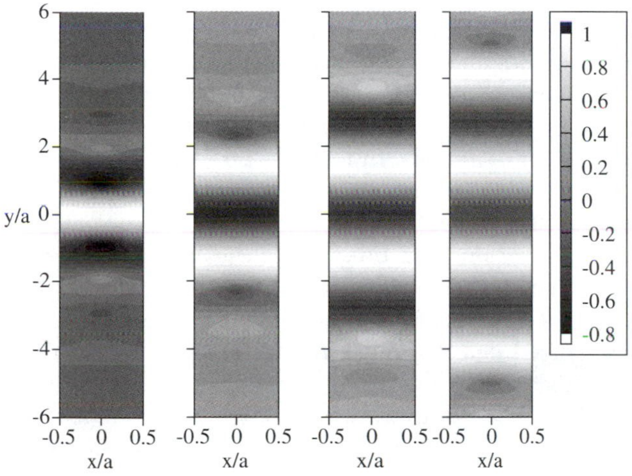

Fig. 6.21. Distribution of the electric fields for the A_1 modes: $A_1(1)$ (*left*) to $A_1(4)$ (*right*). The maximum of each electric field is normalized to unity. (After [70])

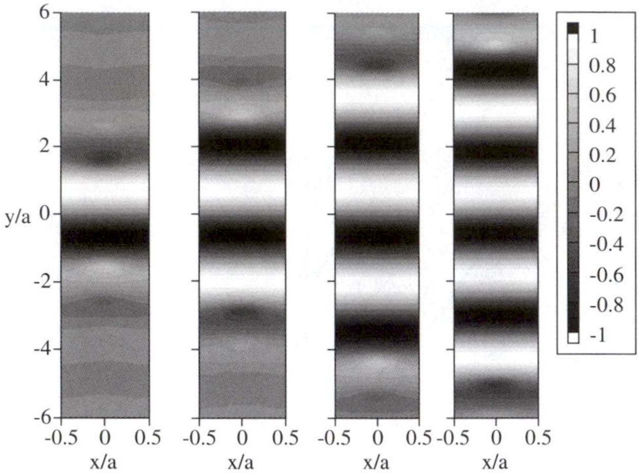

Fig. 6.22. Distribution of the electric fields for the B_2 modes: $B_2(1)$ (*left*) to $B_2(4)$ (*right*). The maximum of each electric field is normalized to unity. (After [70])

6.6 Dielectric Loss and Quality Factor

The high quality factor of the localized defect modes originates from the fact that the amplitude of their wave functions decreases exponentially with the distance from the center of the defect structure, and therefore, the coupling strength between the defect modes and the radiation field outside the crystal can be extremely small. The characteristic length of the exponential decrease is of the order of the relevant wavelength, and therefore we can expect quite large quality factors even with thin samples. In fact, the large quality factor of Lin et al. mentioned above was observed for a 2D crystal with a thickness less than seven times the wavelength. Actually, the quality factor was not limited by coupling to the external radiation field but by the lifetime of the defect mode brought about by the small dielectric loss in the components. Then, evaluation of the dielectric loss is necessary and important for the design and the practical use of the localized defect modes. In this section, we examine the quality factors of the defect modes described in the last section.

In what follows, we will use the notation of Lin et al. to distinguish the structure of the specimen: an m-(l)-n photonic crystal has l rows of empty lattice sandwiched by m and n layers of the dielectric rods. Then $b/a = (l+1)/2$. The structure shown in Fig. 6.17 is, for example, denoted by 4-(2)-4, since it has four layers of the dielectric rods in both sides and it was drawn for $b = 1.5a$. In the actual numerical calculation, we assumed that $m = n$ for simplicity and n was varied from 1 to 9.

We evaluated the lifetime of the defect modes due to the dielectric loss, τ, by analyzing the temporal attenuation of their electromagnetic energy numerically. We now briefly describe the method. In previous sections, we calculated the eigenfrequencies and eigenfunctions of the localized defect modes by means of the numerical simulation of their excitation process by a virtual oscillating dipole moment embedded in the photonic crystals. A similar method can be utilized for the present purpose. Namely, we excited the defect mode by a point dipole oscillating at its eigenfrequency for 200 cycles, and analyzed the temporal attenuation of the accumulated electromagnetic energy after switching off the oscillation in the presence of the imaginary part of the dielectric constant. We can thus obtain the decay rate, and hence the lifetime. The spectral width due to the dielectric loss (FWHM), Γ_{d}, is given by $\Gamma_{\mathrm{d}} = \tau^{-1}$.

First, we examine the spectral widths of the defect modes when the dielectric loss is absent. Figure 6.23 shows the transmission spectra of the n-(2)-n structures calculated by the layer-doubling method based on the cylindrical-wave expansion [72]. As a typical example, the transmission spectrum for the 3-(2)-3 structure in a wide frequency range is presented in Fig. 6.23(a). The abscissa is the normalized frequency, where ω and c denote the angular frequency of the incident microwave and the light velocity in vacuum, respectively. An extremely sharp transmission peak due to the defect mode is observed at $\omega a/2\pi c \simeq 0.36$ in the opaque range between $\omega a/2\pi c \simeq 0.26$ and 0.44 that is brought about by the photonic bandgap. The spectral width of the defect mode decreases rapidly with increasing n as is shown in Figs. 6.23(b) and 6.23(c). The normalized eigenfrequency of the $B_2(1)$ mode of the 5-(2)-5 structure was calculated in the last section to be 0.3602 by means of the numerical simulation of the dipole radiation. The difference between the two calculations is as small as 0.17%. The agreement between the two independent calculations provides clear evidence for the accuracy of our numerical methods.

The n-dependence of the spectral width (FWHM) of the defect modes is presented in Fig. 6.24. In this figure, $\bullet$ represents the value obtained from the calculated transmission spectrum for $\varepsilon'' = 0$, and $\square$ denotes the experimental result of Lin et al. In their experiment, n was not necessarily equal to m. Here, the averaged value, i.e., $(n+m)/2$ is plotted as n. On the other hand, $\circ$ represents the spectral width for $\varepsilon'' = 0.001$, which will be discussed later. It is obvious that a much smaller spectral width, and hence, a much larger quality factor, $Q = \omega_{\mathrm{d}}/(\text{spectral width})$, can be attained for large n if the dielectric loss is absent, and the spectral widths observed by Lin et al. were limited by the dielectric loss. A quality factor larger than 10^7 is expected for $n = 9$ if $\varepsilon'' = 0$.

Now, we examine the spectral width Γ_{d} brought about by the dielectric loss. In Fig. 6.25, the temporal attenuation of the electromagnetic energy U after switching off the oscillating dipole moment is shown for the B_2 mode of the 5-(2)-5 structure as an example. In this figure, the abscissa is the normalized time and U_0 is the accumulated energy just after 200 cycles of the oscillation of the dipole at the eigenfrequency of the defect mode. $\bullet$, $\circ$,

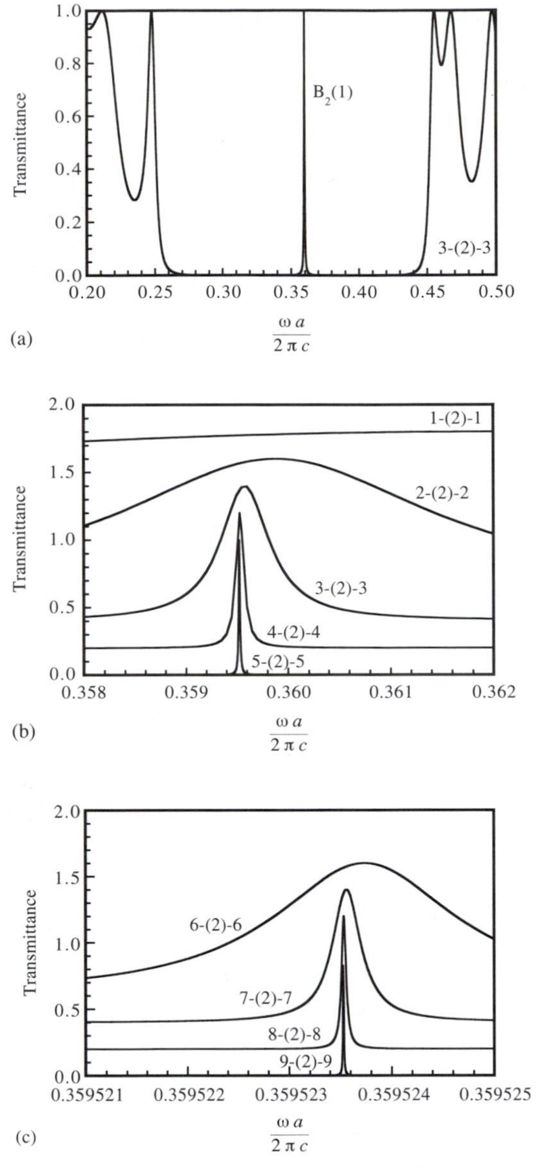

Fig. 6.23. Calculated transmission spectra of the n-(2)-n structures: (**a**) the spectrum of the 3-(2)-3 structure in a wide frequency range, (**b**)(**c**) the spectra of the n-(2)-n structures in the vicinity of the transmission peak due to the localized defect mode with the B_2 spatial symmetry. According to the experimental condition of Lin et al. [77], the following values were assumed for the numerical calculation: $a = 1.27$ mm, $r_\mathrm{a} = 0.255$ mm, $\varepsilon_\mathrm{a} = 10.0$, and $b = 1.5a$. The sharp transmission peaks due to the defect mode are observed at $\omega a/2\pi c \simeq 0.36$ in the opaque range between $\omega a/2\pi c \simeq 0.26$ and 0.44 that is brought about by a photonic bandgap. Note the rapid decrease of the width of the transmission peak with increasing n. (After [72])

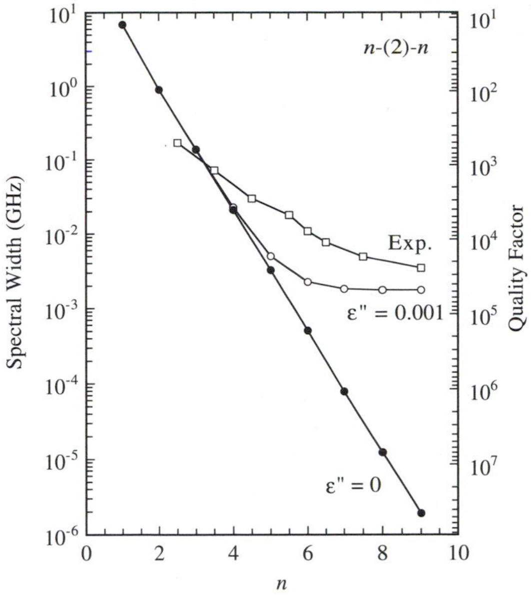

Fig. 6.24. Spectral width (FWHM) of the transmission peak due to the localized defect mode at the Γ point of the 1D impurity band with the B_2 spatial symmetry of the n-(2)-n structure. □ represents the experimental results of Lin et al. [77], whereas • and ∘ represent the calculated values of the present work for $\varepsilon'' = 0$ and $\varepsilon'' = 0.001$, respectively. Because the structure used in the experiment of Lin et al. was not necessarily symmetric about the x axis in Fig. 6.17, the experimental result was plotted for the averaged n. (After [72])

□, and ◇ represent the temporal attenuation for $\varepsilon'' = 0.001$, 0.01, 0.03, and 0.1, respectively. Each curve shows an exponential attenuation, and we can obtain the lifetime of the localized modes from its slope.

We should note that Γ_d is proportional to ε''. This can be easily understood, since the dielectric loss in a unit time, U_loss, is proportional to ε''. It is given by

$$U_\mathrm{loss} = \frac{\omega_\mathrm{d}\varepsilon_0\varepsilon''}{2}\int_\Omega |\boldsymbol{E}_\mathrm{d}(\boldsymbol{r},t)|^2\mathrm{d}V, \tag{6.18}$$

where ω_d and $\boldsymbol{E}_\mathrm{d}$ stand for the eigen-angular frequency and the electric field of the localized defect mode, and Ω denotes the volume that the dielectric rods occupy. Because

$$\Gamma_\mathrm{d}^{-1} = \tau = \frac{U_0}{U_\mathrm{loss}}, \tag{6.19}$$

we can conclude that Γ_d is proportional to ε''.

Returning to Fig. 6.24, ∘ represents the spectral width of each defect mode for $\varepsilon'' = 0.001$, which was obtained as a sum of the spectral width for $\varepsilon'' = 0$

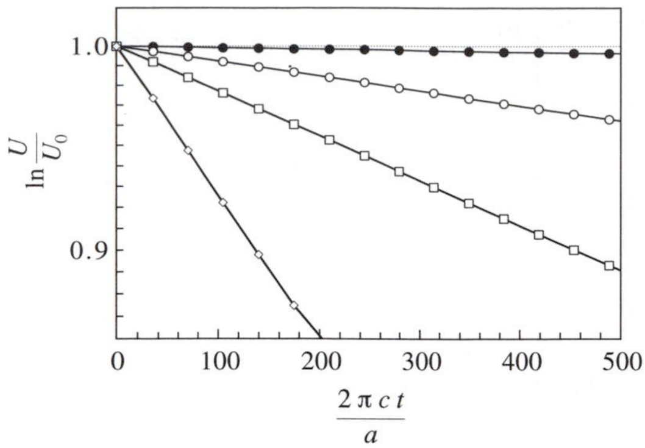

Fig. 6.25. Temporal attenuation of the accumulated electromagnetic energy U of the localized defect mode with the B_2 spatial symmetry with the 5-(2)-5 structure that is caused by the dielectric loss, where U_0 denotes the accumulated energy immediately after 200 cycles of the oscillation of the dipole. The abscissa is the normalized time. •, ○, □, and ◇ represent the calculated results for $\varepsilon'' = 0.001$, 0.01, 0.03, and 0.1, respectively. (After [72])

and Γ_{d}. It is evident that the spectral width is limited by the dielectric loss for $n \geq 6$, and the upper limit of the quality factor is 4.9×10^4. It is clear that a larger ε'' leads to a smaller quality factor. Here we assumed ε'' of 0.001 according to Lin et al. The discrepancy between the theoretical results and the experimental observation in Fig. 6.24 may be attributed to the energy leakage along the direction of the line defect, i.e., along the x axis in Fig. 6.17, as was pointed out by Lin et al. as an additional loss mechanism, the amount of which depends on experimental configuration. Note that we dealt with systems infinite in the x direction in the theoretical analysis, whereas the actual specimens were finite.

7. Band Calculation with Frequency-Dependent Dielectric Constants

The numerical simulation of the dipole radiation that was utilized for the localized defect modes in Chap. 6 will be extended to the band calculation with frequency-dependent dielectric constants [78–80]. Two-dimensional photonic crystals with metallic components will be analized by this method. The dispersion relation, the field distribution, and the lifetime of the radiation modes will be examined. The nature of surface plasmon polaritons will be clarified.

7.1 Principle of Calculation

The dielectric constants of photonic crystals were assumed to be independent of frequency so far. When we deal with transparent materials whose optical transition frequencies are far from those of the relevant radiation field, this assumption is reasonable. However, when we deal with materials whose resonant polarization plays an important role in their optical response, we cannot neglect the frequency depnedence of their dielectric constants.

We begin with the following set of Maxwell's equations.

$$\nabla \times \boldsymbol{E}(\boldsymbol{r},t) = -\mu_0 \frac{\partial}{\partial t}\boldsymbol{H}(\boldsymbol{r},t), \tag{7.1}$$

$$\nabla \times \boldsymbol{H}(\boldsymbol{r},t) = \frac{\partial}{\partial t}\{\boldsymbol{D}_0(\boldsymbol{r},t) + \boldsymbol{P}_\mathrm{d}(\boldsymbol{r},t)\}. \tag{7.2}$$

In (7.2), $\boldsymbol{P}_\mathrm{d}(\boldsymbol{r},t)$ represents the virtual oscillating dipole moment as before and $\boldsymbol{D}_0(\boldsymbol{r},t)$ denotes the electric displacement due to the regular dielectric structure of the photonic crystal. If the dielectric constant does not depend on frequency, the latter is given by the dielectric constant times the electric field. However, if the dielectric constant depends on frequency, a more general treatment is necessary. We assume that the matter system responds linearly to the electric field. Then, a somewhat general form of the displacement field is given by

$$\boldsymbol{D}_0(\boldsymbol{r},t) = \int_{-\infty}^{\infty} \mathrm{d}t' \varPhi(\boldsymbol{r},t-t')\boldsymbol{E}(\boldsymbol{r},t'). \tag{7.3}$$

$\Phi(\boldsymbol{r}, t)$ is referred to as the response function. In this equation, the displacement is given by the convolution of the response function and the electric field. Hence, the former depends not only on the electric field at the same instance but also on those at different times. This is the origin of the frequency dependence of the dielectric constant. Because of the causality principle,

$$\Phi(\boldsymbol{r}, t) = 0 \qquad (t < 0). \tag{7.4}$$

If $\boldsymbol{E}(\boldsymbol{r}, t)$ is monochromatic, i.e., if it has the following form,

$$\boldsymbol{E}(\boldsymbol{r}, t) = \boldsymbol{E}_\omega(\boldsymbol{r}) \mathrm{e}^{-\mathrm{i}\omega t}, \tag{7.5}$$

we have

$$\begin{aligned} \boldsymbol{D}_0(\boldsymbol{r}, t) &= \int_{-\infty}^{\infty} \mathrm{d}t' \Phi(\boldsymbol{r}, t - t') \boldsymbol{E}_\omega(\boldsymbol{r}) \mathrm{e}^{-\mathrm{i}\omega t'} \\ &= \boldsymbol{E}_\omega(\boldsymbol{r}) \mathrm{e}^{-\mathrm{i}\omega t} \int_{-\infty}^{\infty} \mathrm{d}t' \Phi(\boldsymbol{r}, t') \mathrm{e}^{\mathrm{i}\omega t'} . \end{aligned} \tag{7.6}$$

When we define $\varepsilon(\boldsymbol{r}, \omega)$ by

$$\varepsilon(\boldsymbol{r}, \omega) = \frac{1}{\varepsilon_0} \int_{-\infty}^{\infty} \mathrm{d}t' \Phi(\boldsymbol{r}, t') \mathrm{e}^{\mathrm{i}\omega t'}, \tag{7.7}$$

and denote $\boldsymbol{D}_0(\boldsymbol{r}, t)$ by

$$\boldsymbol{D}_0(\boldsymbol{r}, t) = \boldsymbol{D}_\omega(\boldsymbol{r}) \mathrm{e}^{-\mathrm{i}\omega t}, \tag{7.8}$$

we obtain

$$\boldsymbol{D}_\omega(\boldsymbol{r}) = \varepsilon_0 \varepsilon(\boldsymbol{r}, \omega) \boldsymbol{E}_\omega(\boldsymbol{r}). \tag{7.9}$$

Therefore, $\varepsilon(\boldsymbol{r}, \omega)$ is the dielectric function in our previous terminology. However, it may now depend on frequency. Hence, (7.3) gives an extension of our previous treatment of frequency-independent dielectric constants to the frequency-dependent case.

The inverse Fourier transform of (7.7) gives

$$\Phi(\boldsymbol{r}, t) = \frac{\varepsilon_0}{2\pi} \int_{-\infty}^{\infty} \mathrm{d}\omega \varepsilon(\boldsymbol{r}, \omega) \mathrm{e}^{-\mathrm{i}\omega t}. \tag{7.10}$$

For $t < 0$, the path for the integration can be modified to enclose the upper half of the complex ω plane. So, if $\varepsilon(\boldsymbol{r}, \omega)$ has a pole in that region, the integral has a finite value due to the residue theorem, which contradicts the causality. Thus, $\varepsilon(\boldsymbol{r}, \omega)$ cannot have a pole in the upper half of the complex ω plane.

We now show some examples of such dielectric functions. We shall forget the $\boldsymbol{r}$ dependence for a while. For metallic materials, it is known that their dielectric constant is well represented by that of the Drude type in a certain frequency range:

$$\varepsilon(\omega) = \varepsilon_\infty \left\{ 1 - \frac{\omega_\mathrm{p}^2 \tau}{\omega(\omega\tau + \mathrm{i})} \right\}, \tag{7.11}$$

where ω_p is the plasma frequency, τ is the relaxation time, and ε_∞ is the relative dielectric constant at sufficiently high frequencies. Then (7.10) leads to

$$\begin{aligned}\Phi(\boldsymbol{r},t) &= \frac{\varepsilon_0\varepsilon_\infty}{2\pi}\int_{-\infty}^{\infty}\mathrm{d}\omega\left\{1+\mathrm{i}\omega_\mathrm{p}^2\left(\frac{1}{\omega+\mathrm{i}\delta}-\frac{1}{\omega+\mathrm{i}/\tau}\right)\right\}\mathrm{e}^{-\mathrm{i}\omega t}\\ &= \varepsilon_0\varepsilon_\infty\left\{\delta(t)+\omega_\mathrm{p}^2\tau\left(1-\mathrm{e}^{-t/\tau}\right)\right\}\theta(t),\end{aligned} \tag{7.12}$$

where δ is a positive infinitesimal and $\theta(t)$ is the unit step function, i.e.,

$$\theta(t) = \begin{cases} 1 & (t \geq 0) \\ 0 & (t < 0). \end{cases} \tag{7.13}$$

Another example is the dielectric constant appropriate to semiconductors:

$$\varepsilon(\omega) = \varepsilon_\infty \frac{\omega_\mathrm{L}^2-\omega^2}{\omega_\mathrm{T}^2-\omega^2-\mathrm{i}\omega\gamma}, \tag{7.14}$$

where ω_T (ω_L) is the transverse (longitudinal) exciton frequency, and γ is the relaxation rate. The same formula may be used to describe the optical phonons in polar crystals. For this case, we have

$$\begin{aligned}&\Phi(t)\\ &= \varepsilon_0\varepsilon_\infty\left[\delta(t)+\left\{\left(\omega_\mathrm{L}^2-\omega_\mathrm{T}^2+\frac{\gamma^2}{4}\right)\frac{\sin\omega_\mathrm{T}t}{\omega_\mathrm{T}}-\gamma\cos\omega_\mathrm{T}t\right\}\theta(t)\right].\end{aligned} \tag{7.15}$$

Now, we apply our method to a photonic crystal composed of metallic components whose dielectric constant is well represented by that of the Drude type. We locate the oscillating dipole outside the metal. From (7.3) and (7.12), the displacement in the metallic region is given by

$$\boldsymbol{D}_0(\boldsymbol{r},t) = \varepsilon_0\varepsilon_\infty\left\{\boldsymbol{E}(\boldsymbol{r},t)+\omega_\mathrm{p}^2\tau\int_{-\infty}^{t}\mathrm{d}t'\left(1-\mathrm{e}^{-(t-t')/\tau}\right)\boldsymbol{E}(\boldsymbol{r},t')\right\}. \tag{7.16}$$

We thus have

$$\begin{aligned}\frac{\partial^2}{\partial t^2}\boldsymbol{D}_0(\boldsymbol{r},t) = \varepsilon_0\varepsilon_\infty\Bigg[\frac{\partial^2}{\partial t^2}\boldsymbol{E}(\boldsymbol{r},t)+\omega_\mathrm{p}^2\Bigg\{&\boldsymbol{E}(\boldsymbol{r},t)\\ &-\frac{1}{\tau}\int_{-\infty}^{t}\mathrm{d}t'\mathrm{e}^{-(t-t')/\tau}\boldsymbol{E}(\boldsymbol{r},t')\Bigg\}\Bigg].\end{aligned} \tag{7.17}$$

Hence, the wave equation for this region is

$$\begin{aligned}\boldsymbol{\nabla}\times\{\boldsymbol{\nabla}\times\boldsymbol{E}(\boldsymbol{r},t)\} = -\frac{\varepsilon_\infty}{c^2}\Bigg[&\frac{\partial^2}{\partial t^2}\boldsymbol{E}(\boldsymbol{r},t)+\omega_\mathrm{p}^2\boldsymbol{E}(\boldsymbol{r},t)\\ &-\frac{\omega_\mathrm{p}^2}{\tau}\int_0^{\infty}\mathrm{d}t'\exp\left(-\frac{t'}{\tau}\right)\boldsymbol{E}(\boldsymbol{r},t-t')\Bigg].\end{aligned} \tag{7.18}$$

On the other hand, we assume that $\varepsilon(\boldsymbol{r}, \omega) = 1$ (air) outside the metallic region. Then for this region,

$$\nabla \times \{\nabla \times \boldsymbol{E}(\boldsymbol{r}, t)\} = -\frac{1}{c^2}\left[\frac{\partial^2}{\partial t^2}\boldsymbol{E}(\boldsymbol{r}, t) - \omega^2 \mu \delta(\boldsymbol{r} - \boldsymbol{r}_0)\mathrm{e}^{-\mathrm{i}\omega t}\right]. \quad (7.19)$$

We discretize (7.18) and (7.19) to obtain the difference equations and solve them numerically with an initial condition $\boldsymbol{E} = \partial \boldsymbol{E}/\partial t = 0$ and a boundary condition

$$\boldsymbol{E}(\boldsymbol{r} + \boldsymbol{a}, t) = \exp(\mathrm{i}\boldsymbol{k} \cdot \boldsymbol{a})\boldsymbol{E}(\boldsymbol{r}, t), \quad (7.20)$$

where $\boldsymbol{k}$ is a wave vector in the first Brillouin zone and $\boldsymbol{a}$ is the elementary lattice vector. The latter condition extracts the contribution to the radiated electromagnetic field from particular eigenmodes with the specified wave vector. Therefore, we can calculate the resonance frequency as a function of $\boldsymbol{k}$, i.e., we can obtain the dispersion relation.

7.2 Modified Plane Waves in Metallic Crystals

In this section, we will give an example of a numerical calculation. We will deal with a 2D photonic crystal composed of a square array of metallic cylinders for the E polarization. The H polarization will be treated in the next section. The following parameters were assumed: $r_\mathrm{a}/a = 0.472$ (crystal 1) or 0.0564 (crystal 2), where r_a and a denote the radius of the cylinders and the lattice constant, respectively; $\varepsilon_\infty = 1.0$; and $\omega_\mathrm{p} a/2\pi c = 1.0$. Because of the boundary condition, (7.20), it is sufficient to deal with only one unit cell, and therefore, the CPU time necessary for the numerical calculation is small.

The numerical calculation was done for the wave vectors in the (1, 0) and (1, 1) directions in the first Brillouin zone of the 2D square lattice. In the actual calculation, τ^{-1} was set equal to $0.01 \times \omega_\mathrm{p}$. The unit cell was divided into 40×40 parts, and one period of the oscillation was divided into 160 steps. Further decrease of the size of the spatial and temporal meshes did not result in an apparent change in the resonance frequencies. The lifetime of the eigenmodes was evaluated by observing the temporal decrease in the accumulated electromagnetic energy after 200 cycles of the oscillation of the dipole moment. The total electromagnetic energy after switching off the oscillation showed an exponential decrease and we could obtain the lifetime easily.

As an example, Fig. 7.1 shows the electromagnetic energy radiated by the oscillating dipole moment as a function of the oscillation frequency for $\boldsymbol{k} = 0$. The abscissa represents the normalized frequency. □, ○, ◇, and • denote the accumulated electromagnetic energy after 10, 20, 50, and 100 cycles of the oscillation, respectively. A resonance at $\omega a/2\pi c = 0.745$ is clearly observed in this figure. The dispersion relation for crystal 1 thus obtained is presented in Fig. 7.2 where the symmetry of each eigenmode is also shown. In this figure, the number in parentheses is given in order of ascending frequency

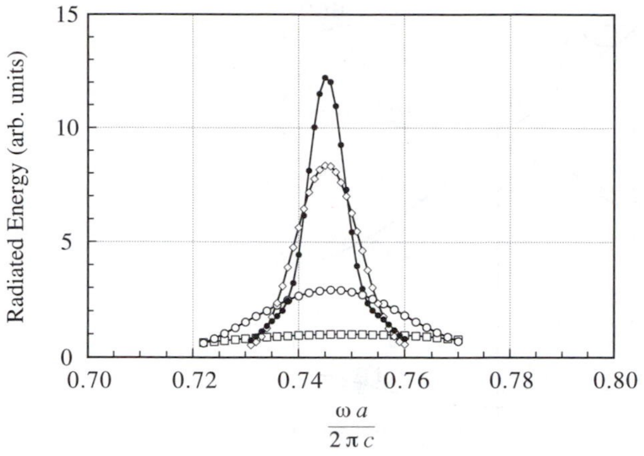

Fig. 7.1. Electromagnetic energy radiated by an oscillating dipole moment located at $\boldsymbol{r}_0/a = (0.3, 0)$ as a function of the oscillation frequency for $\boldsymbol{k} = 0$ for crystal 1. □, ○, ◇, and • denote the accumulated electromagnetic energy after 10, 20, 50, and 100 cycles of the oscillation, respectively. The abscissa represents the normalized frequency. The following parameters were used: $r_\mathrm{a}/a = 0.472$, $\varepsilon_\infty = 1.0$, $\omega_\mathrm{p} a/2\pi c = 1.0$, and $\tau^{-1} = 0.01 \times \omega_\mathrm{p}$. A resonance at $\omega a/2\pi c = 0.745$ is clearly observed. (After [78])

when there are more than one mode of the same symmetry in the analyzed spectral region. The symmetry of each band shown in Fig. 7.2 is consistent with the prediction of the group theory in Table 3.7 that was obtained by the comparison with the irreducible representations of the radiation field in free space. This implies that the radiational eigenmodes for the E polarization in this frequency range are essentially modified plane waves.

We should note that there is no eigenmode for $\omega a/2\pi c < 0.745$. We can show that this cut-off frequency is consistent with the long-wavelength approximation of the Maxwell equations as follows. From (2.47)–(2.49), we have for the E polarization

$$\frac{\partial E_z}{\partial y} = -\mu_0 \frac{\partial H_x}{\partial t}, \tag{7.21}$$

$$\frac{\partial E_z}{\partial x} = \mu_0 \frac{\partial H_y}{\partial t}, \tag{7.22}$$

$$\frac{\partial H_y}{\partial x} - \frac{\partial H_x}{\partial y} = \varepsilon_0 \varepsilon \frac{\partial E_z}{\partial t}. \tag{7.23}$$

If the wavelength is long and the wave vector is small, the eigenmode is not affected by the detail of the spatial variation of the dielectric function. Hence, we may replace the dielectric function by its spatial average over the unit cell. For this case, the spatial variation of the eigenfunctions in the unit cell is also small. Hence, we may approximate the rigorous eigenfunctions by

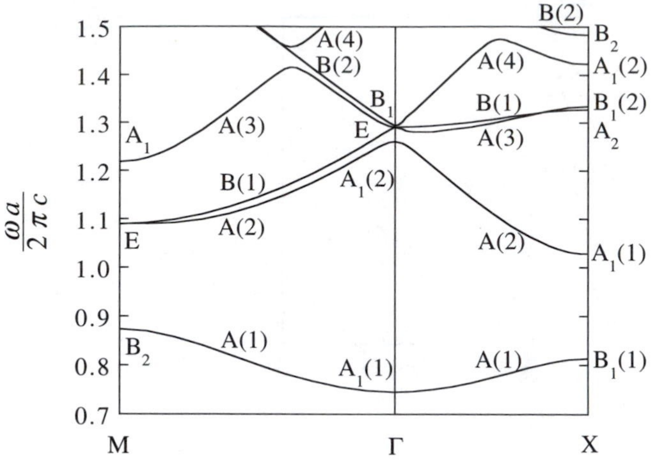

Fig. 7.2. Dispersion relation of the 2D square photonic crystal composed of metallic cylinders (crystal 1) for the E polarization calculated by means of the numerical simulation of the dipole radiation. The ordinate is the normalized frequency where ω, a, and c stand for the angular frequency of the radiation field, the lattice constant of the crystal, and the light velocity in free space. The following parameters were used for numerical calculation: $r_a/a = 0.472$ ($f = 0.7$), where r_a denotes the radius of the metallic cylinders; $\varepsilon_\infty = 1.0$, $\omega_p a/2\pi c = 1.0$, and $\gamma = 0.01 \times \omega_p$ in (7.11). The dispersion relation was drawn for highly symmetric points in the first Brillouin zone of the 2D crystal: $\Gamma(0,0)$, $X(\pi/a, 0)$, and $M(\pi/a, \pi/a)$. The spatial symmetry of each eigenmode is also shown in this figure, where the number in parentheses is given in order of ascending frequency when there are more than one mode of the same symmetry in the analyzed frequency region. (After [79])

$$E_z(\boldsymbol{r}_{/\!/}, t) = u_z(\boldsymbol{r}_{/\!/}) \exp\left\{\mathrm{i}\left(\boldsymbol{k}_{/\!/} \cdot \boldsymbol{r}_{/\!/} - \omega t\right)\right\}$$
$$\simeq \overline{u}_z \exp\left\{\mathrm{i}\left(\boldsymbol{k}_{/\!/} \cdot \boldsymbol{r}_{/\!/} - \omega t\right)\right\}, \quad (7.24)$$

etc., where the overline denotes the spatial average on the unit cell. When we take this spatial averege in (7.21)–(7.23), we obtain

$$k_y \overline{u}_z \simeq \mu_0 \omega \overline{v}_x, \quad (7.25)$$

$$k_x \overline{u}_z \simeq -\mu_0 \omega \overline{v}_y, \quad (7.26)$$

$$k_x \overline{v}_y - k_y \overline{v}_x \simeq -\varepsilon_0 \overline{\varepsilon} \omega \overline{u}_z. \quad (7.27)$$

We should note that we could do the following replacement,

$$\overline{\varepsilon E_z} \simeq \overline{\varepsilon} \overline{E}_z, \quad (7.28)$$

since E_z is a continuous function of $\boldsymbol{r}_{/\!/}$ due to the continuity condition of the tangential component of the electric field on the surface of the metallic cylinder, whereas ε is discontinuous. When both functions were discontinuous, the replacement could not be justified.

From (7.25)–(7.27), we have

$$\frac{\omega^2}{c^2} = \frac{k^2}{\overline{\varepsilon}}, \tag{7.29}$$

where the spatial average of the dielectric function, $\overline{\varepsilon}$, is now given by

$$\overline{\varepsilon} = 1 + (\varepsilon_\infty - 1)f - \frac{f\varepsilon_\infty \omega_\mathrm{p}^2}{\omega^2}. \tag{7.30}$$

In this equation, f is the filling factor of the metallic cylinders: $f = 0.7$ for crystal 1 and $f = 0.01$ for crystal 2. From (7.29) and (7.30), we obtain

$$\omega = \sqrt{\frac{c^2k^2 + f\varepsilon_\infty \omega_\mathrm{p}^2}{1 + (\varepsilon_\infty - 1)f}}. \tag{7.31}$$

Substituting the assumed parameters for crystal 1 into this equation, the cut-off frequency is obtained as 0.84. This value is not very close to the numerical result, but (7.31) tells us that there should be the cut-off and the lowest frequency is realized at the Γ point where $k = 0$.

The distribution of the electric fields of the eigenmodes at the Γ point is presented in Fig. 7.3. The maximum of each electric field is normalized to unity. It is very clear that each eigenmode shows its peculiar symmetry. In particular, Fig. 7.3(c) is a replica of Fig. 7.3(d) given by a 90° rotation, which implies that they are degenerate and attributed to the E representation. Figure 7.4 shows the field distribution of the eigenmodes at the X point. They are consistent with the symmetry assignments given in Fig. 7.2. Figure 7.5 shows the lifetime of the five lowest eigenmodes. This depends strongly on the amount of the field distribution in the metallic region.

The dispersion relation and the lifetime of the lowest band for crystal 2 with a filling factor of 0.01 are shown in Figs. 7.6 and 7.7. In this case, the filling factor is very small and the correspondence with the symmetry assignment for free space is quite good. In the long-wavelength approximation, the cut-off frequency is given by $\omega a/2\pi c = 0.1$, which is very close to the numerical result, 0.097. Figure 7.7 shows a very interesting behavior of the lifetime of the lowest band. It varies by more than three orders of magnitude with $\boldsymbol{k}$. This feature originates, of course, from the variation of the field distribution. For example, the B_2 mode on the M point is antisymmetric about the x and y axes. Its amplitude is thus equal to zero on these axes. Because the electric field is continuous and the radius of the metallic cylinders is small for crystal 2, the electric field is small everywhere in the metal. This is why the dielectric loss is small and the lifetime is long for the B_2 mode. On the other hand, the A_1 mode on the Γ point is totally symmetric and it may have a large amplitude in the metallic region. This is why its lifetime is much shorter than that of the B_2 mode.

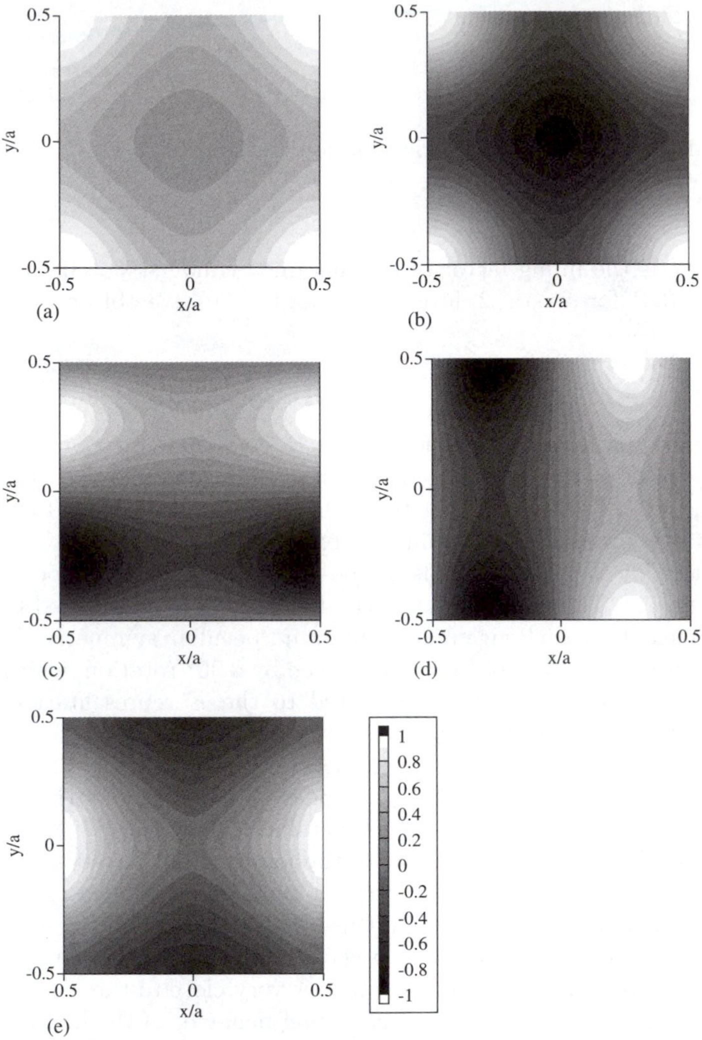

Fig. 7.3. Distribution of the electric fields of the (**a**) $A_1(1)$, (**b**) $A_1(2)$, (**c**) (**d**) E, and (**e**) B_1 modes at the Γ point that has the C_{4v} symmetry. The maximum of each electric field is normalized to unity. For all eigenmodes, the eigenfunctions show their peculiar symmetries. For the doubly degenerate E mode, one eigenfunction is a replica of the other given by a 90° rotation. (After [78])

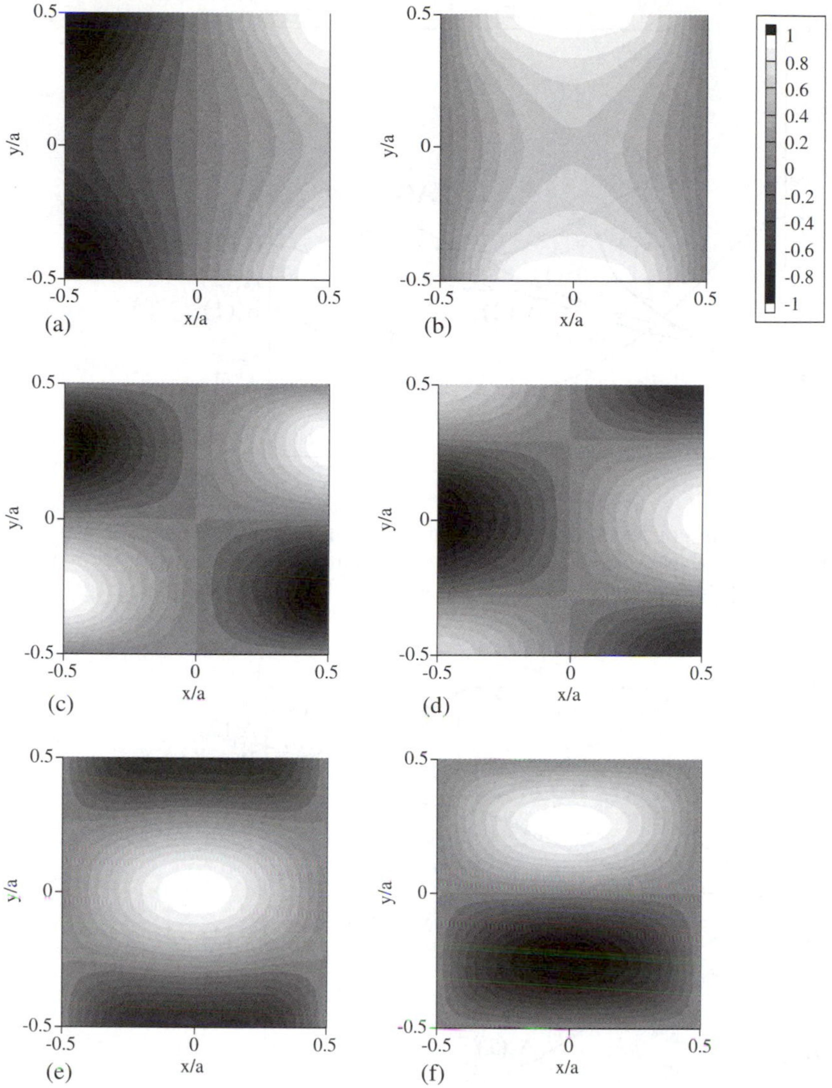

Fig. 7.4. Distribution of the electric fields of the (**a**) $B_1(1)$, (**b**) $A_1(1)$, (**c**) A_2, (**d**) $B_1(2)$, (**e**) $A_1(2)$, and (**f**) B_2 modes at the X point that has the C_{2v} symmetry. The maximum of each electric field is normalized to unity as before. (After [78])

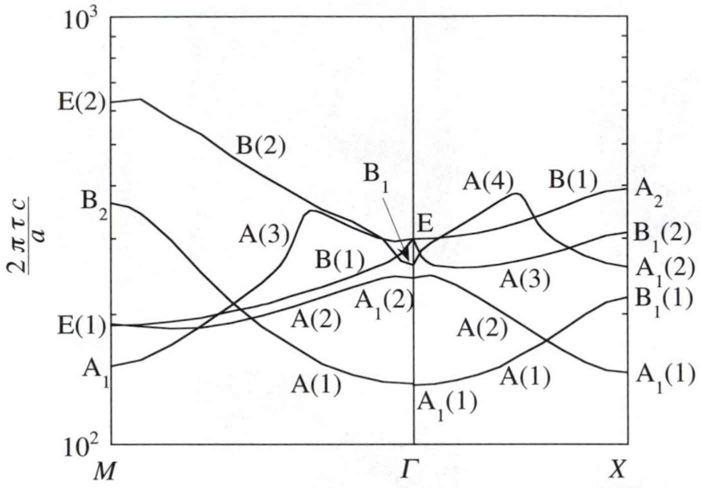

Fig. 7.5. Lifetime of the lowest five eigenmodes of crystal 1. The same parameters as for Fig. 7.1 were used for numerical calculation. The ordinate denotes the normalized lifetime on a logarithmic scale. (After [79])

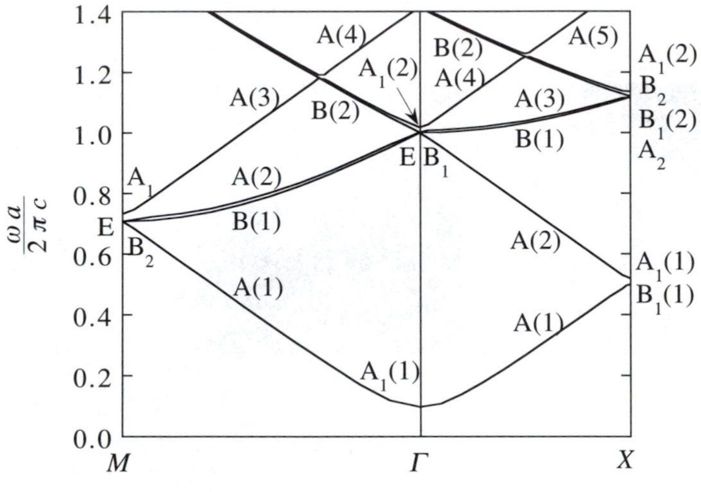

Fig. 7.6. Dispersion relation of the two-dimensional square photonic crystal composed of metallic cylinders (crystal 2) for the E polarization calculated by means of the numerical simulation of the dipole radiation. The same parameters as for Fig. 7.1 were used for numerical calculation except that $r_{\mathrm{a}}/a = 0.0564$ ($f = 0.01$). (After [79])

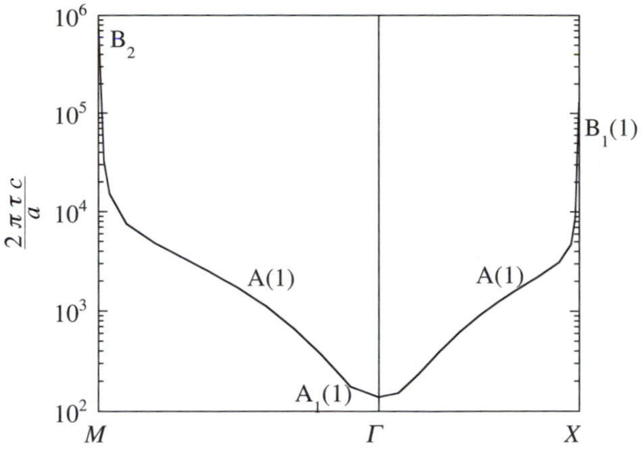

Fig. 7.7. Lifetime of the lowest eigenmode of crystal 2. The same parameters as for Fig. 7.6 were used for numerical calculation. The ordinate denotes the normalized lifetime in a logarithmic scale. (After [79])

We assumed certain analytical forms for the frequency-dependent dielectric constants, such as (7.11) and (7.14). Since the exact analytical forms for actual materials are not known, the present method may seem not to be applicable to actual photonic crystals. However, it is *applicable* in fact. This is a very important and practical feature of the present method. The key idea is that the use of the response function given in (7.10) is necessary to describe the non-steady state of the radiation field after the abrupt introduction of the oscillating dipole at $t = 0$ and that the radiation field after a long period that can be regarded as being in a steady state is well described by the dielectric constant at the oscillation frequency alone. This implies that the choice of the analytical form of the dielectric constant is not important when we calculate the radiation field at a given frequency with a given dielectric constant. Therefore, we can calculate the dipole radiation spectrum using dielectirc constants obtained by experimental observation. The reader my consult [79] for details.

7.3 Surface Plasmon Polaritons

In this section, we will treat the H polarization in the same metallic system as we did in the last section. In this case, a qualitatively different feature appears. That is the excitation of surface plasmons. Before we discuss the surface plasmons in photonic crystals, we will derive their fundamental properties in uniform metals.

7.3.1 Plasmon Polaritons on Flat Surface

First, we consider radiational eigenmodes in a uniform bulk metal with a dielectric constant of the Drude type:

$$\varepsilon_{\mathrm{m}}(\omega) = \varepsilon_{\infty}\left(1 - \frac{\omega_{\mathrm{p}}^2}{\omega^2}\right). \tag{7.32}$$

We omit its imaginary part for simplicity. Since the bulk metal has translational symmetry, its eigenmodes are plane waves and should have the following form:

$$\boldsymbol{E}(\boldsymbol{r}, t) = \boldsymbol{E}_0 \mathrm{e}^{\mathrm{i}(\boldsymbol{k}\cdot\boldsymbol{r} - \omega t)}. \tag{7.33}$$

On the other hand, from one of the Maxwell equations, (2.1), we have

$$\boldsymbol{\nabla} \cdot \boldsymbol{D} \equiv \boldsymbol{\nabla} \cdot \{\varepsilon_{\infty}\varepsilon_{\mathrm{m}}(\omega)\boldsymbol{E}\} = 0. \tag{7.34}$$

From these equations we obtain

$$\varepsilon_{\mathrm{m}}(\omega)\boldsymbol{k} \cdot \boldsymbol{E}_0 = 0. \tag{7.35}$$

Hence, there are two cases:

$$(A) \qquad \boldsymbol{k} \cdot \boldsymbol{E}_0 = 0, \tag{7.36}$$

$$(B) \qquad \varepsilon_{\mathrm{m}}(\omega) = 0. \tag{7.37}$$

For $\omega > \omega_{\mathrm{p}}$, the dielectric constant is positive, and we have only case (A). The eigenmodes are thus transverse waves, i.e., their electric field is perpendicular to their wave vector. Using the vector identity

$$\boldsymbol{\nabla} \times (\boldsymbol{\nabla} \times) = \boldsymbol{\nabla}\boldsymbol{\nabla} \cdot - \nabla^2, \tag{7.38}$$

we obtain the following relation from (2.15) and (7.33):

$$\omega^2 = \frac{c^2 k^2}{\varepsilon_{\mathrm{m}}(\omega)}, \tag{7.39}$$

where $k = |\boldsymbol{k}|$. When we substitute (7.32) into (7.39), we obtain

$$\omega = \sqrt{\omega_{\mathrm{p}}^2 + \frac{c^2 k^2}{\varepsilon_{\infty}}}. \tag{7.40}$$

For $\omega < \omega_{\mathrm{p}}$, we have the same relation as (7.39). However, $\varepsilon_{\mathrm{m}}(\omega)$ is negative in this frequency range, and (7.39) does not have any real solution. For $\omega = \omega_{\mathrm{p}}$, case (B) holds. It is easy to verify that the electric field has only a longitudinal component, i.e., $\boldsymbol{E}_0 /\!/ \boldsymbol{k}$, in this case. The eigen-angular frequency is thus equal to ω_{p} irrespective of $\boldsymbol{k}$. These results are illustrated by curves (a) and (b) in Fig. 7.8.

If we consider a flat interface between a metal and a dielectric material, surface waves localized on the interface appear. These waves are referred to as *surface plasmon polaritons*. Polariton generally means a coupled wave of

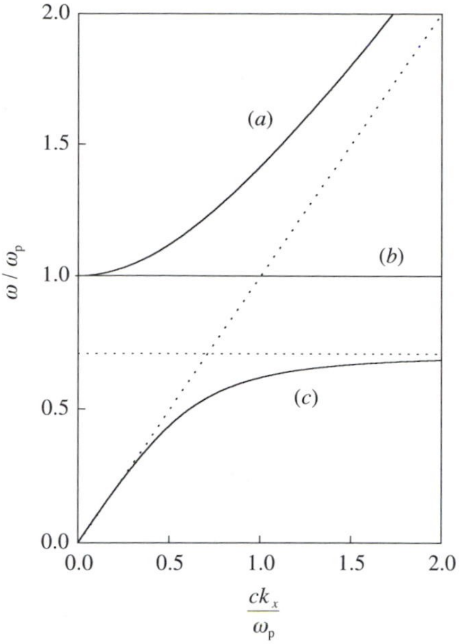

Fig. 7.8. Dispersion relation of (a) transverse plane waves in a bulk metal, (b) longitudinal plasmons in a bulk metal, and (c) surface plasmon polaritons on a flat interface between a metal and a dielectric material. The abscissa represents $c|\boldsymbol{k}|/\omega_{\mathrm{p}}$ for the former two and ck_x/ω_{p} for the latter. We assumed $\varepsilon_\infty = \varepsilon_{\mathrm{b}} = 1$

the radiation field and a resonant polarization in the matter system. Their dispersion relation is obtained as follows. We take the x and z axes as shown in Fig. 7.9, and the y axis in the direction perpendicular to them. We also assume without the loss of generality that the surface wave propagates in the x direction. We denote its wave vector by k_x. Since we can easily show that the electric field of the surface wave does not have a y component, we assume it to have the following form:

$$\boldsymbol{E}(\boldsymbol{r}, t) = \begin{pmatrix} E_{jx} \\ 0 \\ E_{jz} \end{pmatrix} \times \exp\left\{\mathrm{i}\left(k_x x + k_{jz} z - \omega t\right)\right\}, \tag{7.41}$$

where $j = 1$ for the metallic region and 2 for the dielectric region. Substituting this expression into the wave equation, we obtain

$$\omega^2 = \frac{c^2}{\varepsilon_j}\left(k_x^2 + k_{jz}^2\right), \tag{7.42}$$

where ε_j denotes the dielectric constant in region j:

$$\varepsilon_1 = \varepsilon_{\mathrm{m}}(\omega), \tag{7.43}$$

$$\varepsilon_2 = \varepsilon_{\mathrm{b}} \quad \text{(const.)}. \tag{7.44}$$

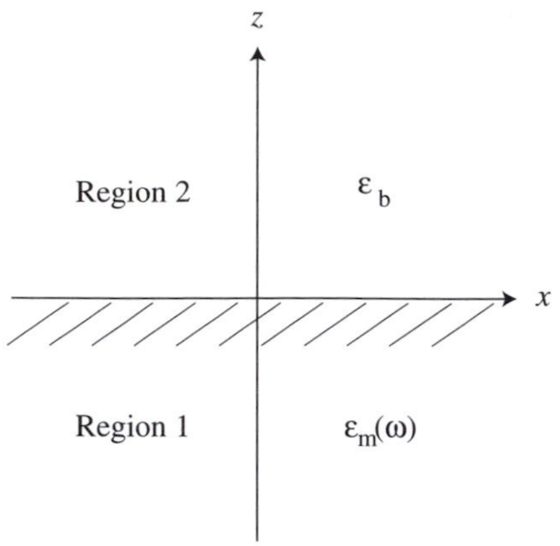

Fig. 7.9. Illustration of a flat interface between a metal and a dielectric material

From $\boldsymbol{\nabla} \cdot \boldsymbol{D} = 0$, we have

$$k_x E_{1x} + k_{1z} E_{1z} = 0, \tag{7.45}$$

$$k_x E_{2x} + k_{2z} E_{2z} = 0. \tag{7.46}$$

The boundary conditions are given by the continuity of E_x and εE_z:

$$E_{1x} = E_{2x}, \tag{7.47}$$

$$\varepsilon_{\mathrm{m}}(\omega) E_{1z} = \varepsilon_{\mathrm{b}} E_{2z}. \tag{7.48}$$

The following determinant should vanish so that these four equations, (7.45)–(7.48), may have a non-trivial solution:

$$\begin{vmatrix} k_x, & k_{1z}, & 0, & 0 \\ 0, & 0, & k_x, & k_{2z} \\ 1, & 0, & -1, & 0 \\ 0, & \varepsilon_{\mathrm{m}}(\omega), & 0, & -\varepsilon_{\mathrm{b}} \end{vmatrix} = 0. \tag{7.49}$$

This equation reduces to

$$\frac{\varepsilon_{\mathrm{m}}(\omega)}{\varepsilon_{\mathrm{b}}} = \frac{k_{1z}}{k_{2z}}. \tag{7.50}$$

We can obtain the dispersion relation of the surface plasmon polariton by solving (7.50) combined with (7.42). We examine some limiting cases here.
(1) $ck_x,\ \omega \ll \omega_{\mathrm{p}}$

We can easily show that the dispersion curve in this range is linear:

$$\omega = \frac{ck_x}{\sqrt{\varepsilon_{\mathrm{b}}}}. \tag{7.51}$$

(2) $\omega \sim \omega_{\mathrm{p}},\ k_x \to \infty$

In this case, the eigen-angular frequency converges to

$$\omega = \sqrt{\frac{\varepsilon_\infty}{\varepsilon_\infty + \varepsilon_{\mathrm{b}}}}\omega_{\mathrm{p}}. \tag{7.52}$$

(3) $\varepsilon_\infty = \varepsilon_{\mathrm{b}} = 1$

The dispersion relation is given by

$$\omega = \sqrt{\frac{2c^2k_x^2 + \omega_{\mathrm{p}}^2 - \sqrt{4c^4k_x^4 + \omega_{\mathrm{p}}^4}}{2}} \tag{7.53}$$

The dispersion relation of the surface plasmon polariton in case (3) is shown by curve (c) in Fig. 7.8. It approaches $\omega_{\mathrm{p}}/\sqrt{2}$ with increasing k_x. The eigenmodes are localized around the interface, since the k_{1z} and k_{2z} are purely imaginary.

7.3.2 Plasmon Resonance on a Metallic Cylinder

As we did in Sect. 7.2, we assume a two-dimensional square array of metallic cylinders. As will be shown in the following sections, the electromagnetic field of the eigenmodes of surface plasmon polaritons are considerably localized around the surface of the metallic cylinders and the overlap of the field between adjacent cylinders is small. This situation resembles the electronic bands composed of well localized atomic orbitals. For the latter case, the tight-binding approximation based on the linear combination of atomic orbitals (LCAO) describes the actual bands well. We thus examine the *atomic orbitals* for the present problem, i.e., the electromagnetic resonant states that originate from the excitation of surface plasmons on a single metallic cylinder that appear for $\omega < \omega_{\mathrm{p}}$.

For this purpose, we assume a plane wave incident on a single cylinder with the magnetic field parallel to the cylinder (z) axis, and study the amplitude of the scattered wave. First we express the magnetic field of the incident plane wave with partial waves using polar coordinates, (r, θ). We take the direction of $\theta = 0$ as the direction of the wave vector $\boldsymbol{k}$ of the incident plane wave. We thus have

$$\begin{aligned} & H_0 \mathrm{e}^{\mathrm{i}(\boldsymbol{k}\cdot\boldsymbol{r}-\omega t)} \\ &= H_0 \mathrm{e}^{\mathrm{i}(kr\cos\theta - \omega t)} \\ &= H_0 \mathrm{e}^{-\mathrm{i}\omega t} \sum_{l=-\infty}^{\infty} J_l(kr) \exp\left\{\mathrm{i}l\left(\theta + \frac{\pi}{2}\right)\right\} \\ &= H_0 \mathrm{e}^{-\mathrm{i}\omega t} \left[J_0(kr) + 2\sum_{l=1}^{\infty} \mathrm{i}^l J_l(kr) \cos l\theta \right], \end{aligned} \tag{7.54}$$

where H_0 is the amplitude of the incident wave, $k = |\boldsymbol{k}|$, and J_l is the Bessel function of the lth order. We used (2.94) to derive (7.54). The eigenvalue equation in the polar coordinate system that describes the magnetic field H_z is given by

$$\frac{1}{\varepsilon}\left(\frac{\partial^2}{\partial r^2} + \frac{1}{r}\frac{\partial}{\partial r} + \frac{1}{r^2}\frac{\partial^2}{\partial \theta^2}\right) H_z = -\frac{\omega^2}{c^2} H_z, \tag{7.55}$$

where ε denotes the position-dependent dielectric constant.

As usual, we express H_z by a product of two functions $R(r)$ and $\Theta(\theta)$. We then obtain a pair of equations for R and Θ:

$$\frac{\partial^2 \Theta}{\partial \theta^2} = -\varepsilon\lambda\Theta, \tag{7.56}$$

$$\left(\frac{\partial^2}{\partial r^2} + \frac{1}{r}\frac{\partial}{\partial r} + \frac{\varepsilon\omega^2}{c^2} - \frac{\varepsilon\lambda}{r^2}\right) R = 0, \tag{7.57}$$

where λ is the parameter for decoupling. Because $\Theta(\theta)$ should be a single-valued function, $\sqrt{\varepsilon\lambda}$ is an integer, which we denote by n. The solutions of (7.56) are thus given by

$$\Theta(\theta) \propto \mathrm{e}^{\mathrm{i}n\theta}, \quad \mathrm{e}^{-\mathrm{i}n\theta}. \tag{7.58}$$

For $r > r_\mathrm{a}$, we define a new variable s by

$$s = kr. \tag{7.59}$$

Then (7.57) is modified to

$$\left(\frac{\partial^2}{\partial s^2} + \frac{1}{s}\frac{\partial}{\partial s} + 1 - \frac{n^2}{s^2}\right) R = 0. \tag{7.60}$$

This is the Bessel equation and its solution is given by the Bessel and Neumann functions, both of which are analytic for $r > r_\mathrm{a}$:

$$R(r) \propto J_n(kr), \quad N_n(kr). \tag{7.61}$$

On the other hand, for $r < r_\mathrm{a}$, the dielectric constant $\varepsilon_\mathrm{m}(\omega)$ is negative for $\omega < \omega_\mathrm{p}$. In this case, we have to take

$$s = \kappa r, \tag{7.62}$$

where

$$\kappa = \sqrt{-\varepsilon_\mathrm{m}(\omega)}\,\frac{\omega}{c}. \tag{7.63}$$

We thus obtain for $r < r_\mathrm{a}$

$$\left[\frac{\partial^2}{\partial s^2} + \frac{1}{s}\frac{\partial}{\partial s} - \left(1 + \frac{n^2}{s^2}\right)\right] R = 0. \tag{7.64}$$

This is the modified Bessel equation and its solution is given by the modified Bessel function of the first kind,

$$R(r) \propto I_n(\kappa r), \tag{7.65}$$

since the modified Bessel function of the second kind is not analytic at $r = 0$. When we take into account that the geometry is symmetric with respect to θ, the total field is given by

$$H_z = \sum_{n=0}^{\infty} A_n I_n(\kappa r) \cos n\theta \quad (r < r_{\mathrm{a}}), \tag{7.66}$$

$$H_z = \sum_{n=0}^{\infty} [B_n J_n(kr) + C_n N_n(kr)] \cos n\theta \quad (r > r_{\mathrm{a}}), \tag{7.67}$$

where A_n, B_n, and C_n are constants that should be determined by the boundary conditions.

The independent boundary conditions at $r = r_{\mathrm{a}}$ are the continuity of H_z and $\varepsilon^{-1}\partial H_z/\partial r$. The latter is equivalent to the continuity of the tangential component of the electric field. From these conditions, we obtain

$$\frac{A_n}{B_n} = \frac{k\varepsilon_{\mathrm{m}}[N_n'(kR)J_n(kR) - N_n(kR)J_n'(kR)]}{k\varepsilon_{\mathrm{m}}N_n'(kR)I_n(\kappa R) - \kappa\varepsilon_{\mathrm{b}}N_n(kR)I_n'(\kappa R)}, \tag{7.68}$$

$$\frac{C_n}{B_n} = -\frac{k\varepsilon_{\mathrm{m}}J_n'(kR)I_n(\kappa R) - \kappa\varepsilon_{\mathrm{b}}J_n(kR)I_n'(\kappa R)}{k\varepsilon_{\mathrm{m}}N_n'(kR)I_n(\kappa R) - \kappa\varepsilon_{\mathrm{b}}N_n(kR)I_n'(\kappa R)}. \tag{7.69}$$

Finally, by comparing (7.54) and (7.67), we obtain

$$B_0 = H_0 \qquad \text{and} \qquad B_n = 2\mathrm{i}^n H_0. \tag{7.70}$$

When we calculate A_n/B_n and C_n/B_n as functions of ω, we observe sharp resonances, which imply the excitation of surface plasmons by the incident plane wave. As an example, Fig. 7.10 shows the ω dependence of A_n/B_n for $n = 1$–5, for which we assumed $\varepsilon_\infty = \varepsilon_{\mathrm{b}} = 1$. A sharp resonance peak is observed for each curve. No resonance is observed with $n = 0$ for this case. The resonance frequencies are listed in Table 7.1. The resonance frequency converges to $\omega_{\mathrm{p}}/\sqrt{2}$ with increasing n irrespective of the ratio $r_{\mathrm{a}} : a$ as is shown in Fig. 7.11. This value is the same as that of the surface plasmon on a flat interface between a metal and a dielectric. These resonant states have finite radiative lifetime and they are not eigenmodes of the system. However, when the metallic cylinders are arrayed in the photonic crystal, the resonant states can be propagated from one cylinder to another to become a Bloch eigenstate yielding an infinite lifetime. Because the resonant states are localized at each cylinder, they may be regarded as atomic orbitals in the theory of LCAO. This feature can be used to classify and predict the symmetry of eigenmodes that originate from the surface plasmons.

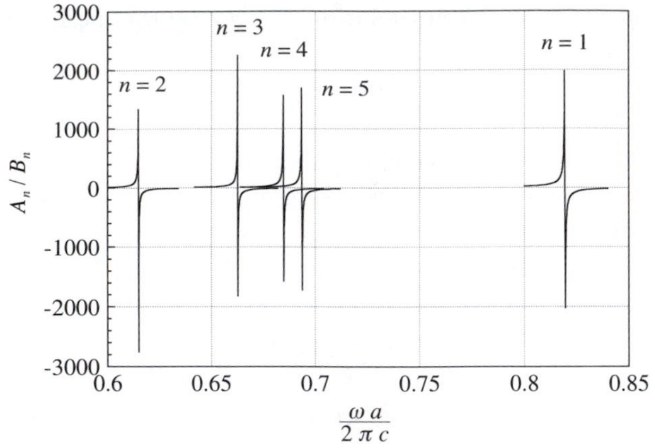

Fig. 7.10. ω dependence of A_n/B_n for small n. The following parameters were assumed: $\varepsilon_\infty = 1.0$, $\omega_\mathrm{p} a/2\pi c = 1.0$, $r_\mathrm{a} = 0.3a$, and $\varepsilon_\mathrm{b} = 1.0$ (air). A sharp resonance that originates from the excitation of the surface plasmon is observed in each curve. (After [80])

Table 7.1. Comparison between the resonance frequencies in the spectra of A_n/B_n and the eigenfrequencies on the Γ point in the unit of $2\pi c/a$. The same parameters as Fig. 7.10 were used for the numerical calculation. (After [80])

n	Resonance frequency	Eigenfrequency
1	0.8194	0.7183
2	0.6149	0.5963, 0.6712
3	0.6627	0.6431
4	0.6849	0.6876, 0.7037
5	0.6937	-

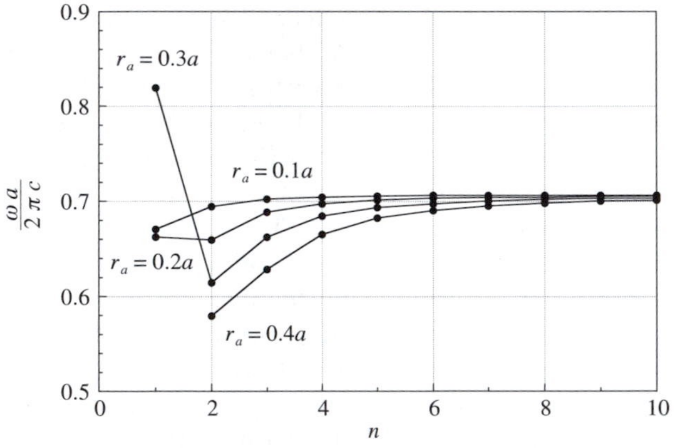

Fig. 7.11. n dependence of the resonance frequency. (After [80])

7.3.3 Symmetry of Plasmon Polaritons

In Sect. 7.2, we examined the photonic bands of two-dimensional metallic lattices for the E polarization. For that case, each dispersion curve in the photonic crystal had its replica in free space. This feature was demonstrated by the comparison of the symmetry of the eigenmodes between the photonic crystal and free space. For the H polarization, the situation is different. Here we also have the modes that originate from the surface plasmons and which do not have replicas in free space. We can classify the symmetry of these eigenmodes by examining the symmetry of the resonant states found in the last section.

First, we examine the Δ point in the two-dimensional Brillouin zone, i.e., $\boldsymbol{k} = (k, 0)$ where $0 < k < \pi/a$. The k group is given by $C_i = \{E, \sigma_y\}$ where E is the identity operation and σ_y is the mirror reflection about the x axis. The eigenfunction at the Δ point is thus either symmetric or antisymmetric about the x axis. On the other hand, the angular (θ) dependence of the resonant state, whose radial variation is described by $I_n(\kappa r)$ for $r < r_\mathrm{a}$, and $J_n(kr)$ and $N_n(kr)$ for $r > r_\mathrm{a}$, is given by $\mathrm{e}^{\pm in\theta}$ for each $n \geq 1$, where we measure θ counterclockwise from the x axis. The symmetric and antisymmetric combinations of these two functions, i.e., $\cos n\theta$ and $\sin n\theta$, possess the appropriate symmetry that the exact eigenfunctions should have. We can thus conclude that the resonant states characterized by index $n \leq 1$ give one symmetric and one antisymmetric mode for the Δ point.

Next, we examine the Γ point, which has the symmetry of the C_{4v} point group. The spatial variation of the resonant states is illustrated schematically in Fig. 7.12. When we compare them with the symmetry of the irreducible representations of C_{4v}, which are shown in Fig. 7.13, the symmetry of the former is easily assigned. The results are also shown in Fig. 7.12. Similar assignments can be made for the X point, which has the symmetry of the

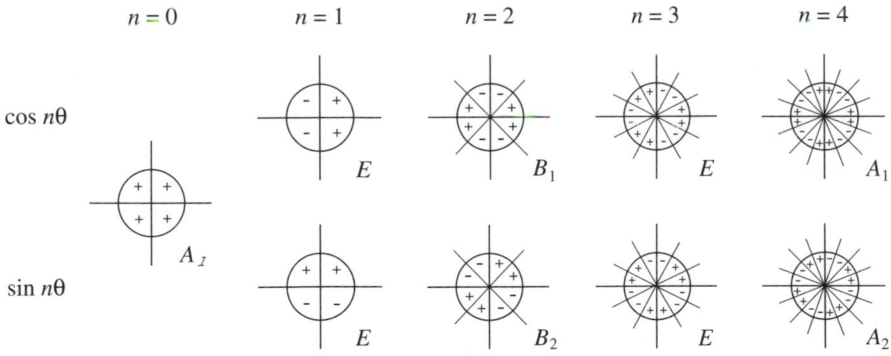

Fig. 7.12. Angular variation of the magnetic field and the symmetry assignment of the resonant states according to the irreducible representations of the C_{4v} point group. (After [80])

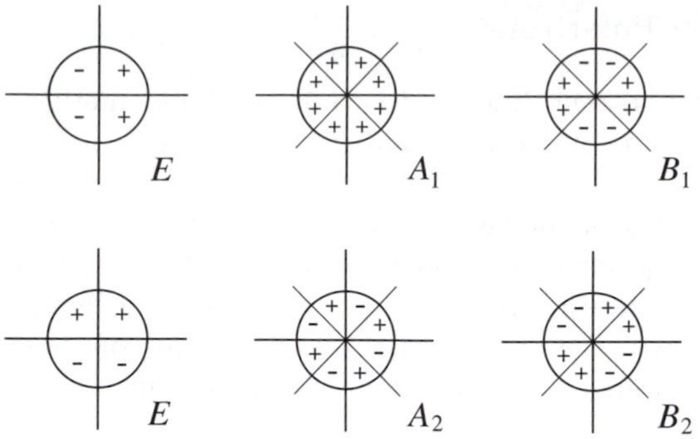

Fig. 7.13. Symmetry of irreducible representations of the C_{4v} point group. (After [80])

C_{2v} point group. All the results are summarized in Table 7.2. Symmetry assignments are presented for $n = 1$–4 in this table. Those for $n' > 4$ can be obtained from the simple relation that the same irreducible representations appear if

$$n' = n + 4j, \tag{7.71}$$

where j is an integer.

Table 7.2. Symmetry of the eigenmodes expected from the LCAO approximation and group theory. (After [80])

n	$\Gamma(C_{4v})$	$\Delta(C_i)$	$X(C_{2v})$
1	E	$A + B$	$B_1 + B_2$
2	$B_1 + B_2$	$A + B$	$A_1 + A_2$
3	E	$A + B$	$B_1 + B_2$
4	$A_1 + A_2$	$A + B$	$A_1 + A_2$

7.3.4 Plasmon Bands in a Square Lattice

The numerical evaluation of the electromagnetic field radiated by an oscillating dipole moment is similar to that for the E polarization. However, in order to obtain a good convergence and accuracy, we have to deal with both the electric and magnetic fields in the discretized Maxwell equations. The reason for this difference between the E polarization and the H polarization is the presence of the spatial differentiation of the discontinuous dielectric function in (2.58), which is absent in (2.57). We thus perform the following replacement in the Maxwell equations.

$$\frac{\partial H_z(x,y,t)}{\partial x} \simeq \frac{H_z(x+\Delta x/2,y,t)-H_z(x-\Delta x/2,y,t)}{\Delta x}, \tag{7.72}$$

$$\frac{\partial H_z(x,y,t)}{\partial y} \simeq \frac{H_z(x,y+\Delta y/2,t)-H_z(x,y-\Delta y/2,t)}{\Delta y}, \tag{7.73}$$

$$\frac{\partial H_z(x,y,t)}{\partial t} \simeq \frac{H_z(x,y,t+\Delta t/2)-H_z(x,y,t-\Delta t/2)}{\Delta t}, \tag{7.74}$$

etc. For the general technique used in numerical calculations by the FDTD method, the reader may consult the text book [59].

Now, Figs. 7.14 and 7.15 shows the calculated band diagram for the square lattice composed of the metallic cylinders. r_a/a was assumed to be 0.3. This figure shows two distinct features: (A) There is no cut-off frequency, i.e., the dispersion curve increases linearly with increasing wave number from $\omega=0$; (B) There are extremely flat bands around $\omega=\omega_\mathrm{p}/\sqrt{2}$.

We first examine item (A). For the H polarization, we have

$$\frac{\partial H_z}{\partial y} = \varepsilon_0\varepsilon\frac{\partial E_x}{\partial t}, \tag{7.75}$$

$$\frac{\partial H_z}{\partial x} = -\varepsilon_0\varepsilon\frac{\partial E_y}{\partial t}, \tag{7.76}$$

$$\frac{\partial E_y}{\partial x} - \frac{\partial E_x}{\partial y} = -\mu_0\frac{\partial H_z}{\partial t}. \tag{7.77}$$

Like the E polarization, we have the following relation in the long-wavelength approximation.

$$\begin{aligned} H_z(\boldsymbol{r}_{/\!/},t) &= v_z(\boldsymbol{r}_{/\!/})\exp\left\{\mathrm{i}\left(\boldsymbol{k}_{/\!/}\cdot\boldsymbol{r}_{/\!/}-\omega t\right)\right\} \\ &\simeq \overline{v}_z\exp\left\{\mathrm{i}\left(\boldsymbol{k}_{/\!/}\cdot\boldsymbol{r}_{/\!/}-\omega t\right)\right\}, \end{aligned} \tag{7.78}$$

etc., where the bar denotes the spatial average over the unit cell as before. When we take the spatial average in (7.75)–(7.77), we have to note that both ε and E_x (E_y) are generally discontinuous on the surface of the metallic cylinder. Hence, we cannot approximate $\overline{\varepsilon E_x}$ by $\overline{\varepsilon}\overline{E}_x$. Instead, we have the following equations.

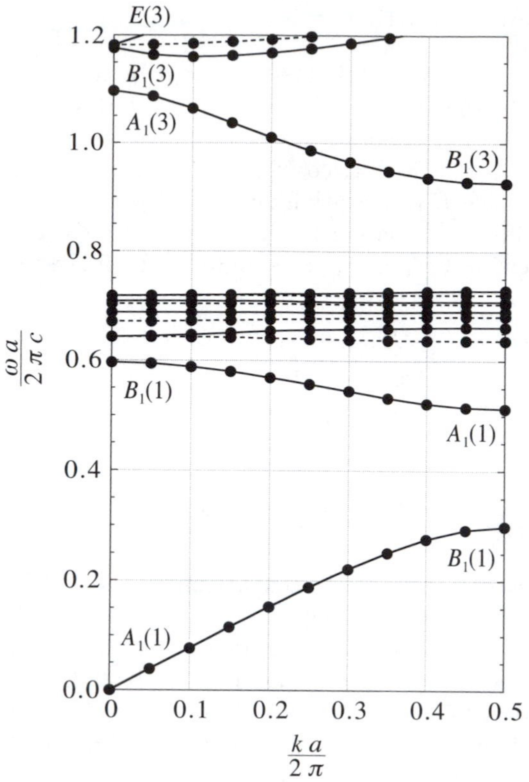

Fig. 7.14. Dispersion relation from the Γ point to the X point for the radiational eigenmodes in the two-dimensional square lattice composed of metallic cylinders calculated for the H polarization. Symmetric and antisymmetric modes are denoted by *solid* and *dashed* lines, respectively. The following parameters were assumed: $r_\mathrm{a}/a = 0.3$, $\omega_\mathrm{p}a/2\pi c = 1.0$, $\gamma = 0.01\omega_\mathrm{p}$, $\varepsilon_\infty = \varepsilon_\mathrm{b} = 1.0$. (After [80])

$$k_y \overline{\left(\frac{1}{\varepsilon}\right)} \overline{v}_z \simeq -\varepsilon_0 \omega \overline{u}_x, \tag{7.79}$$

$$k_x \overline{\left(\frac{1}{\varepsilon}\right)} \overline{v}_z \simeq \varepsilon_0 \omega \overline{u}_y, \tag{7.80}$$

$$k_x \overline{u}_y - k_y \overline{u}_x \simeq \mu_0 \omega \overline{v}_z, \tag{7.81}$$

since H_z is continuous. From these equations, we obtain

$$\frac{\omega^2}{c^2} = \overline{\left(\frac{1}{\varepsilon}\right)} k^2. \tag{7.82}$$

For the dielectric constant of the Drude type, we have

$$\overline{\left(\frac{1}{\varepsilon}\right)} = 1 + \left(\frac{1}{\varepsilon_\infty} - 1\right) f + \frac{f\omega_\mathrm{p}^2}{\varepsilon_\infty(\omega^2 - \omega_\mathrm{p}^2)}. \tag{7.83}$$

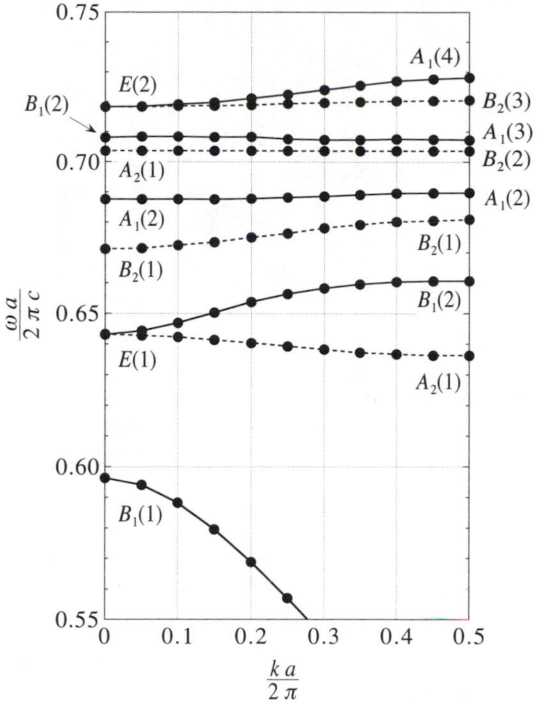

Fig. 7.15. Dispersion relation from the Γ point to the X point. Spectral range $\omega a/2\pi c = 0.55$–0.75 in 7.14 is magnified. (After [80])

From (7.82) and (7.83), we obtain for $k \ll \omega_{\mathrm{p}}/c$

$$\omega \simeq \sqrt{1-f}ck, \tag{7.84}$$

which agrees well with the dispersion relation given in Fig. 7.14.

Proceeding to item (B), the flat bands around $\omega_{\mathrm{p}}/\sqrt{2}$ have no counterparts in free space. They are characteristic of the metallic photonic crystals. Based on the discussion given in Sects. 7.3.1–7.3.3, we may summarize their nature as follows. When an external electric field is applied to the metallic cylinder in the direction perpendicular to its cylinder axis, the positive charges in it move in the direction of the external electric field whereas the negative charges move in the opposite direction. The separation of the positive and negative charges may be regarded as the motion of the solid state plasma, whose characteristic frequency is the plasma frequency ω_{p}. As a result, the induced polarization screens the applied electric field. Thus, the electric field cannot penetrate deeply into the metal as long as the frequency of the applied field is less than ω_{p}. When we consider the eigenmodes of the system, we have to take into account both the genuine electromagnetic field and the polarization field due to the solid state plasma. Their coupled oscillation is called the surface plasmon polariton. When the interface of the metallic region

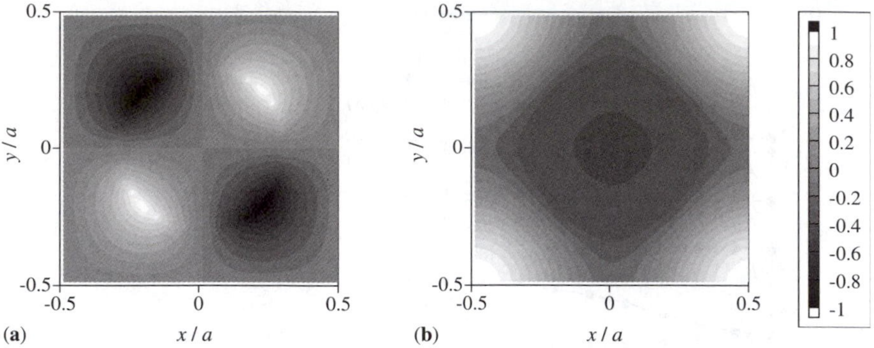

Fig. 7.16. Distribution of the magnetic field for (**a**) the $B_2(1)$ mode at the Γ point, which has an extremely small bandwidth, and (**b**) the $A_1(3)$ mode at the Γ point. The magnetic field, and hence the electric field of the former, is also localized at the surface of the metallic cylinder. These two features are characteristic to the surface plasmon polaritons

and air is an infinite flat plane, we can show that the dispersion curve of the surface plasmon polariton converges from the low frequency side to $\omega_p/\sqrt{2}$ when the magnitude of the wave vector increases. As for the case of one metallic cylinder, we have no genuine eigenmodes. However, we have a series of resonant states characterized by an index n that describes their *rotation* or their *angular momentum*. Strictly speaking, n is the order of the Bessel and the modified Bessel functions that describe the radial variation of the resonant state out of or in the metallic cylinder. Their resonance frequencies also approach $\omega_p/\sqrt{2}$ when n tends to infinity. When the individual cylinders gather to form the regular array, the resonant states can be propagated from one cylinder to another, and thus form the Bloch states, which are eigenmodes of the system and have infinite lifetime if we neglect the imaginary part of the metallic dielectric constant.

The electromagnetic field accompanied by the surface plasmon polariton is localized at the surface of the cylinder, which is inherent in the plasma oscillation. For example, Fig. 7.16(a) shows the magnetic field of the $B_2(1)$ mode at the Γ point that has the flat dispersion and is regarded as the surface plasmon polariton with $n = 2$. Its magnetic field is really localized at the surface of the metallic cylinder, which is apparent when we compare it with that of the $A_1(3)$ mode (Fig. 7.16(b)), which is an ordinary band and has a counterpart band in free space. Because of their localized nature, the overlapping of the electromagnetic field of the resonant states on adjacent cylinders is small, which results in the small bandwidth of the polariton bands. Most of the polariton bands are characterized by a single n and the hybridization between resonant states with different n is relatively small. As a result, the eigenfrequencies of the surface plasmon polaritons are fairly close to the corresponding resonance frequencies, as shown in Table 7.1.

It is expected that the number of polariton bands is infinite, since there are an infinite number of resonant states characterized by n. However, the separation of the eigenfrequencies becomes small when n tends to infinity, and we cannot distinguish them in the spectrum of the dipole radiation. Hence, we plotted only the first nine polariton bands in Figs. 7.14 and 7.15. In the actual metallic system, its exact dielectric constant is not of the Drude type. For example, we have to take into account the spatial dispersion, i.e., the $\boldsymbol{k}$ dependence of the dielectric constant when $\boldsymbol{k}$, and hence n, is large. So in practice, the dispersion curves with small n are relevant to the real system, and we have to go beyond the present treatment if we want to discuss other features.

8. Photonic Crystal Slabs

The radiational eigenmodes in photonic crystal slabs are classified into guided and leaky modes. The former are genuine eigenmodes with infinite lifetimes, while the latter are quasi-eigenmodes with finite lifetimes. We will show that the dispersion curves of photonic crystal slabs are well represented by the folding of those in uniform dielectric slabs into the two-dimensional Brillouin zone. This feature enables us to predict the symmetries of the eigenmodes by group theory. The optical transmission due to the leaky modes in the direction parallel to the slab surface is mainly governed by their diffraction loss, which is a particular feature of photonic crystal slabs. We will show that the diffraction loss is suppressed by symmetry mismatching between internal eigenmodes and the external radiation field [81, 82].

8.1 Eigenmodes of Uniform Slabs

Before we treat photonic crystal slabs, we examine the radiational eigenmodes in uniform dielectric slabs. We assume a geometry shown in Fig. 8.1, where ε_b and d denote the dielectric constant and the thickness of the slab. We assume that the structure is infinite in the x and y directions and surrounded by air in the z direction. We take the x-y plane in the middle of the slab. Then the structure has mirror symmetry about the x-y plane, $\hat{\sigma}_z$.[1]

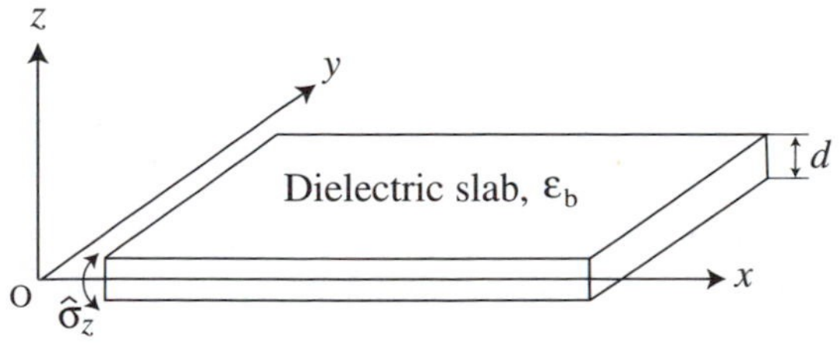

Fig. 8.1. Illustration of a uniform dielectric slab

[1] We denote symmetry operations by hats to distinguish them from their eigenvalues in this chapter.

Guided radiational modes that are confined in a uniform slab are classified into four categories according to the symmetry for the mirror reflection $\widehat{\sigma}_z$ and to their polarizations. Those modes whose electric fields lie in the x-y plane are referred to as transverse electric (TE) modes, whereas those modes whose magnetic fields lie in the x-y plane are referred to as transverse magnetic (TM) modes. Each mode is also characterized by the wave number $k_{/\!/}$ in the x-y plane.

Now, we derive their dispersion relation. We begin with the symmetric TE mode. We assume without the loss of generality that the mode propagates in the x direction. Hence, we have the following form for the electric field in the slab:

$$\boldsymbol{E}_1 = \begin{pmatrix} 0 \\ E_{1y} \\ 0 \end{pmatrix} \exp\left\{\mathrm{i}\left(k_{/\!/}x - \omega t\right)\right\} \cos\left(k_z z\right). \tag{8.1}$$

The z component of the wave vector in the slab, k_z, is related to $k_{/\!/}$ and ω by

$$\omega^2 = \frac{c^2}{\varepsilon_{\mathrm{b}}}\left(k_{/\!/}^2 + k_z^2\right), \tag{8.2}$$

which is derived from the wave equation. From $\boldsymbol{\nabla} \times \boldsymbol{E} = -\partial \boldsymbol{B}/\partial t$, the magnetic field is given by

$$\boldsymbol{H}_1 = \frac{E_{1y}}{\mathrm{i}\omega\mu_0} \begin{pmatrix} k_z \sin k_z z \\ 0 \\ \mathrm{i}k_{/\!/} \cos k_z z \end{pmatrix} \exp\left\{\mathrm{i}\left(k_{/\!/}x - \omega t\right)\right\}. \tag{8.3}$$

The electric field for $z > d/2$ is expressed as

$$\boldsymbol{E}_2 = \begin{pmatrix} 0 \\ E_{2y} \\ 0 \end{pmatrix} \exp\left(\mathrm{i}k_{/\!/}x - \kappa z - \mathrm{i}\omega t\right), \tag{8.4}$$

where κ is related to $k_{/\!/}$ and ω by

$$\omega^2 = c^2\left(k_{/\!/}^2 - \kappa^2\right), \qquad \kappa > 0. \tag{8.5}$$

Since we are dealing with guided modes, we assumed an exponential decrease for the electric field in the air region in (8.4). The magnetic field for $z > d/2$ is given by

$$\boldsymbol{H}_2 = \frac{E_{2y}}{\mathrm{i}\omega\mu_0} \begin{pmatrix} \kappa \\ 0 \\ \mathrm{i}k_{/\!/} \end{pmatrix} \exp\left(\mathrm{i}k_{/\!/}x - \kappa z - \mathrm{i}\omega t\right). \tag{8.6}$$

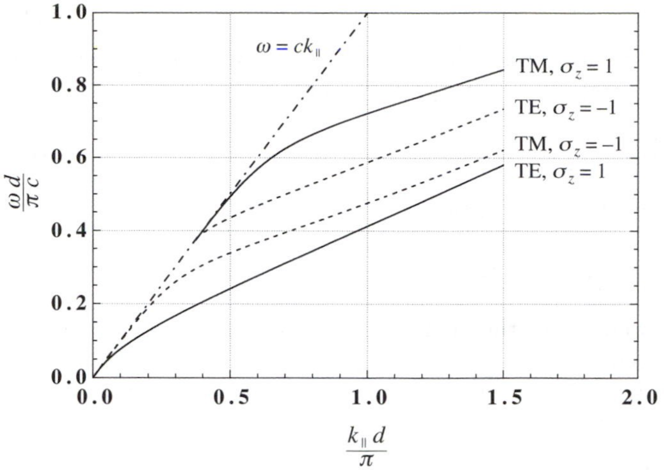

Fig. 8.2. Dispersion curves of the lowest four guided modes in a uniform slab with refractive index 2.86, which is the spatially averaged refractive index of the structure shown in Fig. 8.3. The ordinate is the normalized frequency, where d denotes the thickness of the dielectric slab. The abscissa denotes the normalized in-plane wave number. *Solid lines* represent the dispersion relations of modes with $\sigma_z = 1$ whose in-plane components of the electric field are symmetric about the x-y plane, whereas *broken lines* represent those of modes with $\sigma_z = -1$ whose in-plane electric fields are antisymmetric. The dash-dotted line is referred to as the *light line* and shows the dispersion relation in free space. (After [82])

From the continuity of the tangential components of the electric and magnetic fields at $z = d/2$, we have

$$E_{1y} \cos\left(\frac{k_z z}{2}\right) = E_{2y} \exp\left(-\frac{\kappa z}{2}\right), \tag{8.7}$$

$$k_z E_{1y} \sin\left(\frac{k_z z}{2}\right) = \kappa E_{2y} \exp\left(-\frac{\kappa z}{2}\right). \tag{8.8}$$

As for the continuity of the normal component of the magnetic induction, it does not give an independent boundary condition for this problem. The determinant of coefficients should vanish so that these two equations, (8.7) and (8.8), may have a non-trivial solution:[2]

$$\kappa \cos\left(\frac{k_z d}{2}\right) - k_z \sin\left(\frac{k_z d}{2}\right) = 0 \qquad (\text{TE}, \ \sigma_z = 1). \tag{8.9}$$

The solution of this equation gives the dispersion relation for the symmetric TE modes. The dispersion curve consists of an infinite number of branches. The lowest one is shown in Fig. 8.2.

For the eigenmodes to be localized on the slab, κ must be real and positive. Thus, the limiting case is given by $\kappa = 0$. For this case, from (8.5) we have

[2] σ_z represents the eigenvalue of $\widehat{\sigma}_z$.

$$\omega = ck_{/\!/}. \tag{8.10}$$

This relation is plotted with a dash-dotted line in Fig. 8.2 and is referred to as a *light line*. Substituting $\kappa = 0$ into (8.9), we obtain

$$\sin\left(\frac{k_z d}{2}\right) = 0. \tag{8.11}$$

From this equation and (8.2), we obtain

$$\omega = \frac{2j\pi c}{d\sqrt{\varepsilon_\mathrm{b} - 1}}, \qquad (j = 0,\ 1,\ 2\ \cdots). \tag{8.12}$$

Hence, the dispersion curves of the symmetric TE modes intersect the *light line* at frequencies given above.

Now, we proceed to antisymmetric TE modes. For this case, the electric field in the slab is given by

$$\boldsymbol{E}_1 = \begin{pmatrix} 0 \\ E_{1y} \\ 0 \end{pmatrix} \exp\left\{\mathrm{i}\left(k_{/\!/}x - \omega t\right)\right\} \sin\left(k_z z\right), \tag{8.13}$$

while the magnetic field in the slab is given by

$$\boldsymbol{H}_1 = \frac{E_{1y}}{\mathrm{i}\omega\mu_0} \begin{pmatrix} -k_z \cos k_z z \\ 0 \\ \mathrm{i}k_{/\!/} \sin k_z z \end{pmatrix} \exp\left\{\mathrm{i}\left(k_{/\!/}x - \omega t\right)\right\}. \tag{8.14}$$

The electric and magnetic fields for $z > d/2$ are given by (8.4) and (8.6) as before. From the boundary conditions at $z = d/2$, we obtain the following equation, which gives the dispersion relation for the antisymmetric TE modes:

$$\kappa \sin\left(\frac{k_z d}{2}\right) + k_z \cos\left(\frac{k_z d}{2}\right) = 0 \qquad (\mathrm{TE},\ \sigma_z = -1). \tag{8.15}$$

Now, we examine where the dispersion curves intersect the *light line*. Substituting $\kappa = 0$ into (8.15), we obtain

$$\omega = \frac{(2j+1)\pi c}{d\sqrt{\varepsilon - 1}}, \qquad (j = 0,\ 1,\ 2\ \cdots). \tag{8.16}$$

In particular, the lowest dispersion curve for the antisymmetric TE modes, which is shown in Fig. 8.2, starts from a non-zero value of ω. That is, it has an infrared cut-off.

The dispersion relations of TM modes can be obtained similarly by exchanging the roles of the electric and magnetic fields:

$$\varepsilon\kappa \cos\left(\frac{k_z d}{2}\right) - k_z \sin\left(\frac{k_z d}{2}\right) = 0 \qquad (\mathrm{TM},\ \sigma_z = -1), \tag{8.17}$$

$$\varepsilon\kappa \sin\left(\frac{k_z d}{2}\right) + k_z \cos\left(\frac{k_z d}{2}\right) = 0 \qquad (\mathrm{TM},\ \sigma_z = 1). \tag{8.18}$$

Their lowest branches are also shown in Fig. 8.2.

If the eigenfrequency lies below the light line, the spatial decay constant, κ, is real, and thus, the eigenmode is confined in the slab. In other words, it is a guided mode. If the eigenfrequency lies above the light line, κ is purely imaginary, and the radiation field escapes from the slab. In this case, the radiational mode is not a real eigenmode, but is a resonant state with a complex ω. Only the guided modes are plotted in Fig. 8.2.

One of the important features of the radiational bands in a uniform slab is that only the lowest two bands start from $\omega = 0$, whereas higher bands have infrared cut-off frequencies. Therefore, when we deal with a low frequency region as we will do in the following sections, it is sufficient to take the lowest two bands into account. In particular, this treatment is justified for photonic crystal slabs with small thickness, for which the cut-off frequencies are high. As we will see in the next section, the dispersion relation of modes with $\sigma_z = 1$ can actually be well approximated for $\omega a/2\pi c \leq 0.5$ by the folding of the lowest dispersion curve into the two-dimensional Brillouin zone, provided that we take into consideration the mixing and the frequency splitting where more than one band with the same symmetry crosses each other. We should also note that the mixing of the TE and TM modes with the same σ_z takes place in the photonic crystal slabs.

8.2 Symmetry of Eigenmodes

Figure 8.3 shows the two-dimensional photonic crystal slab that we deal with in the rest of this chapter. We assume that it consists of the regular hexagonal array of cylindrical holes fabricated in a dielectric slab with refractive index 3.4 (GaAs). The right-hand side of the figure shows the top view of the configuration of several cylindrical holes. The lattice constant, the radius of

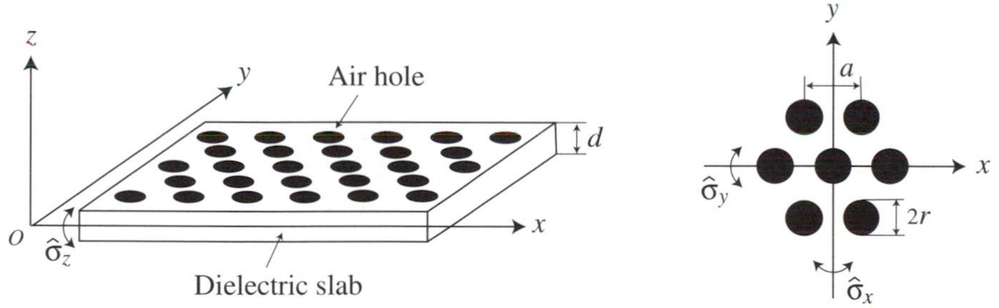

Fig. 8.3. *Left*: Schematic illustration of a two-dimensional photonic crystal slab. *Right*: Top view of the configuration of several air holes, where a and r denote the lattice constant and the radius of the air holes, respectively. For numerical calculation, the following parameters were assumed: The refractive index of the dielectric slab = 3.4, $d = 0.5a$, and $r = 0.25a$. (After [82])

the cylindrical holes, and the thickness of the slab are denoted by a, r, and d, respectively. We assume that the structure is periodic and infinite in the x and y directions, and surrounded by air in the z direction. Photonic crystal slabs of this kind are referred to as the air-bridge type. We take the x-y plane in the middle of the slab as before.

The structure has D_{6h} symmetry, which is a direct product of the C_{6v} and C_{1h} point groups:

$$D_{6h} = C_{6v} \times C_{1h}. \tag{8.19}$$

C_{1h} consists of the identity operation $\widehat{E}$ and the mirror reflection by the x-y plane, $\widehat{\sigma}_z$. Thus any eigenmode of the radiation field should be symmetric ($\sigma_z = 1$) or antisymmetric ($\sigma_z = -1$) about the x-y plane. In order to avoid unnecessary complexity, let us restrict our discussion to the symmetric modes in what follows. The antisymmetric modes can be treated in a similar manner.

We first examine what kind of radiational eigenmodes are expected to appear by the group-theoretical argument. The key idea is the reduction of the reducible representations given by the linear combination of unperturbed eigenfunctions. For the case of two-dimensional photonic crystals with infinite thickness and general three-dimensional photonic crystals, plane waves in free space were used as the unperturbed eigennfunctions in Chap. 3. The group-theoretical prediction was satisfactory in the low frequency range. As for photonic crystals composed of metallic cylinders, the plasmon resonance states were used for this purpose in Chap. 7. For the present problem, the guided modes in a uniform slab with a spatially averaged dielectric constant can be used as the unperturbed eigenfunctions.

Table 8.1. Irreducible representations for the electric field of the guided modes in the uniform slab, whose wave vector, $k_{/\!/}$, is reduced in the two-dimensional first Brillouin zone of the hexagonal lattice that is shown in Fig. 3.6. The irreducible representations were calculated for the lowest TE mode with $\sigma_z = 1$ and the lowest TM mode with $\sigma_z = -1$. (After [82])

Symmetry	Point	TE, $\sigma_z = 1$	TM, $\sigma_z = -1$
C_{6v}	Γ_1	$A_2 + B_1 + E_1 + E_2$	$A_1 + B_2 + E_1 + E_2$
	Γ_2	$A_2 + B_2 + E_1 + E_2$	$A_1 + B_1 + E_1 + E_2$
C_{3v}	K_0	$A_2 + E$	$A_1 + E$
	K_1	$A_2 + E$	$A_1 + E$
	K_2	$A_1 + A_2 + 2E$	$A_1 + A_2 + 2E$
C_{2v}	M_0	$A_2 + B_1$	$A_1 + B_2$
	M_1	$A_2 + B_2$	$A_1 + B_1$
	M_2	$A_1 + A_2 + B_1 + B_2$	$A_1 + A_2 + B_1 + B_2$

Table 8.1 summarizes the results of the symmetry assignment by the folding of the dispersion curves of the lowest TE and TM modes into the first Brillouin zone. The irreducible representations of the electric field obtained by the reduction procedure are listed for three highly symmetric points, that is, the Γ, K, and M points. (For notations of the wave vectors, see Fig. 3.6 and Table 3.11.)

8.3 Photonic Band Structure and Transmission Spectra

The photonic band structure and the transmission spectra for ten lattice layers calculated by the FDTD method are shown in Fig. 8.4 for the Γ-K direction and Fig. 8.5 for the Γ-M direction. The latter was calculated with the lowest TE mode of a uniform slab with refractive index 3.4 as an incident wave.[3] Since the Σ point is invariant under the mirror reflection $\widehat{\sigma}_x$ that is

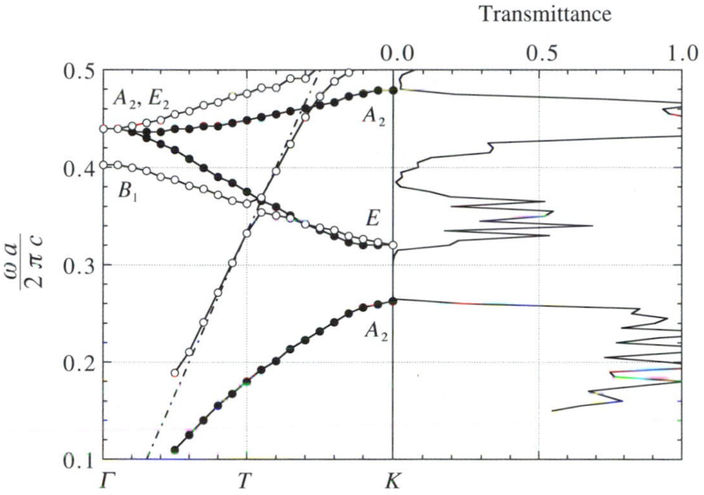

Fig. 8.4. Photonic band structure (*left-hand side*) and optical transmittance (*right-hand side*) in the Γ-K direction. The ordinate is the normalized frequency, where a denotes the lattice constant. For numerical calculation, those parameters that are listed in the caption of Fig. 8.3 were used. In the band diagram, *solid circles* represent the odd (B) modes with $\sigma_y = -1$, whereas *open circles* represent the even (A) modes with $\sigma_y = 1$. The irreducible representations of the k-groups for the electric field are also shown for the Γ and K points. The transmittance was calculated for ten lattice layers with the lowest TE mode in the uniform slab with refractive index 3.4 as an incident wave. Thus, the incident wave has the symmetry of $\sigma_z = 1$ and $\sigma_y = -1$. The even (A) modes do not contribute to the optical transmission, since they do not couple to the incident wave because of the symmetry mismatching. (After [82])

[3] The reader may consult [81, 82] for the details of the numerical calculation.

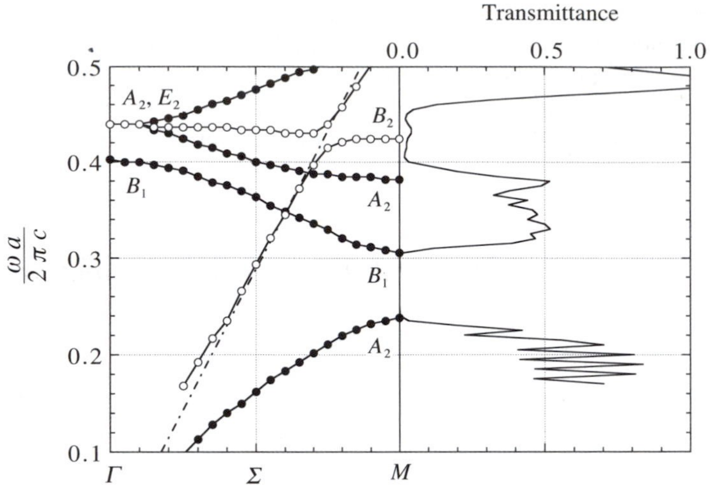

Fig. 8.5. Photonic band structure (*left-hand side*) and optical transmittance (*right-hand side*) in the Γ-M direction. The same parameters as in Fig. 8.4 were assumed for numerical calculation. In the band diagram, *solid circles* represent the odd (B) modes with $\sigma_x = -1$, whereas *open circles* represent the even (A) modes with $\sigma_x = 1$. The irreducible representations of the k-groups for the electric field are also shown for the Γ and M points. The even (A) modes do not contribute to the optical transmission because of the symmetry mismatching. (After [82])

illustrated in Fig. 8.3, its eigenmodes are classified into even ($\sigma_x = 1$) and odd ($\sigma_x = -1$) modes. The former are also referred to as A modes, whereas the latter are referred to as B modes. The same holds for the T point when we take into consideration the symmetry for the mirror reflection $\widehat{\sigma}_y$ instead of $\widehat{\sigma}_x$. In the band diagrams, solid and open circles represent the odd (B) and even (A) modes, respectively. The light line is represented by dash-dotted lines. The symmetries of the eigenmodes on the Γ, K, and M points obtained by the numerical calculation are consistent with the group-theoretical prediction that is given in Table 8.1. They are also consistent, as they should be, with the compatibility relation listed in Table 3.10. Such modes that do not match the group-theoretical prediction based on the folding of the lowest TE mode appear at $\omega a/2\pi c \geq 0.57$.

It is known that modes with $\sigma_z = 1$ tend to have a photonic bandgap below the light line. We actually observe a bandgap at $\omega a/2\pi c = 0.26$–0.31. However, taking into account the leaky A band that exists just above the light line in the same frequency region, this is not a true gap. This leaky band is of TM origin, and the corresponding unperturbed band with complex eigenfrequencies is obtained by the extrapolation of the dispersion curve shown in Fig. 8.2 beyond the infrared cut-off.

Now, we proceed to the transmission spectra. The lowest TE mode of the uniform slab, which was assumed for the incident wave, is odd about the $\widehat{\sigma}_x$ or $\widehat{\sigma}_y$ mirror reflection. Thus the even modes in the photonic crystal do not contribute to the optical transmission, since they do not couple to the incident wave because of the symmetry mismatching. This feature is clearly observed at $\omega a/2\pi c = 0.26$–0.32 and 0.48–0.50 for the Γ-K direction, and $\omega a/2\pi c = 0.24$–0.30 for the Γ-M direction. In these frequency ranges, there is no odd mode and the transmittance is extremely small. The lowest transmittance is less than 10^{-3}. On the other hand, the transmittance is also small even when there is an odd mode if it is leaky, i.e., if it is located above the light line. This is because the incident wave is diffracted into the air region, and the electromagnetic energy transmitted in the Γ-K or Γ-M direction becomes small. This feature is marked when the lifetime and/or the group velocity of the eigenmodes are small. The low transmittance of this kind is observed at $\omega a/2\pi c = 0.38$–0.41 for the Γ-K direction and $\omega a/2\pi c = 0.40$–0.45 for the $\Gamma - M$ direction. On the other hand, the transmittance is high when their lifetime is long. The high transmittance of this kind is observed around $\omega a/2\pi c = 0.45$ for the Γ-K direction and around $\omega a/2\pi c = 0.48$ for the Γ-M direction.

Let us conclude this section with the following. As was mentioned above, the transmittance may be low in two cases: (1) when there is no symmetry-matched mode and (2) when the lifetime of the eigenmode is short. It may also be low in the high-frequency region (3) when Bragg diffraction takes place in the x-y plane. These facts imply that the frequency regions with low transmittance do not necessarily correspond to photonic bandgaps. So, we must be careful when we compare the transmission spectra obtained by experimental observation and the band diagrams.

8.4 Quality Factor

As we saw in the last section, lifetime is an important quantity that characterizes the basic optical properties of leaky modes. It can be evaluated by examining their temporal decay as we did in the previous chapters. When we excite the leaky mode by a dipole moment oscillating at its eigenfrequency and observe its decay after switching off the oscillation, the accumulated electromagnetic energy $U(t)$ decreases with time as

$$U(t) = U(t_0) \exp\left[-\frac{\omega(t-t_0)}{Q}\right], \tag{8.20}$$

where t_0 stands for the switch-off time and Q is the quality factor of the leaky mode. In Fig. 8.6, the calculated Q^{-1} is plotted for the relevant bands. Solid and open circles represent the odd (B) and even (A) modes, respectively. Q^{-1} is equal to zero, or in other words, the lifetime is infinite near the K and M points, since the dispersion curves are located below the light line in these regions. A remarkable feature of this figure is that Q^{-1} is extremely small

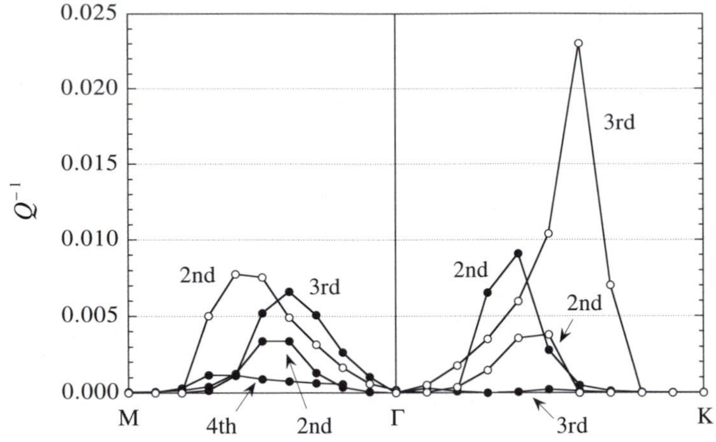

Fig. 8.6. Inverse of the quality factor (Q^{-1}) of several bands. *Solid* and *open circles* represent the odd (B) and even (A) modes, respectively. (After [82])

at the Γ point, where the modes shown in Fig. 8.6 have the B_1, A_2, and E_2 symmetries. As we shall see, this phenomenon originates from the symmetry mismatching with the diffracted radiational field in free space. Namely, the coupling between the internal eigenmodes and the external radiation field is forbidden by symmetry, and the lifetime, and thus Q, are infinite.

We examine the diffraction process of a mode with wave vector $\boldsymbol{k}_{/\!/}$. Because of the conservation of the momentum in the x-y plane, the wave vector of a diffracted wave, $\boldsymbol{k}$, is generally given by

$$\boldsymbol{k} = \boldsymbol{k}_{/\!/} + \boldsymbol{G}_{/\!/} + k_z \boldsymbol{e}_z, \tag{8.21}$$

where $\boldsymbol{G}_{/\!/}$ and $\boldsymbol{e}_z$ are a reciprocal lattice vector of the two-dimensional hexagonal structure and the unit vector in the z direction, respectively. k_z is determined by the dispersion relation in free space and given by

$$k_z = \sqrt{\frac{\omega^2}{c^2} - |\boldsymbol{k}_{/\!/} + \boldsymbol{G}_{/\!/}|^2}. \tag{8.22}$$

It is easy to show that when $\omega a/2\pi c$ is less than $2/\sqrt{3}$, $\boldsymbol{G}_{/\!/}$ should be equal to zero for k_z to be real for the Γ point, i.e., for $\boldsymbol{k}_{/\!/} = 0$. Hence, the diffracted waves are characterized by just one wave vector $\boldsymbol{k}_0$ given by

$$\boldsymbol{k}_0 = \frac{\omega}{c} \boldsymbol{e}_z. \tag{8.23}$$

We shall refer to these waves as the diffracted waves of the zeroth order. They have two polarization components as shown in Fig. 8.7, where $\boldsymbol{e}_x$ and $\boldsymbol{e}_y$ are the unit vectors in the x and y directions. These two waves, of course, have the same frequency. They are a degenerate pair of a two-dimensional irreducible representation of the C_{6v} point group. We can examine how they

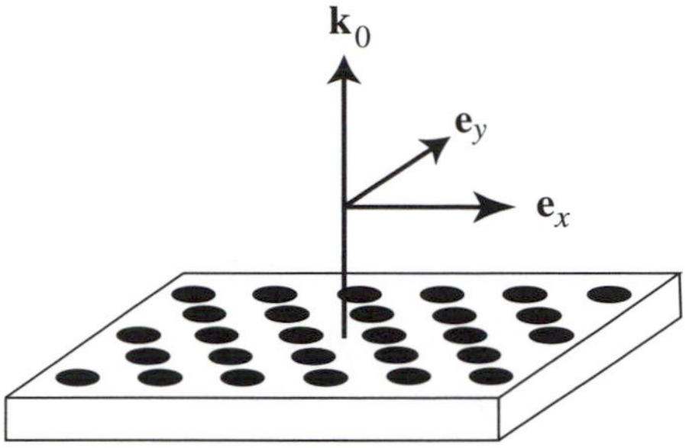

Fig. 8.7. Schematic illustration of the diffracted plane waves of the zeroth order for the Γ point. $\boldsymbol{k}_0$ denotes their wave vector. $\boldsymbol{e}_x$ and $\boldsymbol{e}_y$ denote their two independent polarizations. These two plane waves form the basis of the E_1 irreducible representation of the C_{6v} point group. See text for details. (After [82])

are transformed by symmetry operations of the C_{6v} point group. For example, Their transformation by $\widehat{\sigma}_x$ and $\widehat{\sigma}_y$ is

$$\widehat{\sigma}_x \begin{pmatrix} \boldsymbol{e}_x \\ \boldsymbol{e}_y \end{pmatrix} = \begin{pmatrix} -\boldsymbol{e}_x \\ \boldsymbol{e}_y \end{pmatrix}, \tag{8.24}$$

$$\widehat{\sigma}_y \begin{pmatrix} \boldsymbol{e}_x \\ \boldsymbol{e}_y \end{pmatrix} = \begin{pmatrix} \boldsymbol{e}_x \\ -\boldsymbol{e}_y \end{pmatrix}. \tag{8.25}$$

By comparing their character with the character table for the C_{6v} point group, we can verify that they are a basis set of the E_1 irreducible representation. Because of the mismatching of the spatial symmetry, they can only couple to the E_1 modes in the hexagonal photonic crystal. This is the reason why the B_1, A_2, and E_2 modes at the Γ point shown in Fig. 8.6 do not couple to the radiation field in the air region and have the infinite quality factor. When the wave vector is near the Γ point, the coupling and Q^{-1} are still small although they are not exactly equal to zero.

The absence of diffraction by symmetry mismatching was first reported for a square lattice by Paddon and Young [83]. They found that only the E mode at the Γ point was diffracted and had a finite lifetime. We can show, by a similar argument to the one used for the hexagonal lattice, that the possible diffracted waves in the low-frequency region ($\omega a/2\pi c < 1$) have the E symmetry for the Γ point in the square lattice. Therefore, only the E modes in the photonic crystal slab couple to the external radiation field. The coupling for all other modes is forbidden by symmetry and they have infinite lifetimes. The absence of the coupling discussed here originates from the structural symmetry in the two-dimensional plane. Thus, it can also be observed in those specimens without mirror symmetry about the x-y plane.

Another remarkable feature of Fig. 8.6 is the presence of leaky bands with a very large quality factor. They are the third lowest odd band in the Γ-K direction and the fourth lowest odd band in the Γ-M direction. The quality factor of the former is greater than 3000 everywhere between the Γ and K points. The large quality factors of these two bands result in the high transmittance around $\omega a/2\pi c = 0.45$ in the Γ-K direction and around $\omega a/2\pi c = 0.48$ in the Γ-M direction.

Finally, we give a qualitative estimation of the transmittance. The flow of the radiational energy is described by the group velocity v_{g}. When we denote the propagation length by L, the time necessary for the propagation is equal to L/v_{g}. The damping factor of the leaky mode is thus given by $\exp(-L\omega/Qv_{\mathrm{g}})$, to which the transmittance is proportional. As an example, let us examine the third lowest odd band in the Γ-K direction. Its group velocity is about $c/20$ at the middle of the band. If we assume that $L = 10a$, $Q = 3000$, and $\omega a/2\pi c = 0.45$, its damping factor is as large as 0.83. This is the origin of the high transmittance at this frequency. The third lowest odd mode may travel more than $50a$ in its lifetime. Within this length, the leaky mode may be regarded as a guided mode.

9. Low-Threshold Lasing Due to Group-Velocity Anomaly

The enhancement of stimulated emission due to the small group velocity is evaluated by examining the light amplification spectra for a two-dimensional (2D) crystal. An analytical expression for the lasing threshold is derived and compared with numerical results that show a large reduction of the threshold due to the group-velocity anomaly.

9.1 Enhanced Stimulated Emission

In Chap. 5, we showed that various optical processes in photonic crystals are enhanced by the small group velocity of the eigenmodes. Particularly, we derived the expression for the amplitude amplification factor in a photonic crystal with impurity atoms that have population inversion. As (5.50) shows, the amplitude amplification factor in the unit length is proportional to $v_{\rm g}^{-1}$. Hence, we can expect a large enhancement of stimulated emission at the photonic band edges where $v_{\rm g}$ is equal to zero. Also we can expect the enhancement for those eigenmodes with the group-velocity anomaly in the 2D and 3D crystals. In this chapter, we will show that the group-velocity anomaly is more efficient than the small group velocity at the band edges for crystals with relatively small number of lattice layers. For this purpose, we will examine the light amplification spectra of a 2D crystal in this section and its lasing threshold in the next section.

In Chap. 4, we derived the plane-wave expansion method to calculate the transmission and reflection spectra of 2D photonic crystals and applied it to the square and hexagonal lattices. For those calculations, we assumed real dielectric constants. When impurity atoms with population inversion are distributed in the crystal, the imaginary part of the dielectric constant is no longer equal to zero, and has a negative value, as described in Sect. 5.3. The plane-wave expansion method can be used for this case without any change. Thus, we shall obtain the transmission spectra for the case that the light amplification due to the stimulated emission is present, which we will refer to as the light amplification spectra.

We now examine the light amplification in the 2D square crystal whose band structure was shown in Fig. 3.4(a). In Sect. 4.4.1, we examined its transmission spectrum in the Γ-X direction in the absence of the imaginary part of the dielectric constant (Fig. 4.3). We found that the third (and fourth) lowest band on the Δ point has the group-velocity anomaly and its effective refractive index estimated from the interference patterns in the spectrum was large and consistent with the small group velocity of the band.

Because we are interested in the influence of the photonic band structure on the stimulated emission and the nature of the impurity atoms is irrelevant to the following discussion, we assume that the polarizability of the impurity atoms is independent of ω. In addition, we assume for simplicity that the impurity atoms are uniformly distributed in the dielectric material.

We now proceed to the quantitative evaluation of the enhancement of stimulated emission. In the following calculation, the dielectric constant of the host material was assumed to be 2.1 – 0.01i. Figure 9.1 shows the sum of the transmittance and the reflectance for an incident wave with the E polarization propagated in the Γ-X direction where each solid line with filled circles denotes that of the 2D crystal and each dashed line denotes that of a uniform plate of the same thickness with a spatially averaged dielectric constant. The latter was calculated for comparison. The calculation was performed for crystals with (a) 16, (b) 8, (c) 4, and (d) 2 lattice layers. Note that the sum can be greater than unity because of the stimulated emission that takes place in the crystals and the uniform plates. We first examine Fig. 9.1(a) in detail. The solid line clearly shows a large enhancement of the stimulated emission at $\omega a/2\pi c = 0.788$, which exactly coincides with the upper edge of the third lowest A mode where $v_g = 0$. It also shows a peak at $\omega a/2\pi c = 0.701$ and periodic peaks at $\omega a/2\pi c = 0.733$ to 0.784. The former exactly coincides with the upper edge of the second lowest A mode whereas the latter coincides with the frequency range of the third lowest A mode. Therefore, it is evident that the enhancement of the stimulated emission originates from the low group velocity of the eigenmodes. The enhancement factor, which was calculated as the ratio of the stimulated emission in the photonic crystal and that in the uniform plate was as large as 45 at $\omega a/2\pi c = 0.788$.

When the number of lattice layers is decreased, the spectrum changes considerably. In particular, the peaks at the band edges become small. In Fig. 9.1(c), the peaks that were observed at the upper edges of the second and the third lowest A modes in Fig. 9.1(a) are absent. However, two peaks in the frequency range of the third lowest A mode, where the group velocity is quite small, can be clearly observed. The enhancement factor is large, i.e., 130 at $\omega a/2\pi c = 0.755$ and 68 at $\omega a/2\pi c = 0.776$ even for such a thin geometry. Because the assumed geometry is far from an infinite crystal, it is quite reasonable that the overall correspondence between Fig. 9.1(c) and the photonic band structure in Fig. 3.4 is obscure. In particular, the enhancement just at the band edges where $v_g = 0$ may be difficult to attain with a thin geometry, since the wave vector parallel to the propagation direction is not well defined and the correspondence with the band structure is not assured. However, when the frequency range with a small group velocity is wide

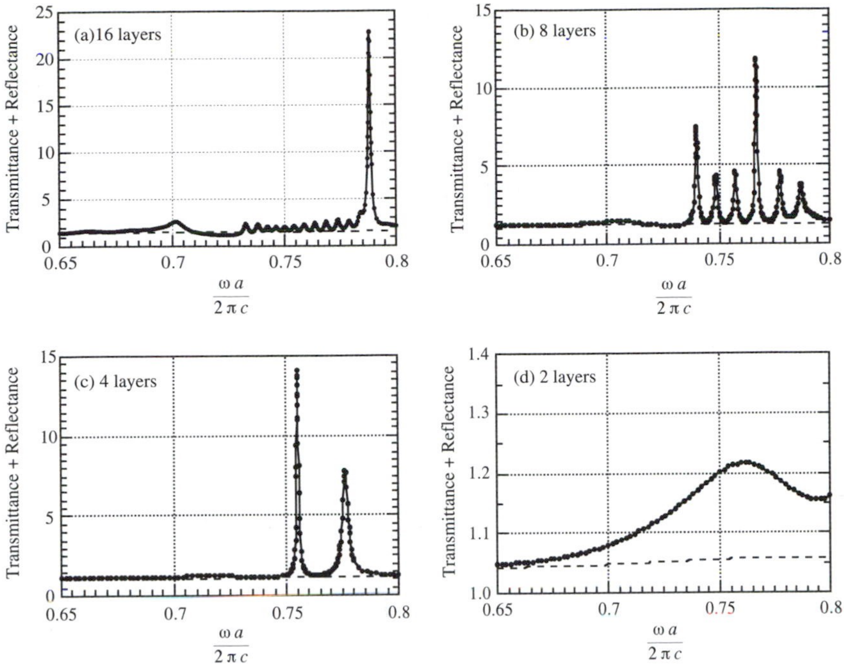

Fig. 9.1. Sum of the transmittance and the reflectance for the E polarization as a function of the normalized frequency calculated for a crystal with (**a**) 16, (**b**) 8, (**c**) 4, and (**d**) 2 layers of air-rods formed in the dielectric material with a dielectric constant of $2.1 - 0.01\mathrm{i}$ (*solid line with filled circles*) and a uniform plate of the same thickness with a spatially averaged dielectric constant (*dashed line*). The negative imaginary part of the dielectric constant stands for the inverted population of the impurity atoms. The incident light was assumed to be propagated in the Γ-X direction. Note that the sum can be greater than unity because of the stimulated emission. (After [84])

enough, as was realized in the third lowest A mode of the present example, the thin geometry has traces of the perfect photonic band structure and we can still expect a large enhancement of the light amplification. A small enhancement is observed even for two layers, as shown in Fig. 9.1(d). Because we do not need a large contrast of the dielectric constant and a very thin crystal is enough to observe the enhancement, the experimental confirmation of this effect may not be difficult.

We now examine the case of the H polarization whose band diagram and transmission spectrum were shown in Figs. 3.4(b) and 4.4, respectively. Figure 9.2 compares them in the frequency range $\omega a/2\pi c = 0.6$–0.8. As mentioned previously, the group-velocity anomaly is observed for the third lowest A mode and the lowest B mode in the Γ-X direction as the E polarization. The bandgap at $\omega a/2\pi c = 0.728$ to 0.742 corresponds very well to a spectral range with low transmittance. The period of interference patterns below 0.720

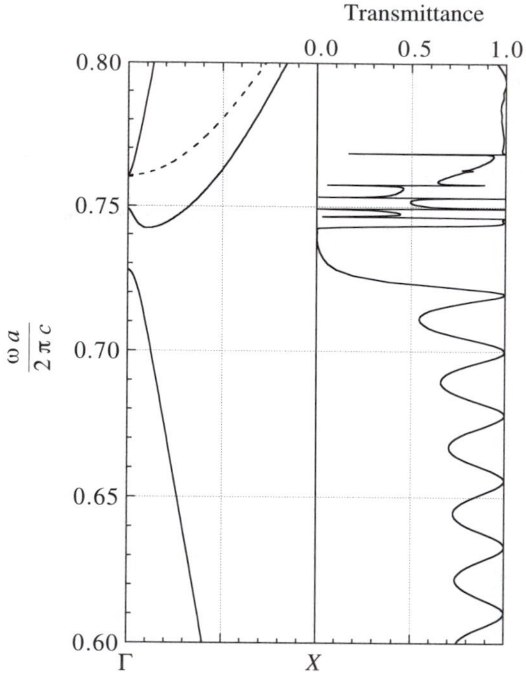

Fig. 9.2. Comparison between the dispersion relation (*left-hand side*) and the transmission spectrum (*right-hand side*) for the H polarization in the Γ-X direction. The same parameters as Fig. 3.4 were assumed for numerical calculation. The *dashed line* in the left-hand side represents a B mode. (After [84])

also corresponds very well to the effective refractive index evaluated from the group velocity of the second lowest A mode (see Table 4.2). The third lowest A mode is convex downward near the Γ point, and most of its frequency range overlaps that of the fourth A mode. This implies that the incident plane wave excites two eigenmodes simultaneously for the most part of the third A branch. This is the origin of quite singular interference patterns observed in that frequency range.

The sum of the transmittance and the reflectance for the H polarization is presented in Fig. 9.3. The same parameters as Fig. 9.1 were used for the numerical calculation. In Fig. 9.3(a), which was calculated for a crystal with 16 lattice layers, the enhancement of stimulated emission can be clearly observed near the upper band edge of the second A mode, $\omega a/2\pi c = 0.728$, and just at the lower band edge of the third A mode, $\omega a/2\pi c = 0.742$. When the number of lattice layers is decreased to four, the peak near the upper band edge of the second A mode disappears completely. However, the peak in the frequency range of the third A mode can be clearly observed even for a crystal with only two lattice layers. In the case of the H polarization

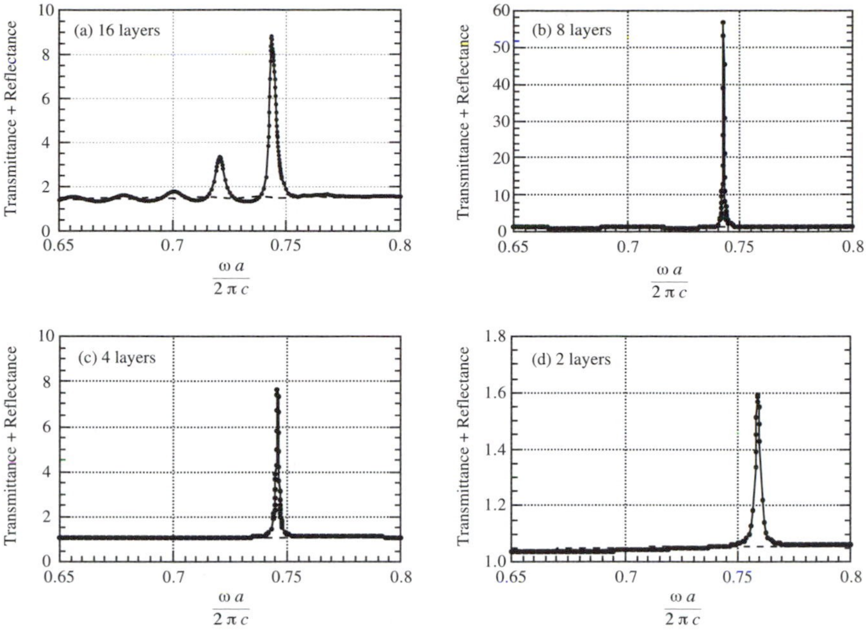

Fig. 9.3. Sum of the transmittance and the reflectance for H polarization as a function of the normalized frequency calculated for a crystal with (**a**) 16, (**b**) 8, (**c**) 4, and (**d**) 2 layers of air-rods formed in the dielectric material with a dielectric constant of $2.1 - 0.01i$ (*solid line with filled circles*) and a uniform plate of the same thickness with a spatially averaged dielectric constant (*dashed line*). The incident light was assumed to be propagated in the Γ-X direction. (After [84])

for this example, the main peak of the light amplification spectrum happens to coincide with the lower band edge of the relevant mode where $v_g = 0$. Therefore, it may be difficult to judge whether the enhancement is caused by the vanishing group velocity at the band edge or the group-velocity anomaly. However, judging from the disappearance of the enhancement at the upper band edge of the second A mode with decreasing lattice layers, we may conclude that the group-velocity anomaly surely contributes to the enhancement of stimulated emission predicted for the 2D crystal with the small contrast of the dielectric constant and the very small number of lattice layers.

9.2 Lasing Threshold

In this section, an analytical expression of the lasing threshold for arbitrary photonic crystals that shows its reduction due to the small group velocities will be derived. The lasing threshold will also be evaluated numerically by examining the divergence of the transmission and reflection coefficients. A large reduction of lasing threshold caused by the group-velocity anomaly will be shown.

9.2.1 Analytical Expression

Before we present the numerical results, we treat the problem analytically. In order to estimate the lasing threshold, we assume a photonic crystal with a thickness L and take a simple model. That is, we assume that the wave function inside the specimen is the same as that of an infinite crystal. Although this is a rough assumption and the field distribution near the surface of the specimen may be considerably different from that in the infinite system, this assumption leads to a qualitatively correct estimation, as will be shown below. When we denote the amplitude reflection coefficient of the relevant eigenmode at each surface by $R_{\boldsymbol{k}n}$, the lasing threshold is given by the balance between the loss at both surfaces and the optical gain in the path of $2L$:

$$R_{\boldsymbol{k}n}^2 \exp\{2(\beta_{\boldsymbol{k}n} + \mathrm{i}k)L\} = 1, \tag{9.1}$$

where $k = |\boldsymbol{k}|$. In this equation, we took into consideration the phase shift of $2kL$, which is consistent with Bloch's theorem. On the other hand, we have neglected additional loss mechanisms such as spontaneous emission and light scattering by imperfections.

As was shown in Sect. 4.4.1, the interference patterns observed in the transmission spectra of the photonic crystals can be described quite well by an effective refractive index defined by $\eta_{\text{eff}} = c/v_{\text{g}}$. Here we assume that the reflection coefficient can also be approximated by that of a uniform material with a refractive index η_{eff}:

$$R \approx \frac{\eta_{\text{eff}} - 1}{\eta_{\text{eff}} + 1}. \tag{9.2}$$

This equation is an empirical one and does not have a logical basis, but the combination of these two equations will show a semi-quantitative agreement with the numerical results. Substituting (9.2) into (9.1), we obtain

$$(\beta_{\boldsymbol{k}n} + \mathrm{i}k)L \approx \log\left(\frac{1 + v_{\text{g}}/c}{1 - v_{\text{g}}/c}\right) + m\pi\mathrm{i}, \tag{9.3}$$

where m is an integer. The explicit expression of $\beta_{\boldsymbol{k}n}$ is given in (5.46). For a rough estimation of the lasing threshold, we first neglect the $(\boldsymbol{k}, n)$ dependence of F_1 which is defined by (5.41). Since the eigenfunction $\boldsymbol{E}_{\boldsymbol{k}n}^{(\mathrm{T})}$ is normalized as

$$\frac{1}{V_0}\int_{V_0} \mathrm{d}\boldsymbol{r}\varepsilon'(\boldsymbol{r}) \left|\boldsymbol{E}_{\boldsymbol{k}n}^{(\mathrm{T})}(\boldsymbol{r})\right|^2 = 1, \tag{9.4}$$

we replace $\left|\boldsymbol{E}_{\boldsymbol{k}n}^{(\mathrm{T})}(\boldsymbol{r})\right|^2$ by $\overline{\varepsilon}'^{-1}$ in (5.41), where the overline denotes the spatial average:

$$\overline{\varepsilon}' = \frac{1}{V_0} \int_{V_0} \mathrm{d}\boldsymbol{r} \varepsilon'(\boldsymbol{r}). \tag{9.5}$$

In (9.4) and (9.5), ε' denotes the real part of the dielectric function:

$$\varepsilon(\boldsymbol{r}) = \varepsilon'(\boldsymbol{r}) + \mathrm{i}\varepsilon''(\boldsymbol{r}). \tag{9.6}$$

Thus, we have

$$F_1 \approx \frac{1}{V_0} \int_{V_0} \mathrm{d}\boldsymbol{r} \frac{\varrho(\boldsymbol{r})}{\overline{\varepsilon}'} = \frac{f\overline{\varrho}}{\overline{\varepsilon}'}, \tag{9.7}$$

where f denotes the filling factor of the dielectric material in which the impurity atoms with population inversion are doped. The real part of (9.3) gives

$$\alpha''_{\mathrm{th}} \approx -\frac{2\varepsilon_0 \overline{\varepsilon}' v_{\mathrm{g}}}{fL\omega\overline{\varrho}} \log\left(\frac{1 + v_{\mathrm{g}}/c}{1 - v_{\mathrm{g}}/c}\right), \tag{9.8}$$

where α''_{th} is the imaginary part of the polarizability of the impurity atoms at the lasing threshold. Since we have the relation that

$$\varepsilon_0 \varepsilon''(\boldsymbol{r}) = \alpha'' \varrho(\boldsymbol{r}), \tag{9.9}$$

(9.8) can be rewritten as

$$\varepsilon''_{\mathrm{th}} \approx -\frac{2\overline{\varepsilon}' v_{\mathrm{g}}}{fL\omega} \log\left(\frac{1 + v_{\mathrm{g}}/c}{1 - v_{\mathrm{g}}/c}\right), \tag{9.10}$$

where $\varepsilon''_{\mathrm{th}}$ is the imaginary part of the dielectric constant at the lasing threshold. If $v_{\mathrm{g}} \ll c$, we have

$$\varepsilon''_{\mathrm{th}} \approx -\frac{4\overline{\varepsilon}' v_{\mathrm{g}}^2}{cfL\omega}. \tag{9.11}$$

The threshold is thus proportional to v_{g}^2, and hence, its large reduction is expected when v_{g} is small. Here, we should note that the reduction of the lasing threshold is brought about by both the enhancement of stimulated emission and the increase of the amplitude reflection coefficient. The imaginary part of (9.3) determines the longitudinal modes, i.e., the wavelength of lasing, which we will not discuss in detail.

9.2.2 Numerical Estimation

We now proceed to the method for the quantitative evaluation of the lasing threshold. Since it is a process of light emission without input signals, which is analogous to the oscillation of electric circuits, the onset of lasing is equivalent to the divergence of the transmittance and/or the reflectance of the assumed specimen [85]. We calculated the transmittance and the reflectance of the assumed specimen as functions of the frequency and the imaginary part of the dielectric constant, and attributed their divergence to the laser oscillation

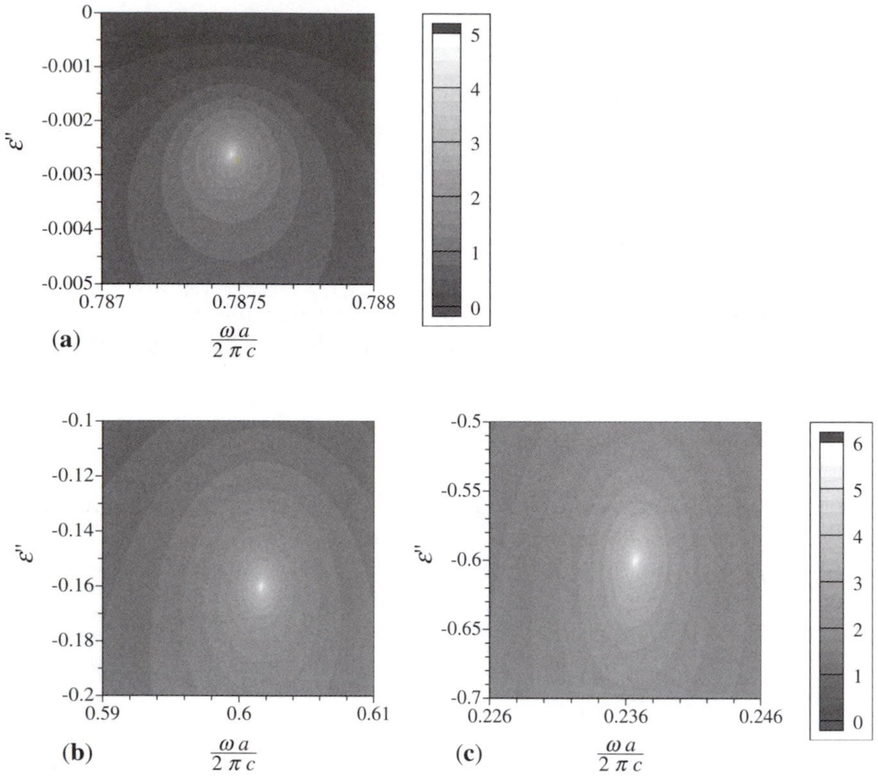

Fig. 9.4. Sum of the transmittance and the reflectance for the E polarization in a logarithmic scale for (**a**) the third, (**b**) the second, and (**c**) the first symmetric bands of the 2D photonic crystal as a function of the normalized frequency, $\omega a/2\pi c$, and the imaginary part of the dielectric constant, ε''. The same parameters as Fig. 9.1 were used for numerical calculation and the incident light was propagated in the Γ-X direction. We assumed that the front and the rear surfaces of the crystal were perpendicular to the propagation direction and that the distance between each surface and the center of the first air cylinder was half a lattice constant. Note that the sum is divergent for certain combinations of $\omega a/2\pi c$ and ε''. (After [61])

[61]. Because this analysis neglects the energy loss caused by spontaneous emission, it only describes the threshold semi-quantitatively. However, it is sufficient, as will be shown below, to clarify the role of the group-velocity anomaly. The number of lattice layers was eight and the incident light was propagated in the Γ-X direction.

Figure 9.4(a) shows an example of the divergence mentioned above, where the sum of the transmittance and the reflectance for the third symmetric band is presented on a logarithmic scale as a function of the normalized frequency and the imaginary part of the dielectric constant. The real part of the dielectric constant was assumed to be 2.1, as before. The numerical

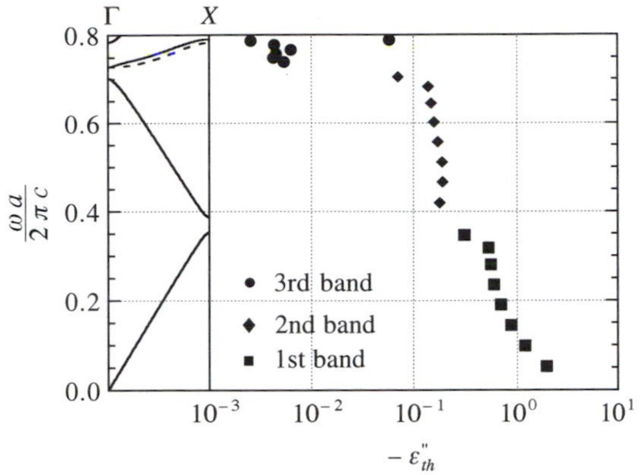

Fig. 9.5. Dispersion relation for the E polarization (*left-hand side*) and the threshold of laser oscillation (*right-hand side*) of a 2D photonic crystal composed of a regular square array of circular air cylinders formed in a dielectric material with a dielectric constant of 2.1. The ordinate is the normalized frequency where ω, a, and c denote the angular frequency of the radiation field, the lattice constant of the 2D crystal, and the light velocity in vacuum, respectively. The radius of the air cylinders was assumed to be 0.28 times the lattice constant. The number of lattice layers was assumed to be eight. The dispersion relation was presented from Γ to X points in the 2D Brillouin zone. The *solid lines* represent the dispersion relation for symmetric modes that can be excited by an incident plane wave, whereas the *dashed line* represents that of an antisymmetric (uncoupled) mode that does not contribute to the light propagation. The threshold of laser oscillation is given by the imaginary part of the dielectric constant of the host material. Note that the third lowest symmetric mode has a small group velocity over its entire spectral range (group-velocity anomaly) and the lasing threshold for this mode is smaller than that for the lowest and the second lowest modes by about two orders of magnitude. Also note the decrease of the threshold at the upper band edges of the latter. (After [61])

calculation was performed for 100×40 sets of (ω, ε''). The maximum of the calculated values was about 10^5 in Fig. 9.4(a). We obtained a larger maximum when we examined the region around the peak with a finer mesh of ω and ε''. Therefore, we could judge that the transmittance and/or reflectance were divergent, and hence, the threshold of laser oscillation was attained. We found six other divergent points in the frequency range of the third symmetric band. These points correspond to the longitudinal modes with different m in (9.3). Figures 9.4(b) and 9.4(c) show other examples of the divergence found in the frequency ranges of the first and the second bands, respectively. Seven other divergent points were found for each band.

Figure 9.5 compares the dispersion relation for the E polarization (left-hand side) and the threshold of laser oscillation (right-hand side) thus obtained. As can be clearly seen, the laser oscillation in the frequency range of the third lowest band takes place with $-\varepsilon''_{\rm th}$ smaller by two orders of magnitude than that necessary for the first and the second bands. Because $-\varepsilon''_{\mathrm{th}}$ is proportional to the pumping rate to create the inverted population, we can conclude that the group-velocity anomaly brings about the reduction of the threshold by two orders of magnitude for the present example. We should also note that the threshold is somewhat small at the upper edges of the first and the second bands compared with that in the middle of both bands. This decrease is caused by the small group velocity at the band edges, which should be equal to zero for a system with infinite thickness. However, for a system with finite thickness, the group-velocity anomaly is much more efficient for the reduction of the lasing threshold even though the relevant group velocity is not exactly equal to zero.

The high threshold at $\omega a/2\pi c = 0.789$ at the upper edge of the third symmetric band is an exception. The incident light excites the third and the fourth symmetric bands simultaneously at this frequency, since the lower edge of the latter is $\omega a/2\pi c = 0.784$. Because the average group velocity of the fourth symmetric band is comparable with that of the second band, a lasing threshold of the same order is expected. Therefore, the lasing at $\omega a/2\pi c = 0.789$ should be attributed to the fourth symmetric band. On the other hand, a low-threshold lasing with $\varepsilon''_{\mathrm{th}} = -2.5 \times 10^{-3}$, which should be attributed to the third symmetric band, is observed at $\omega a/2\pi c = 0.788$ just below the lasing frequency for the fourth symmetric band mentioned above. As this example shows, lasing with different origins can coexist in the frequency ranges where more than one band overlaps each other.

Now we compare the numerical results with the analytical estimation based on (9.10). Because the slope of the dispersion curves varies considerably in the vicinity of the band edges, the estimation of the group velocity, and hence, that of the lasing threshold is difficult in those frequency ranges. Therefore, we compare the numerical and the analytical results in the middle of the frequency range of each band. The effective refractive indices at $\omega a/2\pi c = 0.191$ (1st band), 0.558 (2nd band), and 0.757 (3rd band), which are obtained from the slope of each band, are 1.35, 1.4, and 7.0. Then, the lasing threshold predicted by (9.10) are $\varepsilon''_{\mathrm{th}} = -2.9$, -8.9×10^{-1}, and -2.1×10^{-2}, respectively. These values coincide with the numerical results if we make allowance for an error of a factor of three that may be caused by the rough assumption introduced in the last section. We should note that the fact that the lasing threshold is smaller by two orders of magnitude for the third symmetric band than for the first and the second bands is well reproduced by the analytical estimation, which implies that the reduction of the lasing threshold is really brought about by both the enhancement of stimulated emission and the increase of the amplitude reflection coefficient caused by the small group velocity.

We now examine the case of the H polarization. Figure 9.6 compares the dispersion relation (left-hand side) and the lasing threshold (right-hand side), where the overall features are common with Fig. 9.5. We should note that

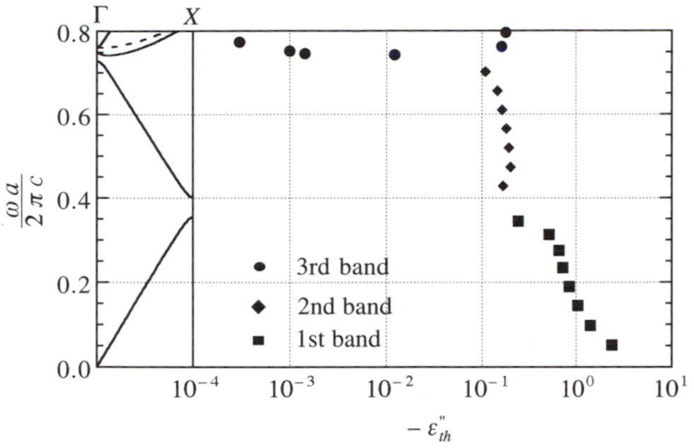

Fig. 9.6. Dispersion relation for H polarization (*left-hand side*) and the threshold of laser oscillation (*right-hand side*) of the 2D photonic crystal. The same parameters as Fig. 9.5 were used for numerical calculation. (After [61])

the third lowest symmetric band shows the group-velocity anomaly like the E polarization. The effective refractive indices at $\omega a/2\pi c = 0.190$ (1st band) and 0.565 (2nd band) are 1.34 and 1.37. Then, the lasing thresholds predicted by (9.10) are $\varepsilon''_{\mathrm{th}} = -2.9$ and -9.3×10^{-1}, which agree with the numerical results qualitatively as the E polarization. On the other hand, the slope of the third band varies considerably with the wave vector and most of its frequency range overlaps that of the fourth symmetric band. Therefore, the comparison with the analytical estimation is not easy for this band. However, it is clearly observed that the lasing thresholds with various magnitude coexist in this frequency range. By analogy with the numerical results for the E polarization, we may regard two longitudinal modes with $\varepsilon''_{\mathrm{th}} \approx -1.7 \times 10^{-1}$ at $\omega a/2\pi c =$ 0.763 and 0.797 as originating from the fourth symmetric band, whereas the other four longitudinal modes with $|\varepsilon''_{\mathrm{th}}| \leq 1.3 \times 10^{-2}$ can be attributed to the third symmetric band. In addition, we should note that the longitudinal mode at $\omega a/2\pi c = 0.774$ with an extremely small threshold of $\varepsilon''_{\mathrm{th}} \approx -3.0 \times 10^{-4}$ does not necessarily correspond to a singular point of the dispersion curves. In Fig. 9.2, quite sharp and irregular patterns were found in the transmission spectrum in this frequency range, which were attributed to the interference between the two relevant bands. Judging from the extraordinary sharpness of the interference patterns, we may conclude that standing waves with extremely high quality factors were realized as a consequence of the interference. Therefore, the amplitude reflection coefficient R for this mode must be particularly high, and we may attribute the longitudinal mode at $\omega a/2\pi c = 0.774$ to this effect.

10. Quantum Optics in Photonic Crystals

The radiation field in photonic crystals is quantized and the Hamiltonian that describes the interaction with the matter system is derived. The theory is applied to quadrature-phase squeezing due to optical parametric amplification and the anomalous Lamb shift due to the peculiar density of states at the photonic band edge.

10.1 Quantization of the Electromagnetic Field

We quantize the free radiation field in the photonic crystal without real charge and current. The radiation field can generally be described by the vector and scalar potentials, $\boldsymbol{A}$ and ϕ:

$$\boldsymbol{B} = \boldsymbol{\nabla} \times \boldsymbol{A}, \tag{10.1}$$

$$\boldsymbol{E} = -\boldsymbol{\nabla}\phi - \frac{\partial \boldsymbol{A}}{\partial t}. \tag{10.2}$$

The radiation field is invariant under the gauge transformation. That is, even when we change $(\phi, \boldsymbol{A})$ in the following manner with an arbitrary function $\xi(\boldsymbol{r}, t)$,

$$\boldsymbol{A} \to \boldsymbol{A} + \boldsymbol{\nabla}\xi, \tag{10.3}$$

$$\phi \to \phi - \frac{\partial \xi}{\partial t}, \tag{10.4}$$

$\boldsymbol{B}$ and $\boldsymbol{E}$ remain as they were. Since the real charge is absent, we have

$$\boldsymbol{\nabla} \cdot (\varepsilon \boldsymbol{E}) = -\boldsymbol{\nabla} \cdot (\varepsilon \boldsymbol{\nabla}\phi) - \frac{\partial}{\partial t}\boldsymbol{\nabla} \cdot (\varepsilon \boldsymbol{A}) = 0, \tag{10.5}$$

where ε is the position-dependent dielectric constant of the photonic crystal. For this case, we can eliminate the scalar potential by the gauge transformation. In order to show this, we assume that we have $(\phi_1, \boldsymbol{A}_1)$ for the first step and $\phi_1 \neq 0$. When we choose ξ such that

$$\frac{\partial \xi}{\partial t} = \phi_1, \tag{10.6}$$

the new scalar potential is equal to zero. From (10.5), the new vector potential should satisfy

$$\nabla \cdot (\varepsilon \boldsymbol{A}) = 0, \tag{10.7}$$

Taking this gauge condition, we have

$$\boldsymbol{B}(\boldsymbol{r}, t) = \nabla \times \boldsymbol{A}(\boldsymbol{r}, t), \tag{10.8}$$

$$\boldsymbol{E}(\boldsymbol{r}, t) = -\frac{\partial}{\partial t}\boldsymbol{A}(\boldsymbol{r}, t). \tag{10.9}$$

The last equation implies that the eigenmode of the vector potential is proportional to that of the electric field, $\boldsymbol{E}_{\boldsymbol{k}n}$.

We can thus expand the vector potential with $\boldsymbol{E}_{\boldsymbol{k}n}$:

$$\boldsymbol{A}(\boldsymbol{r}, t) = \frac{1}{\sqrt{V}} \sum_{\boldsymbol{k}n} \left\{ q_{\boldsymbol{k}n}(t) \boldsymbol{E}_{\boldsymbol{k}n}^{(\mathrm{T})}(\boldsymbol{r}) + q_{\boldsymbol{k}n}^{*}(t) \boldsymbol{E}_{\boldsymbol{k}n}^{(\mathrm{T})*}(\boldsymbol{r}) \right\}, \tag{10.10}$$

where V is the volume of the photonic crystal and $q_{\boldsymbol{k}n}(t)$ is proportional to $\exp(-\mathrm{i}\omega_{\boldsymbol{k}n}^{(\mathrm{T})} t)$. In this equation, only quasi-transverse modes appear since the quasi-longitudinal modes do not satisfy (10.7). We will omit the superscript (T) hereafter. The electric and magnetic fields are thus expressed as

$$\boldsymbol{E}(\boldsymbol{r}, t) = \frac{\mathrm{i}}{\sqrt{V}} \sum_{\boldsymbol{k}n} \omega_{\boldsymbol{k}n} \left\{ q_{\boldsymbol{k}n}(t) \boldsymbol{E}_{\boldsymbol{k}n}(\boldsymbol{r}) - q_{\boldsymbol{k}n}^{*}(t) \boldsymbol{E}_{\boldsymbol{k}n}^{*}(\boldsymbol{r}) \right\}, \tag{10.11}$$

$$\boldsymbol{B}(\boldsymbol{r}, t) = \frac{1}{\sqrt{V}} \sum_{\boldsymbol{k}n} \left\{ q_{\boldsymbol{k}n}(t) \nabla \times \boldsymbol{E}_{\boldsymbol{k}n}(\boldsymbol{r}) + q_{\boldsymbol{k}n}^{*}(t) \nabla \times \boldsymbol{E}_{\boldsymbol{k}n}^{*}(\boldsymbol{r}) \right\}. \tag{10.12}$$

Because $q_{\boldsymbol{k}n}$ ($q_{\boldsymbol{k}n}^{*}$) describes the temporal evolution of a harmonic oscillation, we may replace $q_{\boldsymbol{k}n}$ ($q_{\boldsymbol{k}n}^{*}$) by an annihilation operator $\widehat{a}_{\boldsymbol{k}n}$ (a creation operator $\widehat{a}_{\boldsymbol{k}n}^{+}$) of photons in order to quantize the radiation field:

$$q_{\boldsymbol{k}n}(t) \to \sqrt{\frac{\hbar}{2\varepsilon_0 \omega_{\boldsymbol{k}n}}} \widehat{a}_{\boldsymbol{k}n}(t) \quad \text{and} \quad q_{\boldsymbol{k}n}^{*}(t) \to \sqrt{\frac{\hbar}{2\varepsilon_0 \omega_{\boldsymbol{k}n}}} \widehat{a}_{\boldsymbol{k}n}^{+}(t), \tag{10.13}$$

where the coefficients are determined so that the total electromagnetic energy is reduced to the quantum mechanical Hamiltonian appropriate to an assembly of harmonic oscillators:

$$\widehat{H} = \sum_{\boldsymbol{k}n} \hbar \omega_{\boldsymbol{k}n} \left\{ \widehat{a}_{\boldsymbol{k}n}^{+}(t) \widehat{a}_{\boldsymbol{k}n}(t) + \frac{1}{2} \right\}, \tag{10.14}$$

where $\widehat{a}_{\boldsymbol{k}n}$ and $\widehat{a}_{\boldsymbol{k}n}^{+}$ satisfy the usual commutation relation for a boson:

$$\left[\widehat{a}_{\boldsymbol{k}n}(t), \widehat{a}_{\boldsymbol{k}n}^{+}(t) \right] = 1. \tag{10.15}$$

Actually, the total energy U of the radiation field is given by

$$U = \int_V \mathrm{d}\boldsymbol{r} \left\{ \frac{\varepsilon_0 \varepsilon(\boldsymbol{r}) |\boldsymbol{E}(\boldsymbol{r}, t)|^2}{2} + \frac{|\boldsymbol{B}(\boldsymbol{r}, t)|^2}{2\mu_0} \right\}. \tag{10.16}$$

When we substitute (10.11) and use the orthonormality of the eigenfunctions, (2.145), we obtain

$$\int_V \mathrm{d}\boldsymbol{r} \frac{\varepsilon_0 \varepsilon(\boldsymbol{r}) |\boldsymbol{E}(\boldsymbol{r},t)|^2}{2} = \varepsilon_0 \sum_{\boldsymbol{k}n} \omega_{\boldsymbol{k}n}^2 \left(q_{\boldsymbol{k}n} q_{\boldsymbol{k}n}^* + q_{\boldsymbol{k}n}^* q_{\boldsymbol{k}n} \right) . \tag{10.17}$$

On the other hand,

$$\int_V \mathrm{d}\boldsymbol{r} \frac{|\boldsymbol{B}(\boldsymbol{r},t)|^2}{2\mu_0} = \sum_{\boldsymbol{k}n} \frac{(q_{\boldsymbol{k}n} q_{\boldsymbol{k}n}^* + q_{\boldsymbol{k}n}^* q_{\boldsymbol{k}n})}{\mu_0} \times \int_V \mathrm{d}\boldsymbol{r} \left\{ \boldsymbol{\nabla} \times \boldsymbol{E}_{\boldsymbol{k}n}^*(\boldsymbol{r}) \right\} \cdot \left\{ \boldsymbol{\nabla} \times \boldsymbol{E}_{\boldsymbol{k}n}(\boldsymbol{r}) \right\} . \tag{10.18}$$

When we follow the same procedure as we derived (2.138) and use (2.20), we obtain

$$\begin{aligned} \int_V \mathrm{d}\boldsymbol{r} \frac{|\boldsymbol{B}(\boldsymbol{r},t)|^2}{2\mu_0} &= \sum_{\boldsymbol{k}n} \frac{|q_{\boldsymbol{k}n}|^2}{\mu_0} \int_V \mathrm{d}\boldsymbol{r} \frac{\omega_{\boldsymbol{k}n}^2}{c^2} \varepsilon(\boldsymbol{r}) |\boldsymbol{E}_{\boldsymbol{k}n}|^2 \\ &= \varepsilon_0 \sum_{\boldsymbol{k}n} \omega_{\boldsymbol{k}n}^2 \left(q_{\boldsymbol{k}n} q_{\boldsymbol{k}n}^* + q_{\boldsymbol{k}n}^* q_{\boldsymbol{k}n} \right) . \end{aligned} \tag{10.19}$$

We thus have

$$U = 2\varepsilon_0 \sum_{\boldsymbol{k}n} \omega_{\boldsymbol{k}n}^2 \left(q_{\boldsymbol{k}n} q_{\boldsymbol{k}n}^* + q_{\boldsymbol{k}n}^* q_{\boldsymbol{k}n} \right) . \tag{10.20}$$

When we compare this equation with (10.14), we obtain the coefficients in (10.13). Finally, we have the following expression for the electric field operator:

$$\widehat{\boldsymbol{E}}(\boldsymbol{r},t) = \sum_{\boldsymbol{k}n} \mathrm{i} \sqrt{\frac{\hbar \omega_{\boldsymbol{k}n}}{2\varepsilon_0 V}} \left\{ \widehat{a}_{\boldsymbol{k}n}(t) \boldsymbol{E}_{\boldsymbol{k}n}(\boldsymbol{r}) - \widehat{a}_{\boldsymbol{k}n}^{+}(t) \boldsymbol{E}_{\boldsymbol{k}n}^*(\boldsymbol{r}) \right\} . \tag{10.21}$$

10.2 Quadrature-Phase Squeezing

We consider the quadrature-phase squeezing by means of the degenerate optical parametric amplification. We assume that a pump wave denoted by $\boldsymbol{E}_{\mathrm{p}}(\boldsymbol{r},t)$ and a signal wave denoted by $\boldsymbol{E}_{\mathrm{s}}(\boldsymbol{r},t)$, both of which are eigenmodes of the radiation field, are propagated in the photonic crystal (see Fig. 10.1). We assume that the pump wave can be treated classically because its amplitude is so large that its quantum-mechanical fluctuation can be neglected:

$$\boldsymbol{E}_{\mathrm{p}}(\boldsymbol{r},t) = A\mathrm{i} \left\{ \boldsymbol{E}_{\mathrm{p}}(\boldsymbol{r}) \mathrm{e}^{-2\mathrm{i}\omega_{\mathrm{s}}t + \mathrm{i}\theta} - \boldsymbol{E}_{\mathrm{p}}^*(\boldsymbol{r}) \mathrm{e}^{2\mathrm{i}\omega_{\mathrm{s}}t - \mathrm{i}\theta} \right\} , \tag{10.22}$$

where A and θ stand for the amplitude and the phase, respectively. We assumed that the angular frequency of the pump wave is twice as large as that of the signal wave, ω_{s}. On the other hand, we deal with the electric field of the signal wave as an operator:

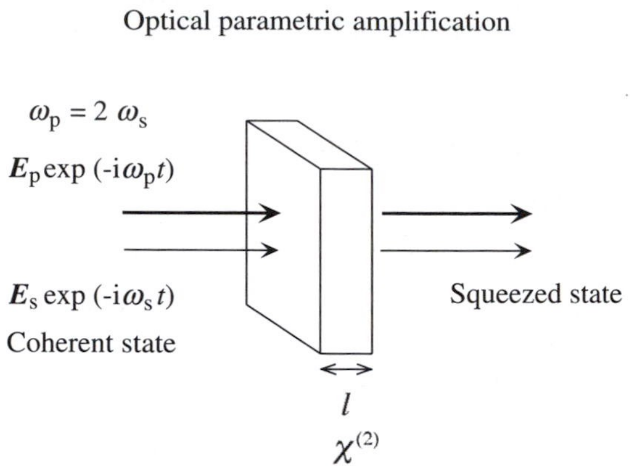

Fig. 10.1. Quadrature-phase squeezing due to degenerate optical parametric amplification

$$\widehat{\boldsymbol{E}}_{\mathrm{s}}(\boldsymbol{r},t) = \mathrm{i}\sqrt{\frac{\hbar\omega_{\mathrm{s}}}{2\varepsilon_0 V}}\left\{\widehat{a}_{\mathrm{s}}(0)\boldsymbol{E}_{\mathrm{s}}(\boldsymbol{r})\mathrm{e}^{-\mathrm{i}\omega_{\mathrm{s}}t} - \widehat{a}_{\mathrm{s}}^{+}(0)\boldsymbol{E}_{\mathrm{s}}^{*}(\boldsymbol{r})\mathrm{e}^{\mathrm{i}\omega_{\mathrm{s}}t}\right\}, \tag{10.23}$$

where we write the time dependence explicitly. We further assume that the signal wave was initially in a coherent state, $|\alpha\rangle$. Light in the coherent state is provided by a laser that oscillates far beyond the lasing threshold.[1] It satisfies the following eigenvalue equation.

$$\widehat{a}_{\mathrm{s}}(0)\,|\alpha\rangle = \alpha\,|\alpha\rangle\,, \tag{10.24}$$

where α is generally a complex number.

When we denote the second order nonlinear susceptibility of the photonic crystal by $\overleftrightarrow{\chi}^{(2)}(\boldsymbol{r})$, the nonlinear polarization induced by these waves is given by

$$\begin{aligned}\widehat{\boldsymbol{P}}^{(0)}(\boldsymbol{r},t) &= \overleftrightarrow{\chi}^{(2)}(\boldsymbol{r}) : \left\{\boldsymbol{E}_{\mathrm{p}}(\boldsymbol{r},t) + \widehat{\boldsymbol{E}}_{\mathrm{s}}(\boldsymbol{r},t)\right\}^2 \\ &\approx 2\overleftrightarrow{\chi}^{(2)}(\boldsymbol{r}) : \boldsymbol{E}_{\mathrm{p}}(\boldsymbol{r},t)\widehat{\boldsymbol{E}}_{\mathrm{s}}(\boldsymbol{r},t) \\ &\approx 2A\sqrt{\frac{\hbar\omega_{\mathrm{s}}}{2\varepsilon_0 V}}\overleftrightarrow{\chi}^{(2)}(\boldsymbol{r}) : \left\{\boldsymbol{E}_{\mathrm{p}}(\boldsymbol{r})\boldsymbol{E}_{\mathrm{s}}^{*}(\boldsymbol{r})\widehat{a}_{\mathrm{s}}^{+}(0)\mathrm{e}^{-\mathrm{i}\omega_{\mathrm{s}}t+\mathrm{i}\theta}\right. \\ &\qquad \left. -\boldsymbol{E}_{\mathrm{p}}^{*}(\boldsymbol{r})\boldsymbol{E}_{\mathrm{s}}(\boldsymbol{r})\widehat{a}_{\mathrm{s}}(0)\mathrm{e}^{\mathrm{i}\omega_{\mathrm{s}}t-\mathrm{i}\theta}\right\},\end{aligned} \tag{10.25}$$

where ":" denotes a product as a tensor. In (10.25), only terms relevant to the parametric amplification were taken. The electric field $\widehat{\boldsymbol{E}}_{\mathrm{ind}}^{(1)}$ induced by $\widehat{\boldsymbol{P}}^{(0)}$ can be calculated according to (5.10). When we substitute (10.25) into (5.10) and follow the same procedure as we derived (5.45), we obtain

$$\widehat{\boldsymbol{E}}_{\mathrm{ind}}^{(1)}(\boldsymbol{r},t)$$

[1] See, for example, [86]

$$= \mathrm{i}l\sqrt{\frac{\hbar\omega_\mathrm{s}}{2\varepsilon_0 V}}\left\{\beta\widehat{a}_\mathrm{s}^+(0)\boldsymbol{E}_\mathrm{s}(\boldsymbol{r})\mathrm{e}^{-\mathrm{i}\omega_\mathrm{s}t+\mathrm{i}\theta} - \beta^*\widehat{a}_\mathrm{s}(0)\boldsymbol{E}_\mathrm{s}^*(\boldsymbol{r})\mathrm{e}^{\mathrm{i}\omega_\mathrm{s}t-\mathrm{i}\theta}\right\}, \quad (10.26)$$

where

$$\beta = \frac{\omega_\mathrm{s}AF}{\varepsilon_0 v_\mathrm{g}}, \quad (10.27)$$

and

$$F = \frac{1}{V}\int_V \mathrm{d}\boldsymbol{r}\boldsymbol{E}_\mathrm{s}^*(\boldsymbol{r})\cdot\overleftrightarrow{\chi}^{(2)}(\boldsymbol{r}) : \boldsymbol{E}_\mathrm{p}(\boldsymbol{r})\boldsymbol{E}_\mathrm{s}^*(\boldsymbol{r}). \quad (10.28)$$

In (10.27), v_g denotes the group velocity of the signal wave and l is the length of the path that the signal wave traveled. When we derived (10.26), we assumed for simplicity that the phase-matching condition for the signal and the pump waves was satisfied. Here, we should note that the factor β becomes large if v_g is small. This is because the interaction time between the signal wave and the photonic crystal is long. Since an extremely small group velocity is easily realized in the photonic crystals as mentioned previously, we can expect a large enhancement.

Now, $\widehat{\boldsymbol{E}}_\mathrm{ind}^{(1)}$ yields another nonlinear polarization, $\boldsymbol{P}^{(1)}$,

$$\widehat{\boldsymbol{P}}^{(1)}(\boldsymbol{r},t) = \overleftrightarrow{\chi}^{(2)}(\boldsymbol{r}) : \left\{\boldsymbol{E}_\mathrm{p}(\boldsymbol{r},t) + \widehat{\boldsymbol{E}}_\mathrm{ind}^{(1)}(\boldsymbol{r},t)\right\}^2$$
$$\approx 2Al\sqrt{\frac{\hbar\omega_\mathrm{s}}{2\varepsilon_0 V}}\overleftrightarrow{\chi}^{(2)}(\boldsymbol{r}) : \{\beta^*\widehat{a}_\mathrm{s}(0)\boldsymbol{E}_\mathrm{p}(\boldsymbol{r})\boldsymbol{E}_\mathrm{s}^*(\boldsymbol{r})\mathrm{e}^{-\mathrm{i}\omega_\mathrm{s}t}$$
$$+\beta\widehat{a}_\mathrm{s}^+(0)\boldsymbol{E}_\mathrm{p}^*(\boldsymbol{r})\boldsymbol{E}_\mathrm{s}(\boldsymbol{r})\mathrm{e}^{\mathrm{i}\omega_\mathrm{s}t}\}, \quad (10.29)$$

which then induces yet another electric field $\widehat{\boldsymbol{E}}_\mathrm{ind}^{(2)}$. We can calculate this using (5.10) again:

$$\widehat{\boldsymbol{E}}_\mathrm{ind}^{(2)}(\boldsymbol{r},t)$$
$$= \frac{\mathrm{i}|\beta|^2 l^2}{2!}\sqrt{\frac{\hbar\omega_\mathrm{s}}{2\varepsilon_0 V}}\left\{\widehat{a}_\mathrm{s}(0)\boldsymbol{E}_\mathrm{s}(\boldsymbol{r})\mathrm{e}^{-\mathrm{i}\omega_\mathrm{s}t} - \widehat{a}_\mathrm{s}^+(0)\boldsymbol{E}_\mathrm{s}^*(\boldsymbol{r})\mathrm{e}^{\mathrm{i}\omega_\mathrm{s}t}\right\}. \quad (10.30)$$

We can carry out this perturbative calculation to the infinite order and the result is given as follows:

$$\widehat{\boldsymbol{E}}(\boldsymbol{r},t) = \widehat{\boldsymbol{E}}_\mathrm{s}(\boldsymbol{r},t) + \widehat{\boldsymbol{E}}_\mathrm{ind}^{(1)}(\boldsymbol{r},t) + \widehat{\boldsymbol{E}}_\mathrm{ind}^{(2)}(\boldsymbol{r},t) + \cdots$$
$$= \mathrm{i}\sqrt{\frac{\hbar\omega_\mathrm{s}}{2\varepsilon_0 V}}\left\{\widehat{b}\boldsymbol{E}_\mathrm{s}(\boldsymbol{r})\mathrm{e}^{-\mathrm{i}\omega_\mathrm{s}t} - \widehat{b}^+\boldsymbol{E}_\mathrm{s}^*(\boldsymbol{r})\mathrm{e}^{\mathrm{i}\omega_\mathrm{s}t}\right\}, \quad (10.31)$$

where the new operators $\widehat{b}$ and $\widehat{b}^+$ are defined by

$$\widehat{b} = \left\{1 + \frac{|\beta|^2 l^2}{2!} + \frac{|\beta|^4 l^4}{4!} + \cdots\right\}\widehat{a}_\mathrm{s}(0)$$
$$+\mathrm{e}^{\mathrm{i}(\theta+\phi)}\left\{|\beta|l + \frac{|\beta|^3 l^3}{3!} + \cdots\right\}\widehat{a}_\mathrm{s}^+(0)$$

$$= \cosh|\beta|l \cdot \widehat{a}_{\rm s}(0) + {\rm e}^{{\rm i}(\theta+\phi)} \sinh|\beta|l \cdot \widehat{a}_{\rm s}^{+}(0), \tag{10.32}$$

and

$$\widehat{b}^{+} = \cosh|\beta|l \cdot \widehat{a}_{\rm s}^{+}(0) + {\rm e}^{-{\rm i}(\theta+\phi)} \sinh|\beta|l \cdot \widehat{a}_{\rm s}(0), \tag{10.33}$$

where we denoted β by

$$\beta = |\beta|{\rm e}^{{\rm i}\phi}. \tag{10.34}$$

Equations (10.32) and (10.33) describe a Bogoliubov transformation and we can verify easily that $[\widehat{b}, \widehat{b}^{+}] = 1$.

Two quadrature-phase components of the signal wave are given by the real and imaginary parts of $\widehat{\boldsymbol{E}}_{\rm s}(\boldsymbol{r}){\rm e}^{-{\rm i}\omega_{\rm s}t}$. We introduce the following two functions:

$$\boldsymbol{E}_1(\boldsymbol{r},t) = -\frac{1}{2{\rm i}}\left\{\boldsymbol{E}_{\rm s}(\boldsymbol{r}){\rm e}^{-{\rm i}\omega_{\rm s}t} - \boldsymbol{E}_{\rm s}^{*}(\boldsymbol{r}){\rm e}^{{\rm i}\omega_{\rm s}t}\right\}, \tag{10.35}$$

$$\boldsymbol{E}_2(\boldsymbol{r},t) = \frac{1}{2}\left\{\boldsymbol{E}_{\rm s}(\boldsymbol{r}){\rm e}^{-{\rm i}\omega_{\rm s}t} + \boldsymbol{E}_{\rm s}^{*}(\boldsymbol{r}){\rm e}^{{\rm i}\omega_{\rm s}t}\right\}. \tag{10.36}$$

Then,

$$\widehat{\boldsymbol{E}}(\boldsymbol{r},t) = \sqrt{\frac{2\hbar\omega_{\rm s}}{\varepsilon_0 V}}\left\{\widehat{q}\boldsymbol{E}_1(\boldsymbol{r},t) + \widehat{p}\boldsymbol{E}_2(\boldsymbol{r},t)\right\}, \tag{10.37}$$

where

$$\widehat{q} = \frac{1}{2}\left(\widehat{b} + \widehat{b}^{+}\right) \qquad \text{and} \qquad \widehat{p} = \frac{\rm i}{2}\left(\widehat{b} - \widehat{b}^{+}\right). \tag{10.38}$$

The commutation and uncertainty relations for $\widehat{q}$ and $\widehat{p}$ are given by

$$[\widehat{q}, \widehat{p}] = -\frac{\rm i}{2} \qquad \text{and} \qquad \Delta q \Delta p \geq \frac{1}{4}. \tag{10.39}$$

Now we assume that the incident signal wave is in a coherent state $|\alpha\rangle$:

$$\widehat{a}_{\rm s}|\alpha\rangle = \alpha|\alpha\rangle. \tag{10.40}$$

We can then easily verify that Δq and Δp for this state are given by

$$(\Delta q)^2 = \frac{1}{4}\Big\{\cosh^2|\beta|l + \sinh^2|\beta|l + 2\cos\left(\theta + \phi - \frac{\pi}{2}\right)\cosh|\beta|l \sinh|\beta|l\Big\} \tag{10.41}$$

and

$$(\Delta p)^2 = \frac{1}{4}\Big\{\cosh^2|\beta|l + \sinh^2|\beta|l - 2\cos\left(\theta + \phi - \frac{\pi}{2}\right)\cosh|\beta|l \sinh|\beta|l\Big\}, \tag{10.42}$$

irrespective of α. When $\theta + \phi = \pi/2$,

$$\Delta q = \frac{1}{2}{\rm e}^{|\beta|l}, \qquad \Delta p = \frac{1}{2}{\rm e}^{-|\beta|l}, \tag{10.43}$$

and when $\theta + \phi = -\pi/2$,

$$\Delta q = \frac{1}{2}\mathrm{e}^{-|\beta| l}, \qquad \Delta p = \frac{1}{2}\mathrm{e}^{|\beta| l}. \tag{10.44}$$

Therefore, the uncertainty of one quadrature-phase component is diminished at the sacrifice of the other component, i.e., the quadrature-phase squeezing is attained. We should note that the squeezed states in photonic crystals are minimum uncertainty states as usual, i.e., $\Delta q \Delta p = 1/4$ [86]. We should also note that there is an enhancement of the quadrature-phase squeezing due to the small group velocity of the electromagnetic eigenmode, as (10.27) shows.

We would like to make three remarks before concluding this section. First, extension of the present calculation to the case without phase-matching is straightforward, since we can do that following the procedure given in Sect. 5.4 that treated sum-frequency generation with and without phase-matching. Second, the perturbation theory presented here is, of course, not limited to the quadrature-phase squeezing, but can be applied to other quantum optical processes in the photonic crystals. Third, the enhancement due to the small group velocity is not limited to the quadrature-phase squeezing either, since it is brought about by the long interaction time between the radiation field and the matter system, which is common to all optical processes.

10.3 Interaction Hamiltonian

Next, we derive the quantized interaction Hamiltonian $\widehat{H}'$ of the radiation field and an electron. It is given by the difference between the Hamiltonian for the electron in the presence of the radiation field and in its absence. When we denote the electric charge and the mass of the electron by e and m_e, $\widehat{H}'$ is thus given by

$$\widehat{H}' = \frac{\left(\widehat{\boldsymbol{p}} - e\widehat{\boldsymbol{A}}\right)^2}{2m_\mathrm{e}} - \frac{\widehat{\boldsymbol{p}}^2}{2m_\mathrm{e}} \approx -\frac{e}{2m_\mathrm{e}}\left(\widehat{\boldsymbol{p}} \cdot \widehat{\boldsymbol{A}} + \widehat{\boldsymbol{A}} \cdot \widehat{\boldsymbol{p}}\right), \tag{10.45}$$

where $\widehat{\boldsymbol{p}}$ is the momentum operator of the electron and the term proportional to $\widehat{\boldsymbol{A}}^2$ was omitted since it is not relevant to the following discussion. If we note that operator $\widehat{\boldsymbol{p}}$ and operator $\widehat{\boldsymbol{A}}$ do not commute and the following equation holds,

$$\widehat{\boldsymbol{p}} \cdot \widehat{\boldsymbol{A}} = \frac{\hbar}{\mathrm{i}} \boldsymbol{\nabla} \cdot \widehat{\boldsymbol{A}} + \widehat{\boldsymbol{A}} \cdot \widehat{\boldsymbol{p}}, \tag{10.46}$$

(10.45) is modified to

$$\widehat{H}' = -\frac{e}{m_\mathrm{e}} \widehat{\boldsymbol{A}} \cdot \widehat{\boldsymbol{p}} - \frac{\mathrm{i}e\hbar}{2m_\mathrm{e}} \widehat{\boldsymbol{A}} \cdot (\boldsymbol{\nabla} \log \varepsilon), \tag{10.47}$$

where we used (10.7). The second term on the right-hand side of (10.47), which is peculiar to the problems of photonic crystals, appears because we adopted the gauge condition (10.7). This term may be important when the electron is located close to the interface between two different components of the photonic crystal. However, in that case, the radiation field and the dielectric constant should be treated microscopically instead of the present macroscopic treatment. Hence, the problem goes beyond the scope of this book. On the other hand, this term is equal to zero when the electron is located in the uniform part of the photonic crystal where the spatial variation of the dielectric function, $\nabla\varepsilon$, vanishes. We assume this case for simplicity hereafter. The final form of the interaction Hamiltonian is thus given by

$$\widehat{H}' = -\frac{e}{m_{\mathrm{e}}}\sum_{\boldsymbol{k}n}\sqrt{\frac{\hbar}{2\varepsilon_0 V\omega_{\boldsymbol{k}n}}}\left\{\widehat{a}_{\boldsymbol{k}n}\boldsymbol{E}_{\boldsymbol{k}n}(\boldsymbol{r}_{\mathrm{e}}) + \widehat{a}^{+}_{\boldsymbol{k}n}\boldsymbol{E}^{*}_{\boldsymbol{k}n}(\boldsymbol{r}_{\mathrm{e}})\right\}\cdot\widehat{\boldsymbol{p}}, \qquad (10.48)$$

where $\boldsymbol{r}_{\mathrm{e}}$ denotes the position of the electron. Here, we took Schrödinger representation and dropped the time dependence of $\widehat{a}_{\boldsymbol{k}n}$ and $\widehat{a}^{+}_{\boldsymbol{k}n}$.

10.4 Lamb Shift

As is well-known, the charge and the mass of the electron change their values because of the interaction with the radiation field. This is termed charge and mass renormalization. What we actually observe are the renormalized quantities. For an electron bound to an atom, a small difference in the renormalized mass may appear according to the character of the bound state. In fact, as Lamb and Retherford [87] reported for the first time, there is an energy difference of about 1.05 GHz between the 2s and 2p states of hydrogen atoms. Later, more accurate mesurements were performed [88, 89]. This energy difference is called the Lamb shift.

As for the Lamb shift of hydrogen atoms in free space, a non-relativistic calculation was given by Bethe [90] for the first time, and a good agreement with the experimental observation was shown. Later, more rigorous relativistic calculations [91–94] were performed, and it was found that the value obtained with the non-relativistic calculation by Bethe gives the main part of the Lamb shift. Hence, in the following, we will calculate the energy shift of the bound electrons in hydrogen-like atoms to the second order of the interaction Hamiltonian after Bethe. That is, we will calculate the self energy of the electron for the Feynman diagram shown in Fig. 10.2, where we subtract the self energy calculated for a free electron. By this procedure, we can cancel the infinite mass renormalization that is present for the latter.

When we calculate the energy shift of the ith eigenstate of a bound electron to the second order of the interaction, we obtain,

$$E_i^{(2)} = \sum_{j}\sum_{\boldsymbol{k}n}\frac{\left\langle i\left|\widehat{H}'\right|j,\boldsymbol{k}n\right\rangle\left\langle j,\boldsymbol{k}n\left|\widehat{H}'\right|i\right\rangle}{E_i - E_j - \hbar\omega_{\boldsymbol{k}n}}$$

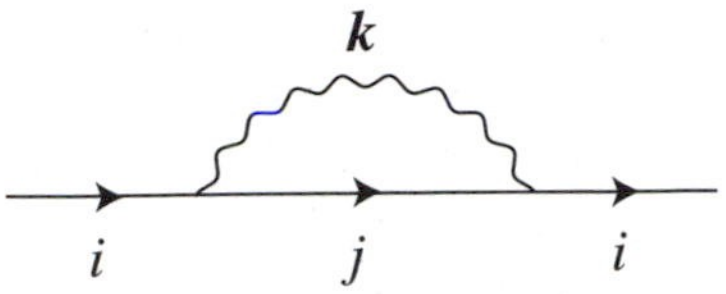

Fig. 10.2. Feynman diagram for the self energy of an electron to the second order of the interaction Hamiltonian

$$= \frac{e^2}{2\varepsilon_0 m_{\mathrm{e}}^2 V} \sum_j \sum_{\boldsymbol{k}n} \frac{\left|\boldsymbol{E}_{\boldsymbol{k}n}(\boldsymbol{r}_{\mathrm{e}}) \cdot \langle i\, |\widehat{\boldsymbol{p}}|\, j\rangle\right|^2}{\omega_{\boldsymbol{k}n}\left(\omega_i - \omega_j - \omega_{\boldsymbol{k}n}\right)}, \tag{10.49}$$

where j denotes the jth eigenstate of the bound electron and $E_j = \hbar\omega_j$ is its eigenenergy. $|j, \boldsymbol{k}n\rangle$ represents the state with the electron in the jth eigenstate and a photon in the $\boldsymbol{k}n$ state. On the other hand, the energy shift of a free electron with momentum $\boldsymbol{p}_{\mathrm{e}}$ is given by

$$E_{\mathrm{free}}^{(2)} = \sum_{\boldsymbol{G}} \sum_{\boldsymbol{k}n} \frac{\left|\left\langle \boldsymbol{p}_{\mathrm{e}} \left| \widehat{H}' \right| \boldsymbol{p}_{\mathrm{e}} - \hbar\boldsymbol{k} - \hbar\boldsymbol{G}, \boldsymbol{k}n \right\rangle\right|^2}{p_{\mathrm{e}}^2/2m - \left\{\left(\boldsymbol{p}_{\mathrm{e}} - \hbar\boldsymbol{k} - \hbar\boldsymbol{G}\right)^2/2m + \hbar\omega_{\boldsymbol{k}n}\right\}}, \tag{10.50}$$

where $\boldsymbol{G}$ is the reciprocal lattice vector of the photonic crystal and

$$|\boldsymbol{p}_{\mathrm{e}} - \hbar\boldsymbol{k} - \hbar\boldsymbol{G}, \boldsymbol{k}n\rangle \tag{10.51}$$

represents the state with an electron with momentum $\boldsymbol{p}_{\mathrm{e}} - \hbar\boldsymbol{k} - \hbar\boldsymbol{G}$ and a photon in the $\boldsymbol{k}n$ state. Here, we should note that $\widehat{H}'$ generally has non-zero matrix elements between states $|\boldsymbol{p}_{\mathrm{e}}\rangle$ and $|\boldsymbol{p}_{\mathrm{e}} - \hbar\boldsymbol{k} - \hbar\boldsymbol{G}, \boldsymbol{k}n\rangle$ with $\boldsymbol{G} \neq 0$ because of the periodic spatial variation of the dielectric constant, although the matrix elements are small for large $\boldsymbol{G}$. Hence, as a rough estimation, we take only one term with $\boldsymbol{G} = 0$ in the summation over $\boldsymbol{G}$.

We will assume later that the momentum of the free electron is equal to that of the bound electron in the ith state to subtract the self energy of the free electron. In (10.50), photons with $\hbar\omega_{\boldsymbol{k}n} \le p_{\mathrm{e}}^2/2m$ make the main contribution. For this case, $\hbar\boldsymbol{k}$ can be neglected compared with $\boldsymbol{p}_{\mathrm{e}}$ for light atoms, since we have

$$\hbar\omega_{\boldsymbol{k}n} \approx \hbar ck \tag{10.52}$$

and

$$\frac{\hbar k}{p_{\mathrm{e}}} \le \frac{p_{\mathrm{e}}}{2m_{\mathrm{e}}c} \ll 1. \tag{10.53}$$

We thus have

$$E_{\mathrm{free}}^{(2)} \approx -\frac{e^2}{2\varepsilon_0 m_{\mathrm{e}}^2 V} \sum_{\boldsymbol{k}n} \frac{\overline{\left|\boldsymbol{E}_{\boldsymbol{k}n}(\boldsymbol{r}_{\mathrm{e}}) \cdot \widehat{\boldsymbol{p}}\right|^2}}{\omega_{\boldsymbol{k}n}^2}, \tag{10.54}$$

where the overline represents the expectation value with a relevant eigenstate.

Now, we will evaluate the Lamb shift for a hydrogen-like atom with an atomic number Z that has only one electron. What is most interesting to us is the case where the transition frequency between the 1s and 2s (or 2p) states lies in a photonic bandgap near the upper edge. For that case, the main contribution from the sum over $\boldsymbol{k}$ and n in (10.49) comes from those eigenfrequencies that are close to the band edge. Hence, we will first calculate the density of states around the upper edge of the bandgap. For simplicity, we assume that the upper edge is located on the Γ point of the first Brillouin zone. We further assume that the effective mass of the radiation field, m_{c}, is isotropic. Then we have

$$\hbar\,\omega = \hbar\omega_{\mathrm{c}} + \frac{\hbar^2 \boldsymbol{k}^2}{2m_{\mathrm{c}}}. \tag{10.55}$$

It is easy to show that the density of states of the radiation field is thus given by

$$D(\omega) = \frac{V}{\pi^2}\sqrt{\frac{m_{\mathrm{c}}^3\,(\omega-\omega_{\mathrm{c}})}{2\hbar^3}} \qquad (\omega \geq \omega_{\mathrm{c}}). \tag{10.56}$$

We approximate the product of $\boldsymbol{E}_{\boldsymbol{k}n}(\boldsymbol{r}_{\mathrm{e}})$ and $\widehat{\boldsymbol{p}}$ by the product of their absolute values times 1/3 and represent $\boldsymbol{E}_{\boldsymbol{k}n}$ by the eigenfunction at the band edge, $\boldsymbol{E}_{\mathrm{c}}$. When we subtract the self energy of the free electron given by (10.54) from that of the bound electron given by (10.49), we obtain the genuine energy shift, $\Delta E_i^{(2)}$:

$$\begin{aligned}
&\Delta E_i^{(2)} \\
&\approx \frac{e^2\,|\boldsymbol{E}_{\mathrm{c}}(\boldsymbol{r}_{\mathrm{e}})|^2}{6\varepsilon_0 m_{\mathrm{e}}^2 V}\left\{\sum_j\sum_{\boldsymbol{k}n}\frac{|\boldsymbol{p}_{ij}|^2}{\omega_{\boldsymbol{k}n}\,(\omega_i-\omega_j-\omega_{\boldsymbol{k}n})}+\sum_{\boldsymbol{k}n}\frac{(\widehat{\boldsymbol{p}}^2)_{ii}}{\omega_{\boldsymbol{k}n}^2}\right\} \\
&= \frac{e^2\,|\boldsymbol{E}_{\mathrm{c}}(\boldsymbol{r}_{\mathrm{e}})|^2}{6\varepsilon_0 m_{\mathrm{e}}^2 V}\sum_j\sum_{\boldsymbol{k}n}\frac{(\omega_i-\omega_j)\,|\boldsymbol{p}_{ij}|^2}{\omega_{\boldsymbol{k}n}^2\,(\omega_i-\omega_j-\omega_{\boldsymbol{k}n})}.
\end{aligned} \tag{10.57}$$

where

$$\boldsymbol{p}_{ij} = \langle i\,|\widehat{\boldsymbol{p}}|\,j\rangle \qquad \text{and} \qquad (\widehat{\boldsymbol{p}}^2)_{ii} = \left\langle i\left|\widehat{\boldsymbol{p}}^2\right|i\right\rangle. \tag{10.58}$$

When we calculate the Lamb shift for the 2s (2p) state (i.e., for $i = 2$), the main contribution on the right-hand side of (10.57) is made by the term with $j = 1$. Hence, we will calculate the summation over $\boldsymbol{k}$ and n for this term first. We set

$$\omega_{21} = \omega_2 - \omega_1, \quad \text{and} \tag{10.59}$$

$$\Omega = \omega_{\mathrm{c}} - \omega_{21}. \tag{10.60}$$

When we transform the summation to an integral over ω, the latter can be calculated analytically:

$$\sum_{\boldsymbol{k}n} \frac{1}{\omega_{\boldsymbol{k}n}^2 (\omega_2 - \omega_1 - \omega_{\boldsymbol{k}n})} = \int_{\omega_c}^{\infty} \frac{D(\omega) d\omega}{\omega^2 (\omega_{21} - \omega)}$$
$$= -\frac{V}{2\pi} \sqrt{\frac{m_c^3}{2\hbar^3 \omega_c}} \frac{1}{\left(\sqrt{\omega_c} + \sqrt{\Omega}\right)^2}. \tag{10.61}$$

For $j > 1$, the integral gives values of the same order. Hence, we use the same expression for $j > 1$ to obtain a rough estimation of the Lamb shift. We thus have

$$\Delta E_2^{(2)} = -\frac{e^2}{12\pi\varepsilon_0 m_e^2} \sqrt{\frac{m_c^3}{2\hbar^3 \omega_c}} \frac{|\boldsymbol{E}_c(\boldsymbol{r}_e)|^2}{\left(\sqrt{\omega_c} + \sqrt{\Omega}\right)^2}$$
$$\times \sum_j (\omega_2 - \omega_j) \,.\, |\boldsymbol{p}_{2j}|^2 \,. \tag{10.62}$$

We denote the Hamiltonian of the electron system by $\widehat{H}_a$.

$$\widehat{H}_a = \frac{\widehat{\boldsymbol{p}}^2}{2m_e} + \phi(\boldsymbol{r}), \tag{10.63}$$

where ϕ is the Coulomb potential due to the atomic nucleus. In general, we have

$$\sum_j (\omega_i - \omega_j) |\boldsymbol{p}_{ij}|^2 = \frac{1}{\hbar} \sum_j \left(\widehat{H}_a \widehat{\boldsymbol{p}} - \widehat{\boldsymbol{p}} \widehat{H}_a\right)_{ij} \cdot \boldsymbol{p}_{ji}$$
$$= \frac{1}{\hbar} \left\{(\phi \widehat{\boldsymbol{p}} - \widehat{\boldsymbol{p}} \phi) \cdot \widehat{\boldsymbol{p}}\right\}_{ii}$$
$$= -\frac{\hbar}{2} \int \mathrm{d}\boldsymbol{r} \, |\psi_i(\boldsymbol{r})|^2 \, \nabla^2 \phi(\boldsymbol{r})$$
$$= -\frac{Z e^2 \hbar \, |\psi_i(0)|^2}{2\varepsilon_0}, \tag{10.64}$$

where ψ_i is the eigenfunction of the ith state. When we derived (10.64), we used the following relation:

$$\nabla^2 \phi(\boldsymbol{r}) = -\frac{Z e^2}{4\pi\varepsilon_0} \nabla^2 \frac{1}{r} = \frac{Z e^2}{\varepsilon_0} \delta(\boldsymbol{r}). \tag{10.65}$$

Finally, we obtain

$$\Delta E_2^{(2)} = \frac{Z e^4}{24\pi\varepsilon_0^2 m_e^2} \sqrt{\frac{m_c^3}{2\hbar \omega_c}} \frac{|\boldsymbol{E}_c(\boldsymbol{r}_e)|^2 \, |\psi_2(0)|^2}{\left(\sqrt{\omega_c} + \sqrt{\Omega}\right)^2}. \tag{10.66}$$

Generally, we have

$$\psi_{ns}(0) = \frac{Z^3}{\pi n^3 a_B^3}, \quad \text{and} \tag{10.67}$$

$$\psi_{n\mathrm{p}}(0) = 0, \tag{10.68}$$

where a_B is the Bohr radius and n is the principal quantum number. Therefore, the Lamb shift between the 2s and 2p states is given by

$$\Delta E^{(2)}_{2\mathrm{s}-2\mathrm{p}} = \frac{e^4 Z^4}{192\pi^2 \varepsilon_0^2 a_\mathrm{B}^3 m_\mathrm{e}^2} \sqrt{\frac{m_\mathrm{c}^3}{2\hbar\omega_\mathrm{c}}} \frac{|\boldsymbol{E}_\mathrm{c}(\boldsymbol{r}_\mathrm{e})|^2}{\left(\sqrt{\omega_\mathrm{c}} + \sqrt{\Omega}\right)^2}. \tag{10.69}$$

Now, we estimate the Lamb shift of hydrogen ($Z = 1$) for a standard photonic crystal. We assume for simplicity that the band edge and the transition frequency coincide:

$$\omega_\mathrm{c} = \omega_{21} = 1.53 \times 10^{16}\ \mathrm{s}^{-1}. \tag{10.70}$$

We also assume a simple cubic crystal with lattice constant a for simplicity. The gap frequency is on the order of $\omega a/2\pi c = 0.25$. Thus we have

$$a \approx 3.08 \times 10^{-8}\ \mathrm{m}. \tag{10.71}$$

On the other hand, we may assume various values for m_c, since the dispersion relation in photonic crystals is diverse. Here, we assume that the dispersion curve starts from $\omega_\mathrm{c} a/2\pi c = 0.25$ on the Γ point and approaches the light line in free space on the X point. We thus have

$$m_\mathrm{c} \approx \frac{\pi^2 \hbar}{2\omega_\mathrm{c} a^2} = 3.59 \times 10^{-35}\ \mathrm{kg}. \tag{10.72}$$

When we further assume that $|\boldsymbol{E}_\mathrm{c}(\boldsymbol{r}_\mathrm{e})| \approx 1$, we obtain

$$\Delta E^{(2)}_{2\mathrm{s}-2\mathrm{p}} = \frac{1}{6}\alpha^5 m_\mathrm{e} c^2 = 0.43\ \mathrm{GHz}, \tag{10.73}$$

where $\alpha = e^2/4\pi\varepsilon_0\hbar c$ is the fine-structure constant.

The anomalous density of states at the band edge mainly contributes to this Lamb shift. $D(\omega)$ assumed for this calculation, however, underestimates the density of states at high frequencies, since it should be approximately proportional to ω^2 whereas $D(\omega)$ is proportional to $\sqrt{\omega}$. Although the accurate calculation should be performed by assuming a more realistic density of states, we may roughly expect that the actual Lamb shift is given by the sum of the anomalous part (0.43 GHz) and the normal part (1.05 GHz). This implies that the Lamb shift can be larger in photonic crystals than in free space by about 40 %. This difference is large enough to detect by an experiment, although a photonic crystal with an extremely small lattice contant, i.e., $a \approx 30$ nm is necessary.

11. Epilogue

The history of photonic crystals is not very long. In 1979, Ohtaka [27, 28] formulated the vector-spherical-wave expansion method, or the vector KKR (Koringa–Kohn–Rostker) method, to calculate the dispersion relation and the transmittance for the regular array of dielectric spheres. To the author's knowledge, this is the first self-consistent treatment of electromagnetic eigenmodes in 3D dielectric systems with large periodic modulation of the dielectric constant. A remarkable step was made by Yablonovitch in 1987 [95] who pointed out the possibility of the realization of photonic bandgaps, localized defect modes, and their applications to various optoelectronic devices. His idea stimulated many researchers, and energetic research activities including his own studies were initiated. The process to realize the photonic bandgaps is described in [96]. In the same year, John [97] discussed the strong localization of electromagnetic waves in disordered photonic crystals. He also predicted many interesting quantum optical phenomena that can be realized in photonic crystals such as the bound state of photons [98] and non-exponential decay of spontaneous emission [99]. The summary of early studies can be found in [100, 101]. Also see [102] and special issues of JOSA [103] and J. Mod. Phys. [104].

Since 1987, many researchers have been engaged in the realization of photonic bandgaps, localized defect modes, and other optical properties peculiar to the photonic crystals. Theoretical studies were indeed useful and indispensable for this purpose. Since the problem of the radiation field is essentially a one-body problem, that is, no photon–photon interaction needs to be taken into account,[1] many accurate numerical calculations could be performed thanks to the development of computing facilities. For example, the FDTD method is now widely used to predict the optical properties of very complicated structures such as 2D crystals of the slab type with line and/or point defects. These calculations were really useful for the design of the crystal structures. However, the essence of the expected phenomena, or we may say, the physics of the photonic crystals is not clarified by numerical

[1] In the case of nonlinear optical processes such as sum-frequency generation, they are usually described by the nonlinear susceptibility tensors, which may be regarded as the *effective* photon–photon interaction brought about by the nonlinear response of the material system. However, this nonlinear coupling is usually small and can be taken into acount in a perturbative manner.

calculation alone. The analytical description and estimation often give us a firm and perspective viewpoint. The symmetry property of the eigenmodes is a good example. The group theory for photonic crystals, that is, the group theory for the vector field with two degrees of freedom, which is indispensable to describe the symmetry property, was formulated by Tanabe and Ohtaka [49] and by this author [42, 50] independently. It was very interesting and exciting to find that the calculated band structures agreed perfectly with the prediction of the group theory. As for the experimental studies, only a few examples have been mentioned in this book. Many researchers in various fields, such as physics, electronics, optics, thin film fabrication, lithography, and chemistry, have been collaborating to make new crystals and measure their properties. As a result, 3D crystals with submicro-meter lattice constants are now available. The number of scientific papers has been increasing rapidly and many researchers in private companies are also interested in this field. Some technological applications of the photonic crystals are being undertaken.

In this book, the fundamental properties of photonic crystals have been described from the theoretical point of view. First, we derived the eigenvalue problem for the electromagnetic field and studied the property of their eigenfunctions. In particular, we introduced an Hermitian operator related to the electric field, and derived the explicit expression for the retarded Green's function. Then, we could obtain the general formula for the optical response of the photonic crystal by means of this Green's function. We applied this method to four typical examples, i.e., (1) dipole radiation, (2) stimulated emission, (3) sum-frequency generation, and (4) free induction decay. The first example was then extended to crystals with the dielectric defect, which enabled us to formulate the numerical method to calculate the eigenfunctions and the eigenfrequencies of the localized defect modes. This method was applied to 2D crystals with structural defects, and excellent agreement with experimental observation was shown. It was also extended to the band calculation with frequency-dependent dielectric constants. The second example showed the enhancement effect due to the small group velocity. In particular, the group-velocity anomaly peculiar to the 2D and 3D crystals led to the large reduction of the lasing threshold. In the third example, we derived the additional selection rule and the phase-matching condition peculiar to the photonic crystals in addition to the enhancement of the induced nonlinear electric field due to the small group velocity. The theoretical treatment given there can readily be extended to other nonlinear optical processes. The last example shows that we can deal with any coherent optical processes in the photonic crystal by means of the method of Green's function

The plane-wave expansion method to calculate the transmission and reflection spectra given in this book was really useful in the early stage of the investigation. The opaque frequency ranges due to uncoupled modes were shown by this method [47] and the vector KKR method [30]. The enhanced stimulated emission and the low-threshold lasing in 2D crystals were also shown by the plane-wave expansion method [61, 84]. However, as was mentioned previously, the convergence is poor when the spatial modulation of the dielectric constant is large, and the plane-wave expansion method is impractical when the contrast of the dielectric constant is, say, larger than four. For that case, more efficient method, such as the transfer matrix method, the vector KKR method, and the FDTD method might be used.

In the last chapter, we quantized the radiation field in the photonic crystal, and the Hamiltonian that describes the interaction between the radiation field and the electron system was derived. Because of the gauge condition that we used, which I believe is the most convenient to deal with the spatial modulation of the dielectric constant, the interaction Hamiltonian includes a term that is absent in free space. Although we did not treat the case where it is important, that term may bring about important and peculiar effects. We examined two quantum optical phenomena, i.e., quadrature-phase squeezing and anomalous Lamb shift. We showed that the method of Green's function can be applied to the former problem and the enhancement effect due to the small group velocity is also present for this case. The latter problem was first discussed by John [98]. The same problem was treated in this book for the case of the density of states described by an isotropic effective mass.

Such items were chosen and described that are really fundamental and indispensable to understand and analyze the optical properties of the photonic crystals. However, there are still many items that are not mentioned in this book. They include solitary waves in the gap frequency region [105, 106], various optical waveguides made of defect structures [107–109], the Smith–Purcell effect for 2D and 3D photonic crystals [110], the idea of sonic crystals, i.e., sound waves in periodic structures [111–113], exciton polaritons in 1D photonic crystal slabs [114], lasing in regular photonic crystal slabs [115-117], the analysis of phase shift in transmission measurements [118], linear and nonlinear optical properties of quasi-periodic systems [119, 120], etc. The theoretical methods given in this book are applicable to these problems as well. The reader may consult original papers.

Photonic crystals, the remarkable invention realized by the combination of the optical physics and the contemporary microfabrication techniques, can indeed control the radiation field and alter the optical properties of embedded atoms and molecules. Thanks to the efforts that people in this field have made in the last ten years, we now have specimens of good quality with submicrometer lattice constants that function in the visible and near infrared frequency ranges where various optical processes are caused by the interaction between the radiation field and matter. We may say that we have come to the stage where we can test the predicted peciliar characteristics of photonic crystals using real specimens and also realize their technological applications.

References

1. E. M. Purcell: Phys. Rev. **69**, 681 (1946)
2. K. Yoshino, Y. Shimoda, Y. Kawagishi, K. Nakayama, M. Ozaki: Appl. Phys. Lett. **75**, 932 (1999)
3. J. G. Fleming, Shawn-Yu Lin: Opt. Lett. **24**,49 (1999)
4. M. Lončar, D. Nedeljković, T. Doll, J. Vučković, A. Scherer, T. P. Pearsall: Appl. Phys. Lett. **77** 1937 (2000)
5. P. L. Gourley, J. R. Wendt, G. A. Vawter, T. M. Brennan, B. E. Hammons: Appl. Phys. Lett. **64**, 687 (1994)
6. M. Kanskar, P. Paddon, V. Pacradouni, R. Morin, A. Busch, J. F. Young, S. R. Johnson, J. Mackenzie, T. Tiedje: Appl. Phys. Lett. **70**, 1438 (1997)
7. S. Fan, P. R. Villeneuve, J. D. Joannopoulos, E. F. Schubert: Phys. Rev. Lett. **78**, 3294 (1997)
8. D. Labilloy, H. Benisty, C. Weisbuch, T. F. Krauss, R. M. De La Rue, V. Bardinal, R. Houdré, U. Oesterle, D. Cassagne, C. Jouanin: Phys. Rev. Lett. **79**, 4147 (1997)
9. B. D'Urso, O. Painter, J. O'Brien, T. Tombrello, A. Yariv, A. Scherer: J. Opt. Soc. Am. B **15**, 1155 (1998)
10. D. Labilloy, H. Benisty, C. Weisbuch, C. J. M. Smith, T. F. Krauss, R. Houdré, U. Oesterle: Phys. Rev. B **59**, 1649 (1999)
11. O. Painter, R. K. Lee, A. Scherer, A. Yariv, J. D. O'Brien, P. D. Dapkus, I. Kim, Science **284**, 1819 (1999)
12. S. G. Johnson, S. Fan, P. R. Villeneuve, J. D. Joannopoulos: Phys. Rev. B **60**, 5751 (1999)
13. S. Kuchinsky, D. C. Allan, N. F. Borrelli, J. C. Cotteverte: Opt. Commun. **175**, 147 (2000)
14. M. Haraguchi, T. Nakai, A. Shinya, T. Okamoto, M. Fukui, T. Koda, R. Shimada, K. Ohtaka, K. Takeda: Jpn. J. Appl. Phys. **39**, 1747 (2000)
15. T. Fujimura, T. Itoh, A. Imada, R. Shimada, T. Koda, N. Chiba, H. Muramatsu, H. Miyazaki, K. Ohtaka: J. Lumin. **87–89**, 954 (2000)
16. H. T. Miyazaki, H. Miyazaki, K. Ohtaka, T. Sato: J. Appl. Phys. **87**, 7152 (2000)
17. See, for example, A. Yariv, P. Yeh: *Optical Waves in Crystals* (Wiley, New York 1984) Chap. 6
18. See, for example, C. Kittel: *Introduction to Solid State Physics* (Wiley, New York 1996) Chap. 7
19. J. P. Dowling, M. Scalora, M. J. Bloemer, C. M. Bowden: J. Appl. Phys. **75**, 1896 (1994)
20. K. Busch, S. John: Phys. Rev. Lett. **83**, 967 (1999)
21. K. M. Leung, Y. F. Liu: Phys. Rev. Lett. **65**, 2646 (1990)
22. Z. Zhang, S. Satpathy: Phys. Rev. Lett. **65**, 2650 (1990)
23. H. S. Sözüer, J. W. Haus, R. Inguva: Phys. Rev. B **45**, 13962 (1992)

24. H. S. Sözüer, J. W. Haus: J. Opt. Soc. Am. B **10**, 296 (1993)
25. K. M. Ho, C. T. Chan, C. M. Soukoulis: Phys. Rev. Lett. **65**, 3152 (1990)
26. P. R. Villeneuve, M. Piché: J. Mod. Opt. **41**, 241 (1994)
27. K. Ohtaka: Phys. Rev. B **19**, 5057 (1979)
28. K. Ohtaka: J. Phys. C **13**, 667 (1980)
29. K. Ohtaka, Y. Tanabe: J. Phys. Soc. Jpn. **65**, 2265 (1996)
30. N. Stefanou, V. Karathanos, A. Modinos: J. Phys. Condens. Matter **4**, 7389 (1992)
31. M. Plihal, A. A. Maradudin: Phys. Rev. B **44**, 8565 (1991)
32. P. R. Villeneuve, M. Piché: Phys. Rev. B **46**, 4969 (1992)
33. R. D. Meade, K. D. Brommer, A. M. Rappe, J. D. Joannopoulos: Appl. Phys. Lett. **61**, 495 (1992)
34. N. Peyghambarian, S. W. Koch, A. Mysyrowicz: *Introduction to Semiconductor Optics* (Prentice Hall, Englewood 1993) Chap. 8
35. S. L. McCall, P. M. Platzman, R. Dalichaouch, D. Smith, S. Schultz: Phys. Rev. Lett. **67**, 2017 (1991)
36. P. Yeh: J. Opt. Soc. Am. **69**, 742 (1979)
37. R. J. Glauber and M. Lewenstein, Phys. Rev. A **43**, 467 (1991).
38. J. P. Dowling, C. M. Bowden: Phys. Rev. A **46**, 612 (1992)
39. K. Sakoda, K. Ohtaka: Phys. Rev. B **54**, 5732 (1996)
40. W. M. Robertson, G. Arjavalingam, R. D. Meade, K. D. Brommer, A. M. Rappe, J. D. Joannopoulos: Phys. Rev. Lett. **68**, 2023 (1992)
41. W. M. Robertson, G. Arjavalingam, R. D. Meade, K. D. Brommer, A. M. Rappe, J. D. Joannopoulos: J. Opt. Soc. Am. B **10**, 322 (1993)
42. K. Sakoda: Phys. Rev. B **52**, 7982 (1995)
43. T. Inui, Y. Tanabe, Y. Onodera: *Group Theory and Its Applications in Physics* (Springer, Berlin 1990)
44. J. Cornwell: *Group Theory and Electronic Energy Bands in Solids* (John Wiley International Publishers, New York 1969)
45. M. Wada, Y. Doi, K. Inoue, J. W. Haus: Phys. Rev. B **55**, 10443 (1997)
46. K. Sakoda: J. Opt. Soc. Am. B **14**, 1961 (1997)
47. K. Sakoda: Phys. Rev. B **51**, 4672 (1995)
48. K. Sakoda, M. Sasada, T. Fukushima, A. Yamanaka, N. Kawai, K. Inoue: J. Opt. Soc. Am. B **16**, 361 (1999)
49. K. Ohtaka, Y. Tanabe: J. Phys. Soc. Jpn. **65**, 2670 (1996)
50. K. Sakoda: Phys. Rev. B **55**, 15345 (1997)
51. S. G. Romanov, N. P. Johnson, A. V. Fokin, V. Y. Butko, H. M. Yates, M. E. Pemble, C. M. Sotomayor Torres: Appl. Phys. Lett. **70**, 2091 (1997)
52. Yu. A. Vlasov, A. K. Luterova, B. I. Pelant, B. Honerlage, V. N. Astratov: Appl. Phys. Lett. **71**, 1616 (1997)
53. H. Miguez, C. Lopez, F. Meseguer, A. Blanco, L. Vazquez, R. Mayoral, M. Ocana, V. Fornes, A. Mifsud: Appl. Phys. Lett. **71**, 1148 (1997)
54. E. Yablonovitch, T. J. Gmitter, K. M. Leung: Phys. Rev. Lett. **67**, 2295 (1991)
55. A. Chelnokov, K. Wang, S. Rowson, P. Garoche, J.-M. Lourtioz: Appl. Phys. Lett. **77**, 2943 (2000)
56. A. Modinos: Physica A **141A** 575 (1987)
57. K. Ohtaka, Y. Tanabe: J. Phys. Soc. Jpn. **65**, 2276 (1996)
58. J. B. Pendry, A. MacKinnon: Phys. Rev. Lett. **69**, 2772 (1992)
59. A. Taflove: *Computational Electrodynamics: The Finite-Difference Time-Domain Method* (Artech House, Boston 1995)
60. K. Sakoda: Phys. Rev. B **52**, 8992 (1995)
61. K. Sakoda, K. Ohtaka, T. Ueta: Opt. Express **4**, 481 (1999)

62. R. Courant, D. Hilbert: *Methods of Mathematical Physics* (Interscience, New York 1953), Vol. 1, pp. 277
63. H. Kosaka, T. Kawashima, A. Tomita, M. Notomi, T. Tamamura, T. Sato, S. Kawakami: Phys. Rev. B **58**, R10096 (1998)
64. H. Kosaka, T. Kawashima, A. Tomita, M. Notomi, T. Tamamura, T. Sato, S. Kawakami: Appl. Phys. Lett. **74**, 1370 (1999)
65. T. Suzuki, P. K. L. Yu: J. Opt. Soc. Am. B **12**, 583 (1995)
66. S. Nojima: Jpn. J. Appl. Phys. Part 2 **37**, L565 (1998)
67. K. Sakoda, K. Ohtaka: Phys. Rev. B **54**, 5742 (1996)
68. Y. R. Shen: *The Principles of Nonlinear Optics* (John Wiley & Sons, New York 1984) pp. 101
69. K. Sakoda, H. Shiroma: Phys. Rev. B **56**, 4830 (1997)
70. K. Sakoda, T. Ueta, K. Ohtaka: Phys. Rev. B **56**, 14905 (1997)
71. K. Sakoda: J. Appl. Phys. **84**, 1210 (1998)
72. T. Ueta, K. Ohtaka, N. Kawai, K. Sakoda: J. Appl. Phys. **84**, 6299 (1998)
73. O. Painter, J. Vučković, A. Scherer: J. Opt. Soc. Am. B **16**, 275 (1999)
74. J-K. Hwang, H-Y. Ryu, Y-H. Lee: Phys. Rev. B **60**, 4688 (1999)
75. D. R. Smith, R. Dalichaouch, N. Kroll, S. Schultz, S. L. McCall, P. M. Platzman: J. Opt. Soc. Am. B **10**, 314 (1993)
76. X. P. Feng, Y. Arakawa: Jpn. J. Appl. Phys. Part 2 **36**, L120 (1997)
77. S.-Y. Lin, V. M. Hietala, S. K. Lyo, A. Zaslavsky: Appl. Phys. Lett. **68**, 3233 (1996)
78. K. Sakoda and J. Kawamata: Opt. Express **3**, 12 (1998)
79. K. Sakoda, N. Kawai, T. Ito, A. Chutinan, S. Noda, T. Mitsuyu, K. Hirao: submitted to Phys. Rev. B
80. T. Ito, K. Sakoda: submitted to Phys. Rev. B
81. T. Ochiai and K. Sakoda: 'Band structure and optical transmittance of leaky modes in two-dimensional photonic crystals of a slab type'. In: *Technical Digest of 2000 Quantum Electronics and Laser Science Conference at San Francisco, May 8–May 12, 2000* (Optical Society of America, Washington D.C. 2000) pp.118
82. T. Ochiai and K. Sakoda: Phys. Rev. B, to be published
83. P. Paddon, J. F. Young: Phys. Rev. B **61**, 2090 (2000)
84. K. Sakoda: Opt. Express **4**, 167 (1999)
85. A. Yariv, *Quantum Electronics* (Wily, New York 1967) Sect. 19.6
86. M. O. Scully, M. S. Zubairy: *Quantum Optics* (Cambridge University Press, Cambridge 1997) Chap. 2
87. W. E. Lamb, Jr., R. C. Retherford: Phys. Rev. **72**, 241 (1947)
88. S. Triebwasser, E. S. Dayhoff, W. E. Lamb, Jr.: Phys. Rev. **89**, 98 (1953)
89. W. H. Parker, B. N. Taylor, D. N. Langenberg: Phys. Rev. Lett. **18**, 287 (1967)
90. H. A. Bethe: Phys. Rev. **72**, 339 (1947)
91. J. B. French, V. F. Weisskopf: Phys. Rev. **75**, 1240 (1949)
92. N. M. Kroll. W. E. Lamb: Phys. Rev. **75**, 338 (1949)
93. R. P. Feynman, Phys. Rev. **74**, 1430 (1948)
94. H. Fukuda, Y. Miyamoto, S. Tomonaga, Prog. Theor. Phys. **4**, 47 (1948)
95. E. Yablonovitch: Phys. Rev. Lett. **58**, 2059 (1987)
96. E. Yablonovitch: J. Opt. Soc. Am. B**10**, 283 (1993)
97. S. John: Phys. Rev. Lett. **58**, 2486 (1987)
98. S. John, J. Wang: Phys. Rev. B **43**, 12772 (1991)
99. S. John, T. Quang: Phys. Rev. A **50**, 1764 (1994)
100. C. M. Soukoulis, ed., *Photonic Band Gaps and Localization* (Plenum, New York 1993)
101. C. M. Soukoulis, ed., *Photonic Band Gap Materials* (Kluwer, Dordrecht 1996)

102. J. D. Joannopoulos, R. D. Meade, J. N. Winn: *Photonic Crystals* (Princeton University Press, Princeton 1995)
103. C. M. Bowden, J. P. Dowling, H. O. Everitt, ed.: 'Development and Applications of Materials Exhibiting Photonic Band Gaps'. In J. Opt. Soc. Am. B, **10** (1993) pp.279–413
104. G. Kurizki, J. W. Haus, ed.: 'Photonic Band Structures'. In J. Mod. Opt. **41** (1994) pp.171–404
105. D. L. Mills, S. E. Trullinger, Phys. Rev. B **36**, 947 (1987)
106. S. John, N. Aközbek: Phys. Rev. Lett. **71**, 1168 (1993)
107. A. Mekis, J. C. Chen, I. Kurland, S. Fan, P. R. Villeneuve, J. D. Joannopoulos: Phys. Rev. Lett. **77**, 3787 (1996)
108. T. Baba, N. Fukaya, J. Yonekura: Electron. Lett. **27**, 654 (1999)
109. S. Noda, A. Chutinan, M. Imada: Nature **407**, 608 (2000)
110. K. Ohtaka, private communication
111. J. V. Sanchez-Perez, D. Caballero, R. Martinez-Sala, C. Rubio, J. Sanchez-Dehesa, F. Meseguer, J. Llinares, F. Galvez: Phys. Rev. Lett. **80**, 5325 (1998)
112. D. Caballero, J. Sanchez-Dehesa, C. Rubio, R. Martinez-Sala, J. V. Sanchez-Perez, F. Meseguer, J. Llinares: Phys. Rev. E **60**, R6316 (1999)
113. K. Imamura, Y. Tanaka, S. Mizuno, S. Tamura: J. Phys. Condens. Matter **12**, 9843 (2000)
114. T. Fujita, Y. Sato, T. Kuitani, T. Ishihara: Phys. Rev. B **57**, 12428 (1998)
115. P. L. Gourley, M. E. Warren, G. R. Hadley, G. A. Vawter, T. M. Brennan, B. E. Hammons: Appl. Phys. Lett. **58**, 890 (1991)
116. P. L. Gourley, M. E. Warren, G. A. Vawter, T. M. Brennan, B. E. Hammons: Appl. Phys. Lett. **60**, 2714 (1992)
117. M. Imada, S. Noda, A. Chutinan, T. Tokuda, M. Murata, G. Sasaki: Appl. Phys. Lett. **75**, 316 (1999)
118. K. Ohtaka, Y. Suda, S. Nagano, T. Ueta, A. Imada, T. Koda, J. S. Bae, K. Mizuno, S. Yano, Y. Segawa: Phys. Rev. B **61**, 5267 (2000)
119. C. Sibilia, F. Tropea, M. Bertolotti: J. Mod. Opt. **45**, 2255 (1998)
120. R. Shimada, T. Koda, T. Ueta, K. Ohtaka: J. Phys. Soc. Jpn. **67**, 3414 (1998)

Index

Springer Series in

OPTICAL SCIENCES

New editions of volumes prior to volume 60

1 **Solid-State Laser Engineering**
By W. Koechner, 5th revised and updated ed. 1999, 472 figs., 55 tabs., XII, 746 pages

14 **Laser Crystals**
Their Physics and Properties
By A. A. Kaminskii, 2nd ed. 1990, 89 figs., 56 tabs., XVI, 456 pages

15 **X-Ray Spectroscopy**
An Introduction
By B. K. Agarwal, 2nd ed. 1991, 239 figs., XV, 419 pages

36 **Transmission Electron Microscopy**
Physics of Image Formation and Microanalysis
By L. Reimer, 4th ed. 1997, 273 figs. XVI, 584 pages

45 **Scanning Electron Microscopy**
Physics of Image Formation and Microanalysis
By L. Reimer, 2nd completely revised and updated ed. 1998,
260 figs., XIV, 527 pages

Published titles since volume 60

60 **Holographic Interferometry in Experimental Mechanics**
By Yu. I. Ostrovsky, V. P. Shchepinov, V. V. Yakovlev, 1991, 167 figs., IX, 248 pages

61 **Millimetre and Submillimetre Wavelength Lasers**
A Handbook of cw Measurements
By N. G. Douglas, 1989, 15 figs., IX, 278 pages

62 **Photoacoustic and Photothermal Phenomena II**
Proceedings of the 6th International Topical Meeting, Baltimore, Maryland,
July 31 - August 3, 1989
By J. C. Murphy, J. W. Maclachlan Spicer, L. C. Aamodt, B. S. H. Royce (Eds.),
1990, 389 figs., 23 tabs., XXI, 545 pages

63 **Electron Energy Loss Spectrometers**
The Technology of High Performance
By H. Ibach, 1991, 103 figs., VIII, 178 pages

64 **Handbook of Nonlinear Optical Crystals**
By V. G. Dmitriev, G. G. Gurzadyan, D. N. Nikogosyan,
3rd revised ed. 1999, 39 figs., XVIII, 413 pages

65 **High-Power Dye Lasers**
By F. J. Duarte (Ed.), 1991, 93 figs., XIII, 252 pages

66 **Silver-Halide Recording Materials**
for Holography and Their Processing
By H. I. Bjelkhagen, 2nd ed. 1995, 64 figs., XX, 440 pages

67 **X-Ray Microscopy III**
Proceedings of the Third International Conference, London, September 3-7, 1990
By A. G. Michette, G. R. Morrison, C. J. Buckley (Eds.), 1992, 359 figs., XVI, 491 pages

68 **Holographic Interferometry**
Principles and Methods
By P. K. Rastogi (Ed.), 1994, 178 figs., 3 in color, XIII, 328 pages

69 **Photoacoustic and Photothermal Phenomena III**
Proceedings of the 7th International Topical Meeting, Doorwerth, The Netherlands,
August 26-30, 1991
By D. Bicanic (Ed.), 1992, 501 figs., XXVIII, 731 pages

Springer Series in

OPTICAL SCIENCES

---

70 **Electron Holography**
By A. Tonomura, 2nd, enlarged ed. 1999, 127 figs., XII, 162 pages

71 **Energy-Filtering Transmission Electron Microscopy**
By L. Reimer (Ed.), 1995, 199 figs., XIV, 424 pages

72 **Nonlinear Optical Effects and Materials**
By P. Günter (Ed.), 2000, 174 figs., 43 tabs., XIV, 540 pages

73 **Evanescent Waves**
From Newtonian Optics to Atomic Optics
By F. de Fornel, 2001, 277 figs., XVIII, 268 pages

74 **International Trends in Optics and Photonics**
ICO IV
By T. Asakura (Ed.), 1999, 190 figs., 14 tabs., XX, 426 pages

75 **Advanced Optical Imaging Theory**
By M. Gu, 2000, 93 figs., XII, 214 pages

76 **Holographic Data Storage**
By H.J. Coufal, D. Psaltis, G.T. Sincerbox (Eds.), 2000
228 figs., 64 in color, 12 tabs., XXVI, 486 pages

77 **Solid-State Lasers for Materials Processing**
Fundamental Relations and Technical Realizations
By R. Iffländer, 2001, 230 figs., 73 tabs., XVIII, 350 pages

78 **Holography**
The First 50 Years
By J.-M. Fournier (Ed.), 2001, 266 figs., XII, 460 pages

79 **Mathematical Methods of Quantum Optics**
By R.R. Puri, 2001, 13 figs., XIV, 285 pages

80 **Optical Properties of Photonic Crystals**
By K. Sakoda, 2001, 95 figs., 28 tabs., XII, 223 pages

81 **Photonic Analog-to-Digital Conversion**
By B.L. Shoop, 2001, 259 figs., 11 tabs., XIV, 330 pages

Printing (Computer to Film): Saladruck, Berlin
Binding: Stürtz AG, Würzburg